SMART AND RESILIENT INFRASTRUCTURE FOR EMERGING ECONOMIES: PERSPECTIVES ON BUILDING BETTER

Smart and Resilient Infrastructure For Emerging Economies: Perspectives on Building Better is about pressing and multidimensional challenges faced in constructing resilient, sustainable, and smart infrastructure in developing countries. The 32 case studies, literature reviews, comparative analyses and systematic reviews, cover a wide range of topics, including:

- sustainable and resilient infrastructure development
- smart cities
- digital innovation in construction
- infrastructure investment
- construction ergonomics
- socio-environmental sustainability
- gender equity, and
- climate change responses

The contributions present innovative solutions, impactful insights, and substantive contributions to the discourse on sustainable infrastructure development, and illuminate the interplay between infrastructure development, social justice, environmental sustainability, and technological advancement.

Smart and Resilient Infrastructure For Emerging Economies: Perspectives on Building Better is essential reading for academics, researchers, practitioners, policymakers, and students involved in the built environment, infrastructure delivery, investment in infrastructure, civil engineering, architecture, urban planning, environmental science, and other related disciplines.

PROCEEDINGS OF THE 9TH INTERNATIONAL CONFERENCE ON DEVELOPMENT AND INVESTMENT IN INFRASTRUCTURE (DII-2023, 19–21 JULY 2023, ZAMBIA)

Smart and Resilient Infrastructure For Emerging Economies: Perspectives on Building Better

Edited by

Innocent Musonda
University of Johannesburg, Johannesburg, South Africa

Erastus Mwanaumo
University of Zambia, Zambia

Adetayo Onososen and Thembani Moyo
University of Johannesburg, Johannesburg, South Africa

CRC Press is an imprint of the
Taylor & Francis Group, an **informa** business

A BALKEMA BOOK

First published 2023
by CRC Press/Balkema
4 Park Square, Milton Park, Abingdon, Oxon, OX14 4RN

and by CRC Press/Balkema
2385 NW Executive Center Drive, Suite 320, Boca Raton FL 33431

CRC Press/Balkema is an imprint of the Taylor & Francis Group, an informa business

© 2024 selection and editorial matter, Innocent Musonda, Erastus Mwanaumo, Adetayo Onososen & Thembani Moyo; individual chapters, the contributors

The right of Innocent Musonda, Erastus Mwanaumo, Adetayo Onososen & Thembani Moyoto be identified as the authors of the editorial material, and of the authors for their individual chapters, has been asserted in accordance with sections 77 and 78 of the Copyright, Designs and Patents Act 1988.

The Open Access version of this book, available at www.taylorfrancis.com, has been made available under a Creative Commons Attribution-Non Commercial-No Derivatives (CC-BY-NC-ND) 4.0 license.

Although all care is taken to ensure integrity and the quality of this publication and the information herein, no responsibility is assumed by the publishers nor the author for any damage to the property or persons as a result of operation or use of this publication and/or the information contained herein.

British Library Cataloguing-in-Publication Data
A catalogue record for this book is available from the British Library

Library of Congress Cataloging-in-Publication Data
A catalog record has been requested for this book

ISBN: 978-1-032-56461-6 (hbk)
ISBN: 978-1-032-56462-3 (pbk)
ISBN: 978-1-003-43564-8 (ebk)

DOI: 10.1201/9781003435648

Typeset in Times New Roman
by MPS Limited, Chennai, India

Table of Contents

Preface	ix
Disclaimer	xi
Declaration and the peer-review process	xiii
Peer Review Process (PRP) Confirmation	xv
Acknowledgment	xvii
Conference Committees	xix
DII-2023 Conference Sponsor	xxi
DII-2023 Conference Partners	xxiii

Sustainable & resilient infrastructure development

Promoting sustainable development through quality infrastructure in Elukwatini node, South Africa *N.P. Nkambule & K. Ntakana*	3
Developing a sustainable water supply system for rural South African communities *N.N. Ngema, S.L. Mbanga, A.A. Adeniran & E. Kabundu*	12
Sustainable community-driven urban regeneration of recreational infrastructure in former mine townships of the Copperbelt Province *L. Makashini, E.K. Munshifwa & Y. Adewunmi*	21
Public participation and sustainability consideration for mega transport infrastructure projects: Lessons from the Gauteng freeway improvement project, South Africa *L. Morejele-zwane, T. Gumbo & S. Dumba*	29
Integration of environmental sustainability practices in real estate development in emerging markets *J. Mahachi & L. Kumalo*	37
Exploring development control tools for improved sustainability in land use compatibility: Experiences from Vuwani town, Collins Chabane local municipality *R. Mulokwe & B. Risimati*	46
Development of South African structural performance criteria for innovative building systems *N. Bhila & J. Mahachi*	54

Smart infrastructure and cities, digital innovation in construction & digital transition

Municipal free Wi-Fi, governance and service delivery in the city of Tshwane *T.P. Mathane & T. Gumbo*	67

Municipal free Wi-Fi intervention: Mamelodi and Soshanguve comparative analysis 74
T.P. Mathane & T. Gumbo

Advancements in E-mobility: A bibliometric literature review on battery
technology, charging infrastructure, and energy management 82
T. Moyo, A.O. Onososen, I. Musonda & H. Muzioreva

Flood simulation with GeoBIM 90
K. Kangwa & B. Mwiya

Framework for integrated inner-city construction site layout planning 99
F.G. Tsegay, E. Mwanaumo & B. Mwiya

Digital twin technology in health, safety, and wellbeing management in
the built environment 109
A.O. Onososen, I. Musonda, T. Moyo & H. Muzioreva

Construction ergonomics, health, safety and wellbeing

Creating safe spaces: Scaling up public transport infrastructure investments in
the city of Johannesburg 119
R.C. Kalaoane, T. Gumbo, A.Y. Kibangou, W. Musakwa & I. Musonda

Influence of public sector built environment professionals on infrastructure
delivery in the Eastern Cape 125
A. Mntu, K. Kajimo-Shakantu & B. du Toit

Factors affecting public sector infrastructure delivery in the Eastern Cape,
South Africa 134
A. Mntu, K. Kajimo-Shakantu & T. Mogorosi

Importance of proactive road maintenance over reactive road maintenance in
developing countries: A case of Malawi 143
B.F. Mwakatobe, W. Kuotcha, I. Ngoma & S. Zulu

Public procurement: Driver for achieving a circular economy among SME
housing developers 152
C.P. Mukumba & K. Kajimo-Shakantu

An assessment of cost and socio-economic implications of applying innovative
active design principles in construction 161
L. Le Roux & K. Kajimo-Shakantu

Construction ergonomics, health, safety and wellbeing

Psycho-social well-being programs required for construction workers
in Zimbabwe 173
W. Mateza & T. Moyo

Implementing construction risk management methods on private sector
projects in South Africa 182
K. Kajimo-Shakantu, H. Nengovhela & F. Muleya

Causes of job burnout amongst Female Quantity Surveyors (FQS) in the
Nigerian Construction Industry (NCI) 191
F.C. Jayeola, A.J. Ogungbile, A.E. Oke & E.M. Mwanaumo

Infrastructure: Economic, social / environmental sustainability

An exploratory and comparative assessment of cost estimates on
infrastructure projects: A client's perspective   203
F. Muleya, M. Beene, A. Lungu, C.K. Tembo & K. Kajimo-Shakantu

Investigation of cost management factors influencing poor cost performance
on large to medium sized projects in the construction industry   213
C.K. Tembo, A. Kanyembo, F. Muleya & K. Kajimo-Shakantu

Gender equity, social justice & social inequality in construction

Adoption of microgrids as an energy solution to uplift rural communities
in the Eastern Cape, South Africa   225
S. Xulaba & C.J. Allen

A framework for sustainable human settlement building methods for low-cost
housing in Northern Cape province   233
T. Bremer & S.C. Monoametsi

Pathways to meaningful upgrading of urban informal settlements: Towards
adequate housing infrastructure in South Africa   242
M.G. Mndzebele & T. Gumbo

Failures of Small Medium Enterprise (SME) in the Kwazulu-Natal construction
industry: Management, technical, and economic factors   250
A.O. Aiyetan & A.B. David

Equalizing opportunity of female: Kwazulu-Natal construction industry   259
A.O. Aiyetan & A.B. David

Tenurial arrangements to encourage growth and development in urban villages
in Nigeria   268
A.N. Abdullahi, G.O. Udo, F.P. Udoudoh, J. Udo & C.S. Okoro

Environment, climate change and shock events impact and response and water resources

Strategies to respond to climate change effects in the Zimbabwean
construction industry   279
M. James & T. Moyo

Critical factors influencing environmental management best practices in the
developing countries real estate industry: A systematic review   289
H. Adjarko, I.C. Anugwo & A.O. Aiyetan

Author index   299

Preface

On behalf of the Organising Committee, I welcome you to the International Conference on Development and Investment in Infrastructure (DII-2023). The DII-2023 conference is part of the DII Conference series on Infrastructure Development and Investment in Africa. It aims to provide an international forum for leaders, researchers, practitioners and other stakeholders in infrastructure development to discuss and devise ways of maximising benefits from infrastructure development in Africa and achieve outputs that will inform policy.

The 2023 conference, themed "Smart and Resilient Infrastructure For Emerging Economies: Perspectives on Building Better" will address a broad range of topics around infrastructure to evaluate and draw lessons on innovations, empowerment, growth and sustainable development.

The broad topics covered by the conference include:

- Sustainable Infrastructure Development
- Smart Infrastructure and Cities
- Quality and Resilient Infrastructure
- Education, Empowerment, Gender Equity, Wellness and Development
- Environmental and Waste Management/Facilities & Real-Estate Management
- Infrastructure, Investment and Finance- Trends and Forecasts
- Infrastructure: Shock Events, Procurement, Project Management, Health & Safety
- Infrastructure: Economic, Social/Environmental Sustainability
- Digital Innovation and transition in the built environment

Warm gratitude is extended to the authors who have successfully gone through a two-tier peer-review process to have their papers accepted and published in this proceeding. The peer-review process would have been impossible without the support of the Scientific and Technical review Committees (STC) members. The organising committee is thankful for this voluntary service central to the quality of the accepted papers.

Special thank you also goes to all the conference delegates from different continents. Thank you for attending the event.

Prof. Innocent Musonda
For/DII-2023

Disclaimer

Every effort was made to ensure accuracy in this publication. However, the publishers and editors make no representation, express or implied, concerning the information contained in these proceedings and cannot accept any legal responsibility or liability in whole or in part for any errors or omissions that article contributors may have made.

Declaration and the peer-review process

All the papers in these conference proceedings were double-blind peer-reviewed at the abstract and full paper stage by the members of the International Review Committee. The process entailed a detailed review of the abstracts and full papers, reporting comments to authors, modification of articles by authors whose papers were not rejected, and re-evaluation of the revised articles to ensure the quality of content.

THE PEER-REVIEW PROCESS

The need for high-quality conference proceedings, evident in the accepted and published papers, entailed a rigorous two-stage blind peer review process by no less than two acknowledged experts in the subject area. Experts, including industry professionals and academics, were assigned to ensure that high standards of scientific papers were produced and included in the proceedings.

The first stage of the review

Submitted abstracts were twice blind-reviewed. Each abstract was examined to ensure relevance to the conference theme and objectives, academic rigour, contribution to knowledge, originality of material and research methodology. Authors whose abstracts were accepted were provided with anonymous reviewers' comments and requested to develop and submit their full papers considering the abstract review comments.

The second stage of the review

Reviewers were assigned the submitted full papers according to their expertise. The full papers were reviewed to ensure relevance to the conference theme and objectives; originality of material; academic rigour; contribution to knowledge; critical current literature review; research methodology and robustness of analysis of findings; empirical research findings; and overall quality and suitability for inclusion in the conference proceedings.

Third stage review

Authors whose papers were accepted after the second review was provided with additional anonymous reviewers' comments on evaluation forms and requested to submit their revised full papers. Evidence was required relative to specific actions taken by the authors regarding the referees' suggestions. After satisfactory evidence was provided, final papers were only accepted and included in the proceedings. To be eligible for inclusion, these papers were required to receive a unanimous endorsement by all the reviewers that the paper had met all the conditions for publication. Of 80 submissions, 32 papers were finally accepted and included in the DII-2023 conference proceedings.

At no stage was any member of the Scientific Review Panel, the Organising Committee, or the editors of the proceedings involved in the review process related to their own authored or co-authored papers. The role of the editors and the scientific committee was to ensure that the final papers incorporated the reviewers' comments and to arrange the papers into the final sequence as captured in the Proceedings.

Regards
Prof. Innocent Musonda
Chair: Scientific Programme

Peer Review Process (PRP) Confirmation

On behalf of the DII-2023 International Conference on Infrastructure Development and Investment Strategies for Africa, we confirm that the manuscripts accepted for oral presentation and publication in the Conference proceedings were blind peer-reviewed by two (2) or more technical specialists.

The reviewers were selected from the experts in the Scientific and Technical Review Committee. To be eligible for inclusion, the papers, reviewed through a three-stage review process (abstract, full paper and final paper), received a unanimous endorsement by all the reviewers that they had met all the conditions for publication. All accepted manuscripts will be published via the conference proceedings.

Regards,
Prof. Justus Agumba
DII-2023 PRP Manager
justusa@dut.ac.za
Conference website: www.diiconference.org
Email: info@diiconference.org

Acknowledgment

The Organising Committee of the DII-2023 is grateful to the Development Bank of South Africa (DBSA) for sponsoring the conference. Their invaluable contribution supports making the proceeding of the conference open access and available to readers.

We appreciate the University of Zambia, Copperbelt University, Zambia, National Council for Construction (NCC), Zambia, Centre for Applied Research and Innovation in the Built Environment (CARINBE), University of Johannesburg, South Africa, other South African, African and International universities and Institutions for supporting the conference through their valued contributions.

The contributions and exceptional support of the International Advisory and Scientific Committees, who worked tirelessly to prepare refereed and edited papers to produce these published proceedings to satisfy the criteria for subsidy by the South African Department of Higher Education and Training (DHET), is truly treasured.

We are grateful to all the Keynote speakers, authors, poster presenters, the Organizing and Scientific Committees, reviewers, and the session chairs for contributing to the success of DII-2023. Their invaluable contribution helped us achieve insightful discussions at the conference.

Conference Committees

Organising Committee
South Africa
Prof Innocent Musonda (Chairman: Scientific Programme)
Dr Thembani Moyo
Mr Adetayo Onososen
Mr Happison Muzioreva

Zambia
Dr Erastus Mwanaumo (Chairman: Technical Programme)
Mr Steve Kabemba Ngoy

Support Team
Mrs Ethel Mwanaumo – *University of Zambia, Zambia*
Miss Reneiloe Malomane – *University of Johannesburg, South Africa*
Miss Kgothatso Tjebane – *University of Johannesburg, South Africa*
Ms Muller Mildred – *University of Johannesburg, South Africa*

Scientific Committee
This committee ensured that the final papers incorporated the reviewers' comments, were correctly allocated to the appropriate theme and met the requirements set by the organisers in line with international standards for inclusion in the proceedings. They also arranged the papers into their final sequence, as the table of contents captured.
Prof Innocent Musonda (Chairman) – *University of Johannesburg, South Africa*
Prof Trynos Gumbo – *University of Johannesburg, South Africa*
Prof. Franco Muleya – *Copperbelt University, Zambia*
Prof. David Olukanni – *Covenant University, Nigeria*
Prof Justus Agumba – *Tshwane University of Technology, South Africa*
Dr Balimu Mwiya – *University of Zambia, Zambia*

Technical Review Committee
The technical review committee comprised experts from the built environment. The committee ensured that the papers were of the highest standard regarding the originality of material, academic rigour, contribution to knowledge, critical current literature review, research methodology and robustness of findings, empirical research findings, and overall quality and suitability for inclusion in the conference proceedings.
Dr Erastus Mwanaumo (Chairman) – *University of Zambia, Zambia*
Prof Chioma Okoro – *University of Johannesburg, South Africa*
Dr Chabota Kaliba – *University of Zambia, Zambia*
Dr Charles Kahanji – *University of Zambia, Zambia*
Dr Neema Kavishe – *Ardhi University, Tanzania*

The panel of Scientific Reviewers
Prof Sing S Wong – *University College of Technology, Sarawak, Malaysia*
Prof Trynos Gumbo – *University of Johannesburg, South Africa*
Prof Ephraim Munshifwa – *Copperbelt University, Zambia*
Prof. Riza Sunindijo – *University of New South Wales*
Prof Franco Muleya – *Copperbelt University, Zambia*
Prof Mbuyu Sumbwanyambe – *University of South Africa*
Prof Ayodeji O Aiyetan – *Durban University of Technology, South Africa*

Prof. Chioma Okoro – *University of Johannesburg*
Dr Sambo Zulu – *Leeds Beckett University, United Kingdom*
Dr Kenneth Park – *Aston University, England, United Kingdom*
Dr Devon Gwaba – *United States*
Dr David Oloke – *University of Brighton, United Kingdom*
Dr Danstan Chiponde – *Copperbelt University, Zambia*
Dr Charles Kahanji – *University of Zambia*
Dr Bupe Mwanza – *University of Zambia*
Mr Danstan Chiponde – *Copperbelt University, Zambia*
Dr Edoghogho Ogbeifun – *University of Johannesburg, South Africa*
Dr Lovemore Chipungu – *University of Kwazulu-Natal, South Africa*
Dr Nalumino Akakandelwa – *University of West England, United Kingdom*
Dr Mehdi Pourmazaherian – *University Technology Malaysia*
Dr Nuru Gambo – *Abubakar Tafawa Balewa University, Bauchi, Nigeria*
Dr Nadine Ibrahim – *University of Toronto, Canada*
Dr Victor Samwinga – *Northumbria University, United Kingdom*
Dr Walied H Elsaigh – *University of South Africa, South Africa*
Dr Simphiwe Gogo – *University of Johannesburg, South Africa*

DII-2023 Conference Sponsor

The Development Bank of Southern Africa is one of the Leading African Development Finance Institutions wholly owned by the government of South Africa.

DBSA's primary purpose is delivering impactful development finance solutions that ignite transformative change in South Africa and on the rest of African Continent.

Improving the quality of life in Africa is the fundamental focus of our developmental impact. DBSA aims to blend the arc of history towards shared prosperity through multifaceted investments in sustainable infrastructure and capacity development.

*Smart and Resilient Infrastructure For Emerging Economies: Perspectives on Building Better –
Musonda et al. (Eds)
© 2024 the Editor(s), ISBN: 978-1-032-56461-6
Open Access: www.taylorfrancis.com, CC BY-NC-ND 4.0 license*

DII-2023 Conference Partners

Sustainable & resilient infrastructure development

Promoting sustainable development through quality infrastructure in Elukwatini node, South Africa

N.P. Nkambule & K. Ntakana
University of Johannesburg, Johannesburg, Gauteng, South Africa

ABSTRACT: This paper investigated the role of infrastructure investment in promoting economic sustainability in Elukwatini node. The paper adopted a qualitative research approach and data collection methods such as desktop research, observations, and interviews with 22 participants consisting of Elukwatini residents, municipal officials and the node's labour force that were selected through convenience sampling. The findings revealed that the node lacks adequate economic infrastructure such as roads and bridges, telecommunication etc., it is overcrowded with informal economic activities and lacks funding. The study concluded that investing in adequate economic infrastructure in the node will result into a formalised economy, expansion of the node and interconnectedness with neighbouring areas. The study recommends the formation of Public Private Partnerships and the involvement of a development bank to fill the funding gap, investment in quality infrastructure to promote economic sustainability and to attract further economic investment into the node.

Keywords: Economic Development, Elukwatini, Infrastructure, Investment, Sustainable Development

1 INTRODUCTION

Sustainable development has become important globally in all dimensions of human life such as social, economic, and political aspects and it is at the core global policy discourse (Ayodeji & Oluwatayo 2016; Olanipekun *et al.* 2014). Glubokova *et al.* (2021) states that for rural areas sustainable development is a problem of world significance. Sustainable development is based on satisfying the developmental (social and economic) needs of human beings and ensuring that this development promotes economic growth and happens in harmony with the natural environment to conserve natural resources and protect the natural environment from developmental harm (Lordos *et al.* 2011). Sango (2022) asserts that there are three supporting dimensions to sustainable development, namely: environmental, social, and economic sustainability.

This study is based on economic sustainability which is understood to be a search for an economic system that aims to respond to the needs of inhabitants, provide adequate employment and rejuvenate the community to guarantee the continuity of services in the long term (Rodriguez-Serrano *et al.* 2017). Furthermore, Lemetti (2011) adds that economic sustainability is also about managing a community's cumulative capital to preserve it and enhance the availability of opportunities meant to improve quality of life and standards of living. In order to achieve economic sustainability, it is of paramount importance to focus on Sustainable Development Goal (SDG) 8 which aims "to promote sustained, inclusive and sustainable economic growth, full and productive employment and decent work for all" (United Nations 2013). To achieve economic growth and development it is imperative to have sustainable infrastructure in place as economic development largely depends on infrastructure. The study aims to investigate the role of infrastructure investment in promoting economic sustainability in Elukwatini node. Specifically, the study assessed the state of infrastructure in Elukwatini and the need for infrastructure investment. The following section contains a literature review on infrastructure investment and the importance of infrastructure in rural areas.

2 LITERATURE REVIEW

2.1 Infrastructure investment

The need for infrastructure is ever-rising due to population growth, migration and urbanisation which causes pressure on existing infrastructure and a demand for additional one (The Economist Intelligence Unit 2019). On the other hand, changing weather patterns, rising sea levels and natural disasters such as floods, poses threat to infrastructure. In developing countries infrastructure investment tends to favour certain areas over other. Wegren (2016) echoes the same sentiments that state policy tends to prioritise cities/urban areas over rural areas in terms of development. Like most rural areas some parts of Elukwatini also lacks access to basic infrastructure such as piped water, electricity, roads, telecommunication, sanitation, housing etc. The available infrastructure is ageing, and it lacks proper maintenance. The lack of investment in rural areas pushed the labour force out of rural areas and leads to the removal of resources such as money and raw materials to be invested in cities, resulting to a decay of rural areas. This places the sustainability and growth of rural areas at risk. As a result of a lack of economic growth and expansion in Elukwatini, the employment rate is deteriorating leading to out migrations in search of better living conditions and economic opportunities.

2.2 The importance of infrastructure in rural areas

Alonge et al. (2021) states that rural areas are an important role player in the economic development of development of developing countries, they provide raw materials necessary for production and are a source of food especially to urban areas where farming/agriculture is not suitable. Elukwatini node plays an important economic role and has potential to grow. Its growth will improve the quality of life and standards of living for the Elukwatini community and surrounding neighbourhoods. Fox & Porca (2014) opines that if rural areas had infrastructure that is competitive with the one available in urban areas, they would be positioned for rapid economic growth. The availability of quality and resilient infrastructure in rural areas can increase the productivity of businesses and allow them to operate at lower costs. The development of infrastructure is vital because it has a direct impact on the demand and supply of goods (Manggat et al. 2018). Adshead et al. (2019) states that the provision of various basic services and the development of infrastructure profoundly influence development, infrastructure is significant to sustain developmental strives. It is a prerequisite to achieving sustainable development because it influences all the 17 SDGs (Thacker et al. 2019). Infrastructure plays various roles such as creating job opportunities; reducing environmental harm and conserving natural resources (in terms of providing sustainable and resilient infrastructure); and improving quality of life and standards of living by granting the society equitable access to amenities and facilities such as health care (The Economist Unit of Intelligence 2019). Moreover, infrastructure plays a role in reducing inequality, poverty and fostering inclusion (Gurara 2018).

The Elukwatini node needs economic infrastructure. Economic infrastructure constitutes of public facilities that support economic activities, such as electricity, roads, telecommunication, water supply, irrigation systems, transportation etc. (Kaur & Kaur 2018).

Economic infrastructure is necessary to facilitate development and economic growth. It serves as an input into private sector production and increases productivity (Mesagan & Ezeji 2017). The provision of quality and resilient infrastructure in Elukwatini can expand the productive capacity of the economy and improve standards of living.

The available literature tends to only talk about rural development in terms of supporting the agricultural sector as it is deemed to be the main economic driver in rural areas. This is because all rural areas are treated the same whereas they are not the same. Chirisa *et al.* (2021) observes that rural areas vary and constitute of various characteristics, based on their proximity to cities. Remote rural areas tend to be dispersed whereas those that are located closer to cities are more integrated. Therefore, a blanket approach for rural economic development will not work since the economy in rural areas that are located closer to cities is diverse and does not only constitute of agriculture, but also ranges from retail, mining to tourism etc. As a result, it is significant to bridge the gap between infrastructure development in urban and rural areas and to balance infrastructure investment in both areas. Additionally, the economic sustainability and investment in the economic infrastructure of rural areas has been under researched.

3 METHODOLOGY

The paper adopted a qualitative research approach, and qualitative methods of data collection such as desktop research, observations, and interviews. The desktop research included collecting data from Chief Albert Luthuli Local Municipality (CALLM) spatial documents (such as the Integrated Development Plan and the Spatial Development Framework), Journal articles, books and conference papers on infrastructure. The living conditions in Elukwatini and the state of infrastructure was observed. Interviews were conducted with a sample of 22 participants, which includes: six (6) CALLM officials involved in spatial planning, service delivery and infrastructure development; ten (10) key informants who reside in Elukwatini; and six (6) people who work in Elukwatini node. Informed consent was obtained from the participants, they were requested to give written consent by signing a consent form. They were also handed an information sheet containing the details of the study and informing them that they can withdraw from the study at any time should they wish not to participate. A small sample size was selected because the municipality is relatively small and lacks capacity.

The participants were selected through non-probability sampling whereby the convenient sampling method was used to allow for the selection of readily available participants. The non-probability sampling technique may limit the generalisability of the study's findings. The data was analysed manually following inductive reasoning whereby codes were identified during data analysis. The codes were then grouped into themes for thematic analysis. The themes that emerged includes the state of infrastructure in Elukwatini and the need for infrastructure investment.

4 STUDY AREA

This paper paper is based on Elukwatini node which is located along the intersection of a main road, known as the R541 (see Figure 1). The node is located in the Elukwatini area and falls under the jurisdiction of Chief Albert Luthuli Local Municipality (CALLM). CALLM is a rural municipality that consists of one main town known as Carolina. The Elukwatini node is a fast-growing economic node and the second main service node after Carolina. Even though Elukwatini is a rural area it is unlike most rural areas that consists of dispersed development and that are dependent on agriculture to keep their economy running. The Elukwatini node is strategically located closer to tourism destination areas such as Badplaas (which offers a wide range of hotels and guesthouses with natural settings, Songimvelo Nature Reserve, kromdraai Camp Nature Reserve, Ebutsini Tourism (Located in Ekulindeni) and Steynsdorp – an agricultural village. In terms of population, Elukwatini is the most populous area in the municipality, consisting of about 41 780 people which makes 31% of the municipality's population (SDF 2017). The node's retail is one of the main drivers of the economy of CALLM, providing retail services to the residents of the surrounding neighbourhoods, the work force and to tourists which makes it important to promote economic sustainability in the node. The node is significant because it contributes towards job creation and economic growth. It attracts people from other neighbouring communities within the municipality and people from Swaziland (a neighbouring country).

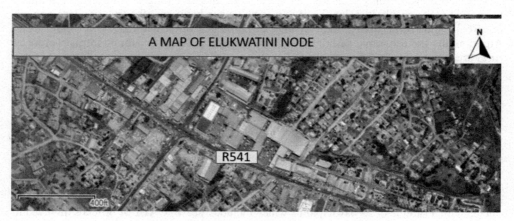

Figure 1. A map of Elukwatini node by RSA property search.

5 FINDINGS AND DISCUSSIONS

While globally the population is urbanising majority of the people in the African continent are living in rural areas where they are faced with several challenges such as the scarcity of land, water, and other resources (Chirisa *et al.* 2021). Lopez-Penabad (2022) observes that the challenges faced in rural areas are coupled with a lack of access to technology, services and infrastructure which causes desertification resulting in poor tax revenues. Additional to infrastructure backlog *"CALLM is faced with challenges such as a relatively low population, low economic growth rate, limited revenue base, informal townships and a negative economic impact resulting from COVID-19"* revealed CALLM officials. In 2021, 2 334 jobs were lost in the municipality due to COVID-19 and the municipality only contributed 2.4% to the economy of the Mpumalanga Province under which it falls (Department of Economic Development and Tourism 2022). However, in 2022 the municipality's contribution into the Province's Gross Domestic Product (GDP) increased to 7.2% (Department of Economic Development and Tourism 2022). The study observes that the limited revenue base in CALLM is due to the existence of numerous informal townships that are not billed for rendered municipal services.

5.1 *The state of infrastructure in Elukwatini*

It was observed that Elukwatini has adequate social infrastructure: there is a district hospital (Embuleni District hospital) that is accessible within a few minutes of driving from Elukwatini node, there are numerous primary and secondary schools, an old age home, and health care clinics. However, like most rural areas the area faces infrastructural challenges such as access to safe drinking water, poor network connectivity, poor road infrastructure, and insufficient trading stalls in the node. Acccording to a CALLM official *"the available infrastructure around Elukwatini node is ageing and results in high water losses and disruptions of water supply"*. There is also a need for additional housing.

Even though the municipality supplies water in the area, in other areas the community only have access to tap water on certain days of the week and others go up to a month without water, such as in Elukwatini G, revealed the residents of Elukwatini. A total of 18.4% households within CALLM lacks access to piped water as of 2016, while 3.6% households have no electricity connection (Department of Economic Development and Tourism 2022). The residents of Elukwatini also complained about network connectivity, which is a big barrier in communication, marketing, and online presence of businesses. Foster and Briceno-Garmendia (2010) states that due to power cuts and outages about 12.5% of production is lost in Africa. Poor connectivity affects businesses such as the wide range of internet cafes available in Elukwatini node and students who hope to do their school assignments from home during tertiary recess.

Elukwatini also lacks an integrated road network that links it to its surrounding neighbourhoods such as Ekulindeni and Steynsdorp. This disadvantages the people of Steynsdorp who are located on an agricultural rich land who are unable to transport their famed goods, and it also affects tourism activities taking place at Ekulindeni. I t was observed that the Elukwatini node experiences a traffic jam that is due to the traffic congestion that occurs during peak hours and at the end of the month. This occurs because the node is located at an intersection of the R541 main road which connects Elukwatini to towns such as Carolina, Ermelo and Barberton/Nelspruit with no alternative routes. The intersection also connects the node to neighbouring areas such as Tjakastaad, Mooiplaas, Aramburg and other extensions of the Elukwatini areas.

The traffic jams also occurs because some of the extensions of Elukwatini are not connected/linked to each other by means of bridges, meaning they can only be accessed by going all the way round the node's intersection, revealed a CALLM official and the residents' of Elukwatini. This presents a need for alternative roads. The road infrastructure is of poor quality, it consists of potholes, invisible to no road markings at all (see Figure 2 below). More especially around the intersection where there are invisible stop markings and no markings for pedestrian crossings, which may cause accidents.

Figure 2. Unmarked intersecting roads.

The Elukwatini node is also overcrowded by an informal economy whereby people trade in informal structure such as shacks and tents along the intersection, due to insufficient formal trading stalls. As put in the municipal IDP (2022), there are undesirable developments happening along the intersection of the main road. This informal economy is mainly retail and ranges from the sale of fast food, vegetables and even clothes, there are Salons and Babershops as well. The informal economy within the node causes informality and makes investors to be reluctant to invest in the areas. Investors tend to look at the infrastructures' ability to cope under challenges such as environmental pollution, growing population, climate change and reduction in non-renewable resources before investing (Beksultanova et al. 2021).

As mentioned previously Elukwatini is the most populous areas and it attracts people from other neighbouring communities and from Swaziland causing pressure on the existing infrastructure which according to a CALLM official currently does not meet the demand of the current population. This presents a need for housing and sustainability in infrastructure development. *"The need for additional housing in the area is depicted by the ongoing land invasions that are happening in Elukwatini and the development of housing in areas that are not suitable for habitation such as agricultural land (as is happening in the Aramburg area), river buffers and flood plains where they are exposed to floods"*, stated a CALLM official.

5.2 The need for infrastructure investment

Investing in resilient infrastructure is key to achieving sustainable development and reducing risks posed by climate change (The Economist Intelligence Unit Limited 2019). In rural areas located in the African continent, sustainable development is undermined by corruption and insufficient infrastructure (Ayodeji & Isaac 2016). Funds meant for infrastructural development are looted and resources are allocated in an unfair manner, resulting in the provision of inadequate infrastructure. Gurara et al. (2018) states that attaining sustainable development requires increased investment in economic infrastructure. Infrastructure investment is a significant determinant of economic growth, and it also plays a critical role in improving quality of life and raising standards of living for a given community (Chaurey & Trung Le 2019). As is *"CALLM is faced with a challenge of having limited funds to tackle infrastructure backlog and lacks investment"* stated a CALLM official.

The availability of basic infrastructure such as roads, clean drinking water, telecommunication and others are the main drivers of economic growth and a key to improving the well-being of a community (Manggat et al. 2018). Significant investments in infrastructure are required to meet the ever-growing need for infrastructure caused by population growth. Beksultanova (2021) states that the development of sustainable and quality infrastructure is significant in promoting economic growth because such infrastructure has limited exposure to climate risks, it has efficient return on investment and is profitable. The development and investment in sustainable and quality infrastructure in Elukwatini node will yield the following benefits: a) it will improve living conditions in Elukwatini and surrounding neighbourhoods; b) improve the competitiveness of businesses; c) reduce overall cost of production because of a reduction of costs such as transportation costs; and d) promote integrated and compact development around the node as investors will be encouraged to invest in the node due to the availability of infrastructure. Additionally, United Nations (2013) notes an increase in the durability of capital goods, higher trade, expansion of demand and supply diversification and economies of scale as contributions that are associated with the development of adequate economic infrastructure. Infrastructure investment in Elukwatini node has potential to diversify the economy of the area and attract industrial land uses, as they are currently lacking. This will ensure that the labour force who migrated out of the area due to the availability of economic opportunities will be retained.

6 IMPLICATIONS

There is a need for infrastructure investment in Elukwatini to enhance development in the node. Investing in infrastructure such as roads will relieve the node from overcrowding and traffic congestions. The infrastructure investment should cater for the development of bridges that will link Elukwatini node to its neighbouring communities, and the communities to each other. This will create job opportunities, reduce transportation costs, lead to the growth and expansion of local businesses and contribute to economic growth. Alternative routes to access the node must also be provided, as well as additional trading stalls to eradicate the existing informality and overcrowding in Elukwatini node. The provision of the road infrastructure will encourage further development in the area and attract investment, more especially in tourism.

Belsultanova *et al.* (2022) states that economic participants such as the government, businesses and the financial sector must be involved and take part in achieving sustainable development, with the government assuming the role of regulating development, the financial sector providing cash injection to economic activities and businesses implementing in practice the achievement of sustainable development. For example, by limiting risks posed by production to the natural environment in order to protect the environment from harm. The municipality currently lacks funding as its available funding is limited. The municipality currently relies on grants from National Treasury to remain operation. As a result, the municipality can consider Public Private Partnerships (PPPs) and the Development Bank of Southern Africa (DBSA) for financing the development of infrastructure.

The formation of PPPs can enhance investment in infrastructure and improve its efficiency by integrating maintenance and operation with the actual construction of infrastructure (Zurich 2017). United Nations (2013) notes that PPPs are value drivers because they provide value for money due to the accountability and transparency associated with the partnership in terms of managing funds. PPPs also presents an opportunity for increased revenue collection since users can be charged for services. Furthermore, capacity constraints in infrastructure development and operation can be resolved through PPPs. Turkey & Uzsoki (2018) reported that the public should be involved in the financing of infrastructure projects as this adds to improved social acceptance of the projects and can prevent deterioration by creating a sense of ownership. This will promote the involvement of the public in taking care of the infrastructure and prevent vandalism. They should be involved throughout all the stages of construction projects, from planning to maintenance. This could also create employment opportunities.

The government should move its focus from the quantity of developed infrastructure into its quality and maintenance. Refurbishment and maintenance of infrastructure should be prioritised in order to increase its life span. For additional revenue, it is recommended for the municipality to start a process of formalising informal townships so that it can be able to collect more revenue, as most townships are currently not paying rates. It is suggested that the municipality use some of its vacant properties to develop sustainable social housing to generate revenue from rentals and fill the housing need gap. Investing in the development of sustainable social housing can transform the area's aesthetic form, attract further investment and reduce out migrations to curb the decay of the area.

The paper also recommends that the municipality device a strategy for infrastructure maintenance around Elukwatini node, and for repairing and rehabilitating existing infrastructure. Such as regular maintenance of the road network which will include ensuring that road markings always stay visible. After additional trading stalls have been provided, a strategy to prevent further development of informal stalls is also needed. To encourage the protection of the natural environment there is a need for the development of recycling station to encourage the reuse of materials needed for production. The municipality can also consider investing in solar energy within the node, since Elukwatini is characterised by a warm and temperate climate.

7 CONCLUSION

In conclusion, the paper has indicated that sustainable development is an important concept globally and is at the core of policy discourse. At the heart of achieving sustainable development is infrastructure which influences all 17 SDGs. Infrastructure plays an important role in promoting economic development, protecting the natural environment from harm (if sustainable infrastructure is developed), improving quality of life and standards of living. In developing countries governments tend to neglect rural areas and prioritise urban areas in terms of infrastructure investment. As a result, rural communities often lack access to adequate infrastructure and other essential basic services. Unlike other rural areas Elukwatini node is well-located and it contributes to the economic development of Chief Albert Luthuli Local Municipality. Therefore, there is a need to invest in the node's infrastructure in order to attain economic sustainability. Infrastructure investment will attract further development into the nide and promote the notable growth of the node and local businesses. When there is improved rural development, capital investment is likely to flow into rural areas (Fox & Porca 2013).

REFERENCES

Adshead, D., Thacker, S., Fuldauer, L.I & Hall, J.W. 2019. Delivering on Sustainable Development Goals Through Long-term Infrastructure Planning. *Global Environmental Change* 59(2019):101975.

African Monitor. 2012. *Rural Infrastructure in Africa, Unlocking the African Moment*. Development support monitor paper series no. 1.

Alonge, O., Lawal, T. & Akindiyo, O. 2021. Addressing the Challenges of Sustainable Rural Infrastructure Delivery in Nigeria: Focus on Ondo North Senatorial District. *International Journal of Academic Research in Business and Social Sciences* 11(2):619–631.

Ayedeio, O.O. & Oluwatayo, I.B. 2016. Drivers and Challenges of Sustainable Development in Africa. *3rd International Conference on African Development Issues (CU-ICADI 2016)*. ISSN 2449-075X.

Beksultanova, A.I., Gaisumova, L.J. & Sdueva, M.A. (2021). The Role of Infrastructure in Sustainable Development. *SHS Web of Conferences*.

Bowen, G.A. 2009. Document Analysis as a Qualitative Research Method. *Qualitative Research Journal*. 9(2): 27–40.

Chaurey, R. & Trung Le, D. 2019. *Rural Infrastructure Development and Economic Activity*.

Chief Albert Luthuli Municipality. 2017. *Spatial Development Framework (SDF)*.

Chief Albert Luthuli Municipality. 2022. *Integrated Development Plan* (IDP).

Chirisa, I., Matai, J. & Mutambisi, T. 2021. *No Sustainability Without Planning for It: Scope and Dimensions for Sustainable Rural Planning in Zimbabwe*. From the edited volume of sustainable rural development by Prof. Dr. Orhan Ozcatalbas.

Department of Economic Development & Tourism. (2022). PopwerPoint Presentation on Relevant Socio-economic Perspectives of Chief Albert Luthuli Municipality. *Presented at a Strategic Spatial Session on the 10th March 2023*.

Foster, V.C. & Briceno-Garmendia. 2010. *Africa's Infrastructure: A Time for Transformation*. A publication of the international bank for reconstruction and development, the World Bank.

Fox, W.F. & Porca, S. 2014. *Investing in Rural Infrastructure*.

Glubokova, L.G., Kokhanenko, D.V. & Lunika, E.V. (2021). Indicators of Sustainable Development Goals in Rural Territories of Russia. *IOP Conference Series: Earth and Environmemtal Science*. 670.

Gurara, D., Klyvuev, V., Mwase, N. & Presbitero, F. 2018. Trends and Challenges in Infrastructure Investment in Developing Countries. *International Development Policy*. Articles, 10.1.

Heard, R., Hendrickson, C. & McMichael, F.C. 2012. Sustainable Development and Physical Infrastructure Materials. *Materials Research Society*. 37:289–394.

Kaur, A. & Kaur, R. (2018). Role of Social and Economic Infrastructure in Economic Development of Punjab. *International Journal of Innovative Knowledge Concepts*. 6(5).

Lemetti, Y.A. (2011). *Strategy for Sustainable Development of Agriculture of the Region: Theory and Practice*. Publishing House of Tver State Agricultural Academy: Russia.

Lopez-Penabad, M.C., Iglesias-Casal, A. & Rey-Ares, L. 2022. Proposal for a Sustainable Development Index for Rural Municipalities. *Journal of Cleaner Production*. 357(2022):1–17.

Lordos, A., Sonan, S. & Solar, A. 2011. *Navigating the Paradigm Shift: Challenges and Opportunities for the Two Communities of Cyprus, in the Search for Sustainable Patterns of Economic and Social Development*. A report by the Cyprus 2015 initiative.

Manggat, Z., Zain, R. & Jamaluddin, Z. 2018. The Impact of Infrastructure Development on Rural Communities: A Literature Review. *International Journal of Academic Research in Business & Social Sciences* 8(1):637–648.

Mesagan, E.P. & Eziji, A. (2017). *The Role of Social and Economic Infrastructure in Manufacturing Sector Performance in Nigeria*. MPRA Paper No. 78310.

Mohamed, H., Judi, M.H., Noor, M.S.F. & Yusof, M.Z. 2012. Bridging Digital Divide: A Study on ICT Literacy Among Students in Malaysian Rural Areas. *Australia Journal of Basic and Applied Sciences* 6 (7):39–45.

Nazeen, S., Hong, X.Ud Din. & Jamil, B. 2021. Infrastructure-driven Development and Sustainable Development Goals: Subjective Analysis of Residents' Perception. *Journal of Environmental Management* 294(2021):112931.

Ojo, A.O. & Oluwatayo, I.B. 2016. Drivers and Challenges of Sustainable Development in Africa. *Third International Conference on African Development Issues.*

Okoli, I.K., Okonkow, S.M. & Chinenye, M.M. 2020. Rural Infrastructure and Sustainable Development in Nigeria: *FUTY Journal of the Environment* 8(1):80–92.

Olanipekun, A.O., Aje, I.O & Awodele, O.A. (2014). Contextualising Sustainable Infrastructure Development in Nigeria. *FUYT Journal of Environment.* 8(1).

Rodriguez-Serrano, I., Caldes, N., De La Rua, C., Lechon, Y. & Garrido, A. (2017). Using the Framework for Integrated Sustainability Assessment (FISA) to Expand the Multiregional Input-output Analysis to Account for the Three Pillars of Sustainability Environmental Development. *Sustainable.* 19: 1981–1997.

Sango, N. (2022). *An Analysis of the Role of Rural Development Initiatives in Promoting Sustainable and Effective Development in the Eastern Cape: A Case Study of Instsika Yethu Municipality*. Master's in public administration Thesis, Stellenbosch University.

Thacker, S., Adshead, D., Fay, M., Hallegatte, S., Harvey, M., Meller, H., O'Regan, N., Rozenberg, J., Watkins, G. & Hall, J.W. 2019. Infrastructure for Sustainable Development. *Nature Sustainability.*

The Economist Intelligence Unit Limited. 2019. *The Critical Role of Infrastructure for Sustainable Development Goals.* Supported by UNOPS.

Turley, L. & Uzsoki, D. 2018. *Financing Rural Infrastructure: Priorities and Pathways for Ending Hunger.* International Institute for Sustainable Development.

United Nations. (2013). Supporting Infrastructure Development to Promote Economic Integration. The role of the Public & Private Sectors. *United Nations Conference on Trade and Development.*

Wegren, S.K. 2016. The Quest for Rural Sustainability in Russia. *Sustainability* 7(8):602.

Zurich. (2017). *Challenges of Our Nation's Aging Infrastructure. The Pressing Need to Maintain Critical Systems.*

Developing a sustainable water supply system for rural South African communities

N.N. Ngema, S.L. Mbanga, A.A. Adeniran & E. Kabundu
Nelson Mandela University, Gqeberha, Eastern Cape, South Africa

ABSTRACT: The study sought to examine shortcomings in the provision of clean water to rural settlements and how these affect attainment of sustainable development goals, and thereby propose a model which integrates indigenous and modern water supply methods to overcome water scarcity in rural areas. Applying a mixed method methodology within a positivist paradigm, data was gathered through document reviews and self-administered survey questionnaires for households, and structured interviews for other stakeholders. Findings indicated that inadequate access to reliable and safe water hinder sustainable development. Water supply in rural areas is affected by lack of resources, lack of funding, illegal connection, political instability, poor operation and maintenance, lack of capacity and skills. It is recommended that the enhancement of innovative technologies and modern approaches to maintain water systems be accepted and blended with traditional knowledge that exists to make them more appealing and valuable for the intended purpose.

Keywords: Rural, Water, Supply, Resources, Model

1 INTRODUCTION

A quarter of the world's population across seventeen (17) countries are living in regions of tremendously high-water stress, with a high measure of the level of competition over water resources (Food and Agriculture Organization of the UN 2013:17). Sub-Saharan Africa suffers from frequently overstrained water systems from rapid growth of urban areas (Jacobsen *et al.* 2012:16). Northern India encounters severe groundwater depletion, envisioned on Aqueduct's maps, and encompassed in calculations of water stress for the first time.

Qatar encounters an enormously high level of water strain owing to its lack of natural renewable water resources and high rates of water consumption (Al-Ansari *et al.* 2014:438). These elements, combined with the absence of permanent rivers in Qatar have led to heavy reliance on groundwater resources, mainly groundwater aquifers, for agricultural irrigation (Blazev 2016:719). In developing nations, inadequate water and sanitation conditions contribute to a staggering 80% of infections (Alirol *et al.* 2011:134). Water strains affect women and girls as they are regularly the primary managers of natural resources, specifically for household use and small-scale agriculture and it alters their health safety and opportunity to engage in economic activities (Prihatiningtyastuti *et al.* 2017:71).

1.1 *Background*

South Africa has a population of 59,197,704 people with 67 percent of the population living in urban environments and 33 percent living in rural settlements (Anyangwe & Mtonga 2007:97). Currently, South Africa has access to surface water (77% of total use), groundwater (9% of total use), and recycled water (14% of total use). Conversely, the population's dependence on water is not equally distributed. Rural Settlements lacks water infrastructure, as a result, 74 percent of all rural people are entirely dependent upon groundwater such as local wells and pumps. Presently, 19 percent of the rural population lacks access to a reliable water supply and 33 percent do not have basic sanitation services, over 26 percent of all schools both in urban and rural settlements, and 45 percent of clinics also have no access to water supply (Hedden & Cilliers 2016:06).

South Africa's policy of Free Basic Water Access, guaranteed by the constitution of the Republic of South African (1996), states that every citizen is entitled to a certain amount of water regardless of that person's ability to pay for it; this policy defines the amount of entitlement be 6000 litres per household per month. Meanwhile rural areas lack water monitoring devices, this has become a challenge for the organization in charge of water allocation, the South African Department of Water Affairs and Forestry (SADWAF), to effectively determine the amount of water people use per month in rural areas. The human right to water incorporates both freedoms and entitlements (Shikwamabane 2017:22).

The objective of this research is to develop a comprehensive model that integrates indigenous and modern water supply systems in rural areas of South Africa. The aim is to enhance rural development and address the water crisis prevalent in poverty-stricken rural regions, ultimately promoting sustainable development. In what way(s) can indigenous and modern water supply system be integrated in rural areas for sustainable development in South Africa?

1.2 *Policies and regulations*

Water Sector in South Africa is led and regulated by the Department of Water and Sanitation (DWS). DWS is regulated by two Acts, the National Water Act (1998) which is part of National Environmental Management Act (NEMA), 107 of 1998 and the Water Services Act (1997), and together with national strategic objectives, governance, and regulatory frameworks, provides an enabling environment for effective water use and management in the country. National Water Act 36 of 1998 ensures that water resources are safeguarded, used, developed, conserved, managed, and controlled in a justifiable, efficient and equitable manner by establishing suitable institutions (Department of Water Affairs 2015:20). Water Services Act 108 of 1997 affords the right of admission to basic water supply and the right to basic sanitation mandatory to secure sufficient water and an environment not harmful to human health and well-being (Department of Water Affairs 2014:02).

National Water Policy Review (2013) aims and focuses on incapacitating the water challenges of faced by the Department of Water and Sanitation and the whole of South Africa in order to improve access to water, efficiency, equity and sustainability (Department of Water Affairs 2015). National Water Resources Strategy Second Edition (2013) clarifies how water supports development eliminates poverty and inequality; the contribution of water to job creation and economy development; and ways in which water is conserved, protected, developed, used, controlled, and managed sustainably and equitably. The fundamental focus of the strategy is to ensure equitable and sustainable access (Department of Water Affairs 2013:12).

2 LITERATURE REVIEW

Water resource systems have benefited both people and their economies for many centuries. However, in many regions of the world, basic drinking water and sanitation needs is still not achieved. Nor do many of the water resource systems support and sustain a resilient biodiverse ecosystem (Lane *et al.* 2017:13). From an African standpoint, water is not merely for economic and social importance, but as well for cultural and spiritual significance (Kapfudzaruwa & Sowman 2009:683).

2.1 *State of water supply infrastructure*

South Africa is extremely dependent on storage reservoirs to preserve dependable water supplies in times of water stress. Dams and rivers serve as a major water storage. Most (60%)

of the large dams in Africa are located in South Africa and Zimbabwe. South Africa has more than 500 large dams, of which 50 have a storage capacity exceeding 100 million m3. The principal purpose of dams in South Africa is for irrigation and urban and industrial water supply (Department of Water Affairs 2015:13). The scarcity of water is compounded by pollution of the surface- and ground-water resources. The state of rivers and dams is an apprehension as some rivers have excessively high levels of pollution (Ali 2011:196). Groundwater is also significantly used in South Africa, predominantly in the rural and more waterless areas. On the other hand, it is restricted by the geology of the country, greatly of which is hard rock, while large porous aquifers occur only in a few areas. South Africa's groundwater resources presently supply about thirteen percent (13%) of the total volume of water consumed nationally (Pietersen *et al.* 2011:26).

2.2 Indigenous knowledge of water management

Indigenous methods of water resource management and irrigation methods differ from canal, pond and well digging to cultivation of low adaptive crops (Mujtaba *et al.* 2018:24).
Examples of indigenous practices for water harvesting and management:

- Rainwater Harvesting.
- Gelesha technique incorporates soil conservation and water harvesting.
- Stone terracing – the enclosure of certain segments of lands by boulders & stones for water-flow management.
- Homestead ponds.
- Contouring or the construction of contour ridges.
- The Saaidamme, which is Afrikaans for 'planting dams'.
- 'Klipplaate en Vanggate' in Afrikaans can be interpreted as 'paved-rock and catchpits'.

2.3 Modern water supply network design and dimension

In modern way, distribution networks are used to supply communities with suitable quality and quantity of water. There are four network types: dead end, gridiron, circular and radial systems. Networks are a system of pipes and trenches supplying the suitable quantity and quality of water to a community. The network construction and layout are cautiously arranged in order to ensure enough pressure and ensure hygienically safe water (Charlesworth & Booth 2016:15).

2.4 Causes of water scarcity

Water scarcity encompasses water stress, water crisis, water shortage or water deficit. Water scarcity can be cause by physical water scarcity and economic water scarcity. Physical water scarcity refers a situation where natural water resources are incapable of meeting a region's water demand. Economic water scarcity is a result of poor water management resources (Chartres & Varma 2010:45). Water scarcity is caused by the following: Drought, Climate change and variability, Overuse and misuse of water, and Pollution of Water.

2.5 Effect of water scarcity on women

In the African setting, African rural women are primarily affected by scarcity of water. Women in rural areas are more vulnerable compared to men because they are more responsible for water related duties (Pahwaringira *et al.* 2015:67). Women must travel longer distances to fetch water when there is water scarcity. In some cases, young girls help their mothers as a result, they must drop out of school, and this has some bearing on low literacy levels among women. Water sources in rural areas are often located far from the reach of most women and an average of 5km is travelled to reach the water source (Ngarava *et al.* 2019:04). This distance increases during the dry season when most springs and wells drying up, leaving women with no other alternative but to walk further distances in search of water. During the dry seasons, women's nights become shortened as they wake up in the early hours of the morning to go and search for water (Sigenu 2006:52). Women have specific hygiene requirements in relative to menstruation, pregnancy and childbirth, and lack of water affords certain discomfort and distress. The lack of clean water and sanitation also escalates the risk of infections and diarrheal diseases for everyone (Daily Maverick 2018).

2.6 *Effect of water scarcity on education in the rural areas*

Education is a crucial tool for breaking the cycle of poverty, however over half of the world's schools lack access to safe water and sanitation facilities. Absence of clean water has severe effects on students' academic performance and attendance rates (Jasper *et al.* 2012:2779). Lack of safe water can cause students to lose momentum as they deal with stomach pains and diarrhea from disease and hunger. Students miss class to go fetch water, or to care for sick parents or siblings. The water scarcity issue is twice hard for girls. With the lack of proper toilets in school, girls drop out once they reach puberty. Further, the society has imposed responsibility of fetching water to women thus limiting their access to both education and business opportunities (Prihatiningtyastuti *et al.* 2017:82). Lifewater (2014) added that, over half of the girls in sub-Saharan Africa who drop out of primary school as an outcome of poor water and sanitation facilities. Education and water are intertwined with the challenges of gender equality. Girls miss school as they spend time gathering water for the family. Water is linked to health; millions of children get sick and die every year from water-borne diseases and also lack of basic sanitation and hygiene. Moreover, students utilise significant class time to fetch water far from the school. Sometimes the roads are dangerous for children (Agol & Harvey 2018:286).

3 RESEARCH METHODOLOGY

This study is primarily motivated by a research problem and is guided by a pragmatic research philosophy that combines positivist and interpretivist approaches. The study encompasses both epistemological and ontological assumptions, which pertain to the understanding of human knowledge and the nature of reality in research. The research problem aims to investigate a model that integrates indigenous and modern water supply systems in rural areas to foster rural development and address water scarcity in impoverished rural regions, with the ultimate goal of achieving sustainable development in South Africa.

To gain a comprehensive understanding of the research question, a mixed research method was employed. Data were collected through the utilization of survey questionnaires and structured interviews as research instruments, with a focus on addressing the sub-problems identified. Field observations were also conducted as part of the data collection process. The target population of the study consisted of households residing in the four selected rural settlements, agriculture officers from the two chosen local municipalities, local business owners, community healthcare executive officials, community school executives, and municipal officials.

A non-probability sampling approach, specifically convenience sampling, was utilized to select the other stakeholders involved in the study. As for the households, a probability convenience sampling method was employed, taking into consideration the availability of adult participants during the data collection period.

4 STUDY FINDINGS AND DISCUSSION

The identified rural areas depend mostly on groundwater as a source of water supply, thus mostly used borehole as a formal water supply (see Table 1 below). Physical evaluations on the rural settlements also identified a number of sources such as wells, dams, springs, rivers and streams, and these were also used by communities for their domestic water needs. Furthermore, the infrastructure identified included reservoir, reticulation networks, communal-taps and boreholes pumps. As per the community participants, they have yard-taps as well but barely have running water (see figures below).

Table 1. Types of water sources and the level to which each rural settlement depends on.

Areas	Ground water	Surface water
Dundonald	>75%	<25%
Betty's Goed	83%	17%
Hlankomo	>87%	13%
Mndeni	91%	9%

Figure 1. Types of water infrastructures in the areas.

Technical assessments as well as discussions with community members also uncovered that springs are the main sources of water supply in some areas as they are mostly reliable and accessible water sources (refer to Figure 1). According to the respondents, rivers and dams are not reliable sources of water supply as they dry out in winter seasons, and water is always dirty on rainy days (Figure 3). The community members also outlined that they use rivers and streams for their livestock, and other household use. Furthermore, discussions further highlighted that while boreholes are available, they are not always operational. Through group discussions it was discovered that existing boreholes are unable to meet the demand in the selected communities due to low yield and lack of servicing. For this reason, communities such as those in Mndeni and Hlankomo are using springs and wells to augment the supply of water for their basic needs, additionally community members have to buy water from the households with private boreholes. Communities such as Dundonald and Betty's Goed also mentioned that they sometimes get water from water tank trucks, which is supplied by the municipality but sometimes they buy it from private supplies.

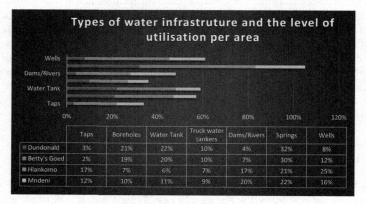

Figure 2. Available and accessible water infrastructure per area.

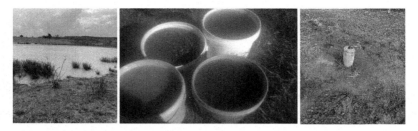

Figure 3. Water from dam/river.

Interview with the schools and health care officials in the selected areas discovered that both the schools and health care centres have taps fixed in their yards, but they mostly rely on the water tanks that are used as back-ups. Some schools use borehole drills to keep the tanks full since they hardly have running tap water, but some utilise water tank trucks to fill up their tanks at least once a week. Dundonald health care centre runs for 24 hours, meaning water scarcity weighs heavily on them. In all the identified areas, local business owners, especially those that depend on water for operation such as salons indicated that water scarcity has negatively impacted on their income. The local business owners utilise alternative means of accessing water, which has become part of their expenses as some have hired people to collect water for them. The agricultural office emphasised that they rely on ground water, which is a common practice in the agricultural sector.

4.1 *Challenges for water supply in the study areas*

Lack of capacity and skills at municipalities: Water supply in rural communities continues to be a major concern. Rural areas such as Dundonald and Betty's Goed majority stated that more than four weeks would pass, and communities would be without running water (see Figure 4 below). This problem is caused by several factors such as the availability of operators and the municipal officials responsible for operation and management of public water supply. In Dundonald the borehole had not been functional for more than four years. Community members also reported that it has been over five years since their taps has been broken. During the interview with the officials, it was outlined that this is caused by lack of maintenance and communication between the officials and the communities. Chief Albert Luthuli Local municipality located in Mpuluzi in which both Betty's Goed and Dundonald falls in, reported that they lack operators onsite that should be available at the time the infrastructure is broken, thus it takes longer to fix and have it operational on time as they have to wait for technicians to be sent from their main office at Nelspruit, and this often lead to inconsistent water supply. Due to this reason, households that could not afford to buy water from water tank trucks turn to use untreated water from rivers, wells and springs.

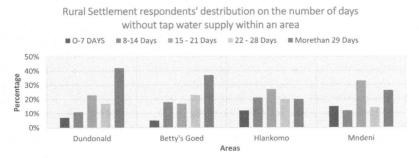

Figure 4. Number of days without tap water (Per respondents' population distribution).

Poor operation and maintenance of infrastructure: In all the rural settlements approached for this study there was some infrastructure in place which ranged from borehole pumps, reservoir, reticulation pipes, street taps and others. However, most of the infrastructure in these settlements are poorly maintained. Some of the infrastructure problems included leakages (reservoirs, pipes, and taps) either due to damage or aging. In some areas tap heads were damaged or stolen (see Figure 5 below). The poorly maintained infrastructure affects the water supply as they result to lot of water losses during the distribution period. Poor operation and maintenance of boreholes is a major problem in the selected areas as boreholes would have a significant reduction in terms of their expected yields. Hence, other boreholes fail to meet the expected demand.

Figure 5. Poorly maintained wate infrastructure.

Illegal connections: Officials outline that they identified illegal connections which affect the supply of water in rural areas. Illegal connections affect the pressure of water in the pipelines, either to or from the reservoir, as some piped water get lost along the way, which enables the reservoir to get full. For example, in some areas under Hlankomo, there are identified illegal connections on the feeder pipes to the reservoir which challenge the ability of the borehole to fill-up the reservoir.

Limited resources: Officials reported that Rural municipalities often have limited resources and lack the financial resources to invest in water infrastructure, making it difficult to provide safe and reliable water services to rural communities. Water resources are not distributed equally across South Africa, with some areas having abundant water resources while others have very limited access to water. This has contributed to disparities in water supply between urban and rural areas. Political interference in municipal operations in the development of infrastructure is a further operational issue that affects delivery of water sector infrastructure. Such interference results in funds allocated for water infrastructure development not being appropriately spent.

5 CONCLUSION AND RECOMMENDATIONS

The inadequate access to sustainable water in the rural areas directly affects people within those settlements and other departments such as health care community centres, schools and agriculture. People in rural areas mostly depend on agriculture to make a living. Lack of water supply directly affects sanitation as well. It is impossible to achieve a clean and sanitary environment without water. The progress of sanitation throughout the world has been closely associated with the availability of water. Amongst developing countries, water insecurity is rising, with the poorest and most vulnerable particularly at risk. Global water demand is increasing at approximately 1% per annum, whilst between 4.8 and 5.7 billion people are estimated to live in areas that are possibly water scarce for one month per year by 2050 (Cooper 2020:02). Regardless of the world's progress, billions of people still lack safe water, sanitation and handwashing facilities. Extra effective utilisation and management of water are crucial to addressing the rising demand for water, pressures to water security and the increasing frequency and harshness of droughts and floods stemming from climate change.

As water demand continues to rise in future, an increasing number of cities will face challenges of managing scarcer and less reliable water resources in an efficient way. Reliable drinking water and suitable sanitation services are fundamental to human health. Water is amongst the critical substances on earth that are required for the survival of plants, animals, and human beings. Benefits of water to society include health and hygiene, clean water for drinking is vital for functioning of human body as the bodies needs to be suitably hydrated so that the organs function properly to keep human in good health.

5.1 *Recommendations*

The integration of indigenous and modern water supply systems offers a range of advantages, such as augmenting water availability, enhancing water quality, and bolstering community resilience in the face of climate change. To achieve this integration effectively, the following recommendations are proposed:

1. Knowledge sharing and collaboration: Promote and facilitate the exchange of knowledge and collaboration between indigenous water practitioners and experts in modern water supply systems. Emphasize the importance of fostering dialogue, joint planning, and inclusive decision-making processes to effectively merge the strengths and best practices of both systems.
2. Incorporate traditional infrastructure: Assess the feasibility of incorporating traditional infrastructure, such as indigenous water storage methods or traditional water harvesting techniques, into modern water supply systems. This may involve retrofitting existing infrastructure or designing new systems that integrate indigenous elements.
3. Adaptation of modern technologies: Categorise modern technologies that can complement and enrich indigenous water systems. For instance, the utilisation of solar-powered pumps, filtration systems, or rainwater harvesting systems can be integrated with existing indigenous practices to improve water availability and quality.
4. Community-led initiatives: Empower local communities to take an active role in managing and maintaining the integrated water systems. Encourage community ownership through training programs, capacity building, and the establishment of community water management committees.
5. Involve local communities: It's essential to involve local communities in the integration process to ensure that their needs and concerns are considered. Engage community leaders, elders, and water experts to provide valuable insights and guidance.
6. Conduct a needs assessment: A needs assessment can help identify the gaps and opportunities for integrating indigenous and modern water supply systems. It's essential to consider factors such as water quality, quantity, accessibility, and cultural preferences.
7. Policy and regulatory support: necessitate the development of policies and regulations that recognize and support the integration of indigenous and modern water systems. This incorporates confirming that legal frameworks deliberate and accommodate indigenous water rights and practices, as well as stimulating the utilisation of sustainable and inclusive water management practices.
8. Monitoring and evaluation: Implement monitoring and evaluation mechanisms to evaluate the performance and impact of the integrated systems. Frequently observe water quality, availability, and community satisfaction to identify any challenges or areas for enhancement.
9. Sustainable financing and investment: Obtain sustainable financing options and secure investments for the integration of indigenous and modern water systems. This can involve leveraging government funding, international support, or exploring innovative financing models that promote long-term sustainability.

By integrating indigenous and modern water supply systems, the departments can create a sustainable and resilient water supply that meets the needs of local communities and the environment. By implementing these practical methods, rural areas can effectively integrate indigenous water systems with modern water supply systems, resulting in improved water access, quality, and sustainability for the communities involved.

REFERENCES

Agol, D. & Harvey, P., 2018. Gender Differences Related to WASH in Schools and Educational Efficiency. *Water Alternatives*, 11(2), pp. 284–296.

Al-Ansari, N., Knutsson, S., & Ali, T. S. (2014). Qatar's Water Resources Management and Sustainability: Status and Future Prospects. *Journal of Water Resource and Protection*, 6(5), pp. 431–461.

Ali, H., 2011. *Practices of Irrigation & On-farm Water Management: Volume 2*. New York: Springer Science & Business Media.

Alirol, E., Getaz, L., Stoll, B., Chappuis, F. and Loutan, L., 2011. Urbanisation and Infectious Diseases in a Globalised World. *The Lancet Infectious Diseases*, 11(2), pp.131–141

Anyangwe, S.C. and Mtonga, C., 2007. Inequities in the Global Health Workforce: The Greatest Impediment to Health in Sub-Saharan Africa. *International Journal of Environmental Research and Public Health*, 4(2), pp.93–100.

Blazev, A. S., 2016. *Global Energy Market Trends*. Lilburn: The Fairmont Press, Inc.

Charlesworth, S. M. & Booth, C. A., 2016. *Sustainable Surface Water Management: A Handbook for SUDS*. United Kingdom: John Wiley & Sons.

Chartres, C. & Varma, S., 2010. *Out of Water: From Abundance to Scarcity and How to Solve the World's Water Problems*. New Jersey: FT Press.

Cooper, R., 2020. *Water Security Beyond Covid-19*, United Kingdom: University of Birmingham.

Daily Maverick, 2018. *Gendered Impact of Water Deprivation Must be Addressed Must be Addressed*. [Online] Available at: https://genderjustice.org.za/article/gendered-impact-of-water-deprivation-must-be-addressed/ [Accessed 23 April 2020].

Department of Water Affairs (DWA), 2013. *National Water Resource Strategy: Water for an Equitable and Sustainable Future*. Pretoria: Department of Water Affairs.

Department of Water Affairs, 2014. *Water Services Act 108 of 1997*. Government Gazette, 390(18522).

Department of Water Affairs, 2015. Strategic Overview of the Water Services Sector in South Africa 2015. *National Water Policy Review (2013)*, 20 January.

Food and Agriculture Organization of the UN, 2013. *The State of the World's Land and Water Resources for Food and Agriculture*. Milton Park: Routledge.

Hedden, S. and Cilliers, J., 2014. Parched Prospects-the Emerging Water Crisis in South Africa. *Institute for Security Studies Papers*, 2014(11), p.16

Jacobsen, M., Webster, M. & Vairavamoorthy, K., 2012. *The Future of Water in African Cities: Why Waste Water?*. Washington: World Bank Publications.

Jasper, C., Le, T. T. & Bartram, J., 2012. Water and Sanitation in Schools: A Systematic Review of the Health and Educational Outcomes. *International Journal of Environmental Research and Public Health*, 9(8), pp. 2772–2787.

Kapfudzaruwa, F. & Sowman, M., 2009. Is There a Role for Traditional Governance Systems in South Africa's New Water Management Regime. *Water SA*, 35(5), pp. 683–692.

Lane, A., Norton, M. & Ryan, S., 2017. *Water Resources: A New Water Architecture*. United Kingdom: John Wiley & Sons.

Lifewater, 2014. *Water and Education: How Safe Water Access Helps Schoolchildren*. [Online] Available at: https://lifewater.org/blog/water-education/ [Accessed 28 April 2020].

Mujtaba, I. M., Majozi, T. & Amosa, M. K., 2018. *Water Management: Social and Technological Perspectives*. Parkway: CRC Press.

Ngarava, S., Zhou, L. & Monde, N., 2019. Gendered Water Insecurity: A Structural Equation Approach for Female Headed Households in South Africa. *Water*, 11(12), pp. 2–19.

Pahwaringira, L., Chaminuka, L. & Muranda-Kaseke, K., 2015. The Impacts of Water Shortages on Women's Time-Space Activities in the High-Density Suburb of Mabvuku in Harare. wH2O: *The Journal of Gender and Water*, 4(1), pp. 65–76.

Pietersen, K., Beekman, H. E. & Holland, M., 2011. *South African Groundwater Governance Case Study, Gezina*: Water Research Commission.

Prihatiningtyastuti, E., Dayaram, K. and Burgess, J., 2020. Women's Role in Water Management: A Tale of Two Villages. In *Developing the Workforce in an Emerging Economy* (pp. 68–82). Routledge.

Sigenu, K., 2006. *The Role of Rural Women in Mitigating Water Scarcity*, Bloemfontein: University of the Free State.

Sustainable community-driven urban regeneration of recreational infrastructure in former mine townships of the Copperbelt Province

L. Makashini & E.K. Munshifwa
Copperbelt University, Kitwe, Zambia

Y. Adewunmi
University of Witwatersrand, Johannesburg, South Africa

ABSTRACT: The aim of this paper is to explore ways in which communities can assist in the redevelopment of formerly mine-owned recreational facilities in the Copperbelt Province. Since the privatization of mining conglomerate Zambia Consolidated Copper Mines (ZCCM) in the early 1990s, this recreational infrastructure has remained in a management quandary resulting in disuse, neglect, disrepair and decay, and contributing to blighted neighborhoods. The study used surveys, focus group discussions and interviews for data collection. The study concluded that communities can lead the process of regeneration, as evidenced by the fact that most of the current activities in these facilities are being run by them, however this first requires a clear institutional and policy framework from government (local or central). It was clear that community-led initiatives lack financial resources and organizational structure. A complete institutional structure of the public, private and communities is therefore necessary for a successful regeneration initiative.

Keywords: sustainable urban regeneration, social cohesion, redevelopment, recreational infrastructure, Zambia

1 INTRODUCTION

Urban decay is a "natural" process for aging towns and cities (Alade *et al.* 2021; Andersen 2019), however remedial actions result in rejuvenation of these areas. Sustainable urban regeneration is therefore a process by which consolidated cities' actions result in renewing the dead parts of the city. Evidence from former mine townships revealed dying recreational facilities which could rightly be referred to as brownfields and up for regeneration. With the change in political administration in 1991, so came the change in economic orientation from communist to capitalist systems of production. The new Movement for Multi-Party (MMD) administration's decision was therefore to privatize all formerly State Owned Enterprises (SOEs) including the Zambia Consolidated Copper Mines (ZCCM) (Fraser and Lungu 2005; Kaonga & Nguvulu 2015); the mining conglomerate at the time and also the owners of the recreational centers in mining towns. However, the new mine owners were mostly interested in productive assets, smelters and copper ore bodies, and less in social assets. The result has been that since the early 1990s, these facilities have fallen into a management quandary with no institution directly responsible for their management. This has resulted in many of them falling into disrepair and decay. The aim of this paper is therefore to explore the role that local communities can play to fill up this management gap. A preliminary survey of the facilities revealed that most of them are now used by churches for their religious functions.

The paper first examines the current state of these recreational centers before exploring the role local communities are playing and how that can be enhanced to ensure sustainability, even in the absence of private investors. It goes without saying that communities are more permanent residents in these areas than private investors, hence their involvement would ensure long term results. This paper asserts that recreational facilities do not just provide social benefits but also contribute to economic and environmental development. It further explores how communities can lead the process of regeneration in the absence of the public and private leadership. This paper examines this issue within the prism of sustainable community-driven urban regeneration.

2 LITERATURE REVIEW

A number of studies recognise the emerging problem of decaying cities, in what is commonly referred to as "brownfields" (Greenland 2018). The problem is more pronounced in the old cities of Europe with studies reporting 120, 000 hectares in Germany, 28,000 hectares in the United Kingdom, 100,000 in France, to mention a few (Environment and Energy Management Agency 2014; Elrahman 2016; Greenland 2018). Studies have also shown that decayed or derelict buildings result in a number of vices, such as environmental degradation, hazards to humans, havens for criminals, and unsightly appearances (Elrahman 2016; Hollander et al. 2010; Tang 2013). This phenomenon is slowly being seen in cities in the developing world, with the former mine-owned recreational facilities in Zambia being a case in point. A combination of literature on community based management and sustainable development helped to frame this paper in the context of understanding how former mine-owned recreational centers can be redeveloped and managed sustainably.

As noted earlier, sustainable urban regeneration is a process by which the cities' action results in renewing the dead parts. Its basis is the 2015 Sustainable Development Goals (SDGs). Sustainability in this sense involves much more than just the environment but also includes physical, social, economic and cultural components (Lee et al. 2019). Urban regeneration has been used as a tool to stimulate economies and solve urban and social problems, with initiatives growing over time (Lee & Chan 2008; Xuili & Maliene 2021). For instance, Li et al. (2016) acknowledge that the last 50 years has witnessed a huge wave of urban regeneration projects in major cities of Shanghai, Hong Kong and London; however these efforts have faced serious implementation challenges. One of the major reasons cited is the dissimilar characteristics of stakeholders with varying interests (Ho et al. 2012). Li et al. (2016) further argued that despite decades of experimenting with public involvement in regeneration projects, successful cases are still few. Therefore, the key stakeholders are the community where these brownfields are located who should thus be included in the redevelopment process right from the planning stage.

The role of stakeholders, and their management, in projects of various types has come to be understood as one of the important success factors (Tseng et al. 2019; Zhuang et al. 2019). Specifically, Zhuang et al. (2019) examined the role of stakeholders in cases involving decision making in urban renewal projects in China. The study argued that for complex urban renewal projects to be successful, it required integrated, coordinated and multifaceted strategies involving a wide range of stakeholders. Studies (cf. Kujala et al. 2016, 2022; Kujala & Sachs 2019) have also argued that stakeholder theory calls for the probing of relationships that exist between organizations and individuals or groups who have an effect on it or may be affected by it. This is critical for this study because for regeneration to take place at the recreational centers of interest, stakeholders must be identified who will be called upon to influence decision making, provide resources as well as be available for management of the facilities beyond the regeneration process as advised by Neville et al. (2011).

Literature also alludes to the need for an institutional framework to support stakeholder engagement (Grootaert & van Bastelaer 2002; Schreiner et al. 2011). Grootaert & van Bastelaer (2002) stated that an active institutional framework is needed for cooperation and respect in communities. Poor policy implementation was identified as a challenge that needs to be addressed.

Community based approaches are closely intertwined with participatory governance in the management of various resources from tourism, coastal and ocean, and generally natural resources (Gurney *et al.* 2016; Harrington *et al.* 2019; Lucero *et al.* 2018). For instance, Kearney *et al.* (2007) in investigating participatory governance in community based management in coastal and ocean resources, argued that community based approaches are some of the best ways of facilitating participatory governance. Social cohesion is also important in any governance system (Chavez-Miguel *et al.* 2022; Fonseca *et al.* 2019). For instance, Chavez-Miguel *et al.* (2022) discussed the importance of social cohesion in the management of natural resources; similar in context to recreational facilities as a "public" good. Kearns & Forest (2000) went further to break down social cohesion into a number of interrelated dimensions, namely: common values and civic culture; social solidarity; social networks and social dimensions; and territorial belonging and identity.

From this literature review, a number of factors have been identified as key to the success of community participation, namely: institutional framework, stakeholders' identification and management; participatory governance and social cohesion.

3 RESEARCH METHODOLOGY

This paper is an extract from a larger research project. Overall, the study used an exploratory sequential mixed-method approach where the qualitative phase of data collection and analysis is then followed by the quantitative part. Thus data was collected using household surveys, focus group discussions (FGDs) and in-depth interviews with key informants. Furthermore, the study commissioned a condition assessment in order to establish the current physical state of the recreational facilities. The facilities were rated from 1- critical (unsafe and high risk) to 6 – Excellent (new or state of the art and meets current and foreseeable future requirements).

Four facilities were selected, two in each town of Kitwe and Mufulira, on the Copperbelt Province, in Zambia. These comprised the main recreation centers and one within the townships, namely, Nkana Main Recreation Center and Chamboli Soccer Grounds in Kitwe and Mufulira Main Recreation Center and Bufuke Club House in Mufulira. A total of 386 questionnaires were distributed in these research sites, four focus group discussions held (FGD 1- 4) and 18 interviews with key informants conducted (Interviewees A-R).

The Krushal-Willis H-test, a rank based nonparametric test, was also used to rank the responses of respondents in relation to their willingness to participate in community activities.

4 FINDINGS AND DISCUSSIONS

4.1 *Condition assessment*

The entry point for this paper was to argue that the current condition of recreational facilities in former mine townships in Kitwe and Mufulira is as a result of a vacuum in management. It was thus important to assess the physical state of the buildings. An independent Structural Engineer was then engaged to prepare a Condition Assessment Report. The Report revealed serious decay of these facilities indicating a lack of maintenance over a prolonged period. For instance, the main recreation center in Kitwe was rated 3 (marginal), meaning it was still generally structurally sound although some of the sections were rated 2 (poor). Only one section (the Rugby Club) was rated 5 (good). A further scrutiny of these ratings revealed that the good areas were still under the sponsorship of the new mine owner, Mopani Copper Mines Plc., while the others were either being used and managed by churches or sports associations or the union for the miners.

This pattern was also seen in Mufulira where the Rugby Club and hockey grounds, also funded/managed by Mopani, were rated 5 (good) while the cricket and squash clubs, being run by volunteers, were rated 3 (marginal). The indication is that the sections being funded/managed by the new mining firm are in better condition than those funded/managed by local community organizations; prompting the question, what role are the community organizations playing in the redevelopment of these facilities? Can this be improved?

4.2 Descriptive information

As noted earlier, a total of 386 questionnaires were administered in the four research sites. Focus group discussions and interviews were also conducted.

In terms of the demographics, the majority of respondents (55%) fell within the ages of 25–45 years. This is considered the most productive and energetic group of society. The employment status of respondents were also examined, which revealed that only 42% were in formal gainful employment while the majority were involved in other activities, such as studying (19%), unemployed (33%) and retired (7%). One of the major reasons for examining these parameters was to understand the depth of social capital within the community, being an important factor in the ability of a community to self-organize. A further analysis of the employed (160 respondents) was also conducted. The analysis revealed that 36% were self-employed or in part-time employment (9%); others included public officers (19%), mine workers (14%) and other private sectors (23%). The logic of this breakdown is to understand how many would have time when called upon to participate in community activities.

McMilan & Chavis (1986) argued that the length of stay in a community was an important factor in determining the sense of belonging to a particular group. Thus, this study examined the length of stay of respondents in the research sites. The study found that at least 61% of respondents had lived in their respective communities for the last 10 years, a sufficiently long time to be able to develop this sense of belonging.

The study further examined the existence of self-organizing groups within the communities based on the thinking that people will be much willing to assist if they have already developed a volunteering spirit. The study revealed that at least 74% of respondents belonged to a religious group while 67% were also in cooperatives; the others included 49% in women's groups and 56% in sports.

4.3 Critical facets supporting community based management

As discussed in the literature review (Section 2. above) community based management systems are anchored on identifiable critical tenets of the community. This paper restricted itself to four such tenets, namely: institutional and policy framework; stakeholders' identification and management; participatory governance; and, social cohesion and willingness to participate.

4.3.1 Institutional and policy framework

It was clear from the perspective of most respondents that they consider the government's direction, with regards to the institutional and policy framework, as the starting point to ensure new investors contributed to the social-wellbeing of their communities. For instance, Interviewee B lamented that "there are no policies to compel the mines to look at social infrastructure. If it is requested from a Corporate Social Responsibility (CSR) point of view, investors reserve the right to accept or reject the request. Thus, they cannot be compelled to act". This was similar to Interviewee K who noted that "although the government provides the policy and legal framework for the sports and recreation industry to survive" it seems to fall short in terms of enactment of a law to support communities. Hence, Grootaert & van Bastelaer (2002) argued that where the institutional framework is neglected, individuals cannot be persuaded to cooperate or respect each other.

4.3.2 Stakeholders' identification and management

It was also important to understand who the respondents thought were the key stakeholders in the management of these facilities. Participants in focus group discussions named various groups in the redevelopment process; these included; ward councilors, banking institutions, government (local and central), sports associations, churches, and various individuals within the community. With such a long list of interested parties, it is no wonder Zhuang *et al.* (2019) argued that the complex process of urban renewal needed an integrated, coordinated and multi-faceted strategy involving a wide range of stakeholders. Conflicts and squabbles often ensue when this process is mismanaged.

4.3.3 *Participatory governance*

Participatory governance was seen to be intertwined with community based management (Gurney *et al.* 2016; Harrington *et al.* 2019; Lucero *et al.* 2018). Participants in focus group discussions were asked to identify people who had influence on decision making within the community. Participants argued that with the right leaders, the community can be involved in those activities that promote empowerment and participation. For instance, in FGD4 it was revealed that "there are a few business men that have influence. When they spearhead something, progress is seen". Participants in FGD1 also agreed that 'implementation is supposed to come from leaders', while those in FGD3 said, 'those in power will always want to champion things'. Therefore, leaders play an important coordinating function for the rest of the community.

4.3.4 *Social cohesion and willingness to participate in community activities*

The study also sought to investigate the social cohesion within the community as a precursor to working together on common projects. Specific parameters for this variable included knowledge of neighbors and sharing of information, giving and receiving help from neighbors and causes and resolution of conflicts (Tables 1–3). These concepts fall within the various definitions of social cohesion discussed by Kearns & Forrest (2000), and Chavez-Miguel *et al.* (2022).

As shown in Table 1, an average of 81.3% of respondents agreed to having knowledge of their neighbors on all sides and 71.2% shared or received information from their neighbors. Table 2 also shows that community members helped each other by giving or receiving help in the form of funds, food/groceries, hospital and funeral assistance, etc.

Table 1. Knowledge of neighbors and sharing of information.

Response	Knowledge of neighbors on all sides		Information sharing with neighbors	
	Frequency	Percent	Frequency	Percent
Yes	314	81.3	274	71.2
No	72	18.7	111	28.8
Total	386	100.0	385	100.0

Table 2. Giving and receiving help from neighbors.

Response	Helping neighbors		Being helped by neighbors	
	Frequency	Percent	Frequency	Percent
Financially	32	8.3	28	7.3
With food/groceries	120	31.2	105	27.3
Funeral assistance	33	8.6	28	7.3
Hospital matters	27	7.0	18	4.7
Transportation and car matters	9	2.3	10	2.6
Yard works	26	6.8	33	8.6
Other	41	10.6	55	14.3
Nothing	97	25.2	108	28.1
Total	385	100.0	385	100.0

The study sought to understand the causes of differences/conflicts and how they are resolved based on the reasoning of social cohesion within social relations (Schiefer & van der Noll 2017). Table 3 revealed that some of the major sources of conflicts include: lack of information; selfish leaders, misunderstandings and inequitable distributions of resources. It was generally acknowledged that the solution to most of these problems lay in people working together and sorting out their problems. It is particularly noteworthy that despite the prominence of church and community leaders in society, the community did not see them as the first call for solving their problems.

Table 3. Causes of conflicts and their resolution.

Causes of conflicts			Handling of differences or conflicts		
Variable	Count	%	Variable	Count	%
Lack of information	232	60.1	People work it out themselves as individuals	275	71.2
Selfishness of leaders	239	61.9	Families/ households intervene	149	38.8
Inequitable distribution of resources	200	51.8	Neighbors intervene	169	43.8
Misunderstandings	240	62.2	Community leaders mediate	99	25.6
Suspiciousness	147	38.1	Religious leaders mediate	175	45.3
Favoritism of leaders	170	44.0	Judicial leaders	184	47.7
Tribalism	68	17.6			

Lack of willingness of members to participate in community projects poses serious challenges in implementation (Abowen-Dake 2013). In the research sites, the lack of willingness to participate was reflected in the levels of enthusiasm exhibited by participants of the focus group discussions. Based on the numbers of participants present, more enthusiasm was displayed in Chamboli and Butondo which are low cost areas. Nkana West and Mufulira High Cost areas seemed less enthusiastic and this was further supported by a statement made during FGD4. Although made in response to sources of conflict in the community, one participant noted that "people normally mind their own business and keep to themselves. They are often indoors", which suggested that they were not bothered by any community activities. This is further supported by the H-statistics in Table 4.

Table 4. Community willingness to participate in projects and the concern for others.

Question	H-statistic	Number of successes	Sample size	p-value	Direction of the responses
If a community project will not benefit you directly you would be willing to participate	2.016	200	383	0.207	Disagreement
Generally, people in this community live in harmony with others	25.681	321	383	0.000	Agreement
Generally, people in this community are concerned about the welfare of others?	6.066	242	384	0.000	Agreement
If the community is not involved in community projects, they are most likely to vandalize the project.	26.446	147	384	1.000	Disagreement

5 CONCLUSION

The paper revealed that stakeholder participation is an important success factor in projects of varying types; including urban regeneration projects. However, despite over 40 years of literature supporting this concept from its narrow understanding within urban renewal to now a broader concept of sustainable urban regeneration, the role of communities, as stakeholders, in this process has remained an issue of debate. This paper isolated institutional framework, stakeholder identification and management, participatory governance, and social cohesion and willingness to participate in community activities. This paper particularly found that social cohesion is the underpinning characteristic of community participation and was therefore examined in more detail.

The paper finds that a number of key components are missing in the study areas to enable successful management of the former mine owned recreational facilities. Besides social cohesion, funding of the management function is also a critical component. The communities consist mainly of people who are either not working or earn very little. It was clear that components of the facilities, such as the Rugby Clubs which are still directly funded by the new mining firm in Kitwe and Mufulira, had better ranking of 5 (good) than the rest of the facilities run by volunteers and church organizations. Furthermore, members of the community are less willing to participate in anything without pay; thus challenging the spirit of volunteerism and sense of belonging. In addition, the community felt that the government had a role to play to ensure that firms operating in these areas are part of the stakeholders. This paper concludes that sustainable regeneration of former mine owned recreational facilities should start with a clear institutional and policy framework from government (central or local) which would then guide the communities in their participation. Local leaders should spearhead the regeneration process since they have the respect and recognition of community members. They can do this by engaging more directly and closely with the self-organising groups that are already functioning within the community.

REFERENCES

Abowen-Dake, R. and Nelson, M.M., 2013. The Applicability of Community-based Facilities Management Approach to Regeneration: A Case Study. *Journal for Facility Management*, 1(7), 20–36.

Alade, W., Ogunkan, D. And Alade, B., 2021. Analysis of Urban Decay In The Core Residential Areas Of Ota, Southwest Nigeria. *Ethiopian Journal of Environmental Studies & Management*, 14.

Andersen, H.S., 2019. *Urban Sores: On The Interaction Between Segregation, Urban Decay and Deprived Neighborhoods.* Routledge.

Chavez-Miguel, G., Bonatti, M., Ácevedo-Osorio, Á., Sieber, S. and Löhr, K., 2022. Agroecology as a Grassroots Approach for Environmental Peace-building: Strengthening Social Cohesion and Resilience in Post-conflict Settings with Community-based Natural Resource Management. *GAIA-Ecological Perspectives for Science and Society*, 31(1), pp.36–45.

Elrahman, A.A., 2016. Redevelopment Aspects for Brownfields Sites in Egypt. *Procedia Environmental Sciences*, 34, pp 25–35.

Environment and Energy Management Agency, 2014. *Reconversion of Urban Wastelands in France. Urban Wastelands in SCoTs and PLUs.* https://www.ademe.fr/collectivites-secteur-public/integrer-lenvironnement-domaines-dintervention/urbanisme-amenagement/dossier/reconversion-friches-urbaines/friches-urbaines-scot-plu. Date accessed 26.2.2019.

Fonseca, X., Lukosch, S. and Brazier, F., 2019. Social Cohesion Revisited: A New Definition and How to Characterize It. Innovation: *The European Journal of Social Science Research*, 32 (2), pp.231–253.

Fraser, A. and Lungu, J., 2007. *For Whom the Windfalls. Winners & Losers in The Privatization of Zambia's Copper Mines.* Retrieved from https://sarpn.org/documents/d0002403/Zambia_copper-mines_Lungu_Fraser.pdf

Greenland, M., 2018. *What are the Barriers to Brownfield Development?* https://developmentfinancetoday.co.uk/article-desc-6233what-are-the-barriers-to-brownfield-development. Date accessed 22.11.2019.

Grootaert, C. and Van Bastelaer, T. eds., 2002. *Understanding and Measuring Social Capital: A Multidisciplinary Tool for Practitioners (Vol. 1).* World Bank Publications.

Gurney, G.G., Cinner, J.E., Sartin, J., Pressey, R.L., Ban, N.C., Marshall, N.A. and Prabuning, D., 2016. Participation in Devolved Commons Management: Multiscale Socioeconomic Factors Related to Individuals' Participation in Community-based Management of Marine Protected Areas in Indonesia. *Environmental Science & Policy*, 61, pp.212–220.

Harrington, C., Erete, S. and Piper, A.M., 2019. Deconstructing Community-based Collaborative Design: Towards More Equitable Participatory Design Engagements. *Proceedings of the ACM on Human-Computer Interaction*, 3(CSCW), pp.1–25.

Ho, D.C.W., Yau, Y., Poon, S.W. and Liusman, E., 2012. Achieving Sustainable Urban Renewal in Hong Kong: Strategy for Dilapidation Assessment of High Rises. *Journal of Urban Planning and Development*, 138(2), pp.153–165.

Hollander, J.B., Kirkwood, N.G. and Gold, J. L., 2010. *Principles of Brownfield Regeneration: Clean Up, Design and Reuse of Derelict Land*. Island Press. Washington.

Kaonga, L. and Nguvulu, A., Investigating Factors Affecting the Sustainability of Corporate Funded Community Based Projects: The Case of Mopani Copper Mine-Mufulira (Kankoyo).

Kearney, J., Berkes, F., Charles, A., Pinkerton, E. and Wiber, M., 2007. The Role of Participatory Governance and Community-based Management in Integrated Coastal and Ocean Management in Canada. *Coastal Management*.

Kearns, A. and Forrest, R., 2000. Social Cohesion and Multilevel Urban Governance. *Urban Studies*, 37(5–6), pp.995–1017.

Kujala, J. and Sachs, S., 2019. The Chapter14 Practice of Stakeholder Engagement. *The Cambridge Handbook of Stakeholder Theory*, p.227.

Kujala, J., Lehtimäki, H. and Myllykangas, P., 2016. Toward a Relational Stakeholder Theory: Attributes of Value-creating Stakeholder Relationships. In *Academy of Management Proceedings* (Vol. 2016, No. 1, p. 13609). Briarcliff Manor, NY 10510: Academy of Management.

Kujala, J., Sachs, S., Leinonen, H., Heikkinen, A. and Laude, D., 2022. Stakeholder Engagement: Past, Present, and Future. *Business & Society*, 61(5), pp.1136–1196.

Lee, G.K. and Chan, E.H., 2008. The Analytic Hierarchy Process (AHP) Approach for Assessment of Urban Renewal Proposals. *Social Indicators Research*, 89, pp.155–168.

Lee, T.H. and Jan, F.H., 2019. Can Community-based Tourism Contribute to Sustainable Development? Evidence from Residents' Perceptions of Sustainability. *Tourism Management*, 70, pp.368–380.

Li, L., Hong, G., Wang, A., Liu, B. and Li, Z., 2016. Evaluating the Performance of Public Involvement for Sustainable Urban Regeneration. *Procedia Engineering*, 145, pp.1493–1500.

Lucero, J., Wallerstein, N., Duran, B., Alegria, M., Greene-Moton, E., Israel, B., Kastelic, S., Magarati, M., Oetzel, J., Pearson, C. and Schulz, A., 2018. Development of a Mixed Methods Investigation of Process and Outcomes of Community-based Participatory Research. *Journal of Mixed Methods Research*, 12(1), pp.55–74.

McMillan, D.W. and Chavis, D.M., 1986. Sense of Community: A Definition and Theory. *Journal of Community Psychology*, 14(1), pp.6–23.

Neville, B.A., Bell, S.J. and Whitwell, G.J., 2011. Stakeholder Salience Revisited: Refining, Redefining, and Refueling an Underdeveloped Conceptual Tool. *Journal of Business Ethics*, 102, pp.357–378.

Schiefer, D. and Van der Noll, J., 2017. The Essentials of Social Cohesion: A Literature Review. *Social Indicators Research*, 132, pp.579–603.

Tang, J. 2013. Public Participation in Brownfield Redevelopment: A Framework for Community Empowerment in Zoning Practices. *Seattle Journal of Environmental Law*. Vol. 3, Iss. 1, Article 9, pp 241–269.

Xuili, G. and Maliene, V., 2021. A Review of Studies on Sustainable Urban Regeneration. *EPiC Series in Built Environment*, 2, pp.615–625.

Zhuang, T., Qian, Q.K., Visscher, H.J., Elsinga, M.G. and Wu, W., 2019. The Role of Stakeholders and Their Participation Network in Decision-making of Urban Renewal in China: The Case of Chongqing. *Cities*, 92, pp.47–58.

Public participation and sustainability consideration for mega transport infrastructure projects: Lessons from the Gauteng freeway improvement project, South Africa

L. Morejele-zwane & T. Gumbo
Department of Urban and Regional Planning, Faculty of Engineering and the Built Environment, University of Johannesburg, South Africa

S. Dumba
Department of Marketing and Logistics, Faculty of Commerce, Human Sciences and Education, Namibia University of Science and Technology, Namibia

ABSTRACT: Public participation has evolved, emphasizing the need for meaningful engagement between authorities and the public in sustainable transport infrastructure projects. However, power imbalances still hinder public participation processes. This paper examines factors influencing public participation in the Gauteng Freeway Improvement Project (GFIP) and explores its impact on project sustainability. Quantitative data, analyzed using statistica, reveals that limited meaningful public participation resulted in public opposition to GFIP tolling and non-compliance with payments. The paper concludes that statutory requirements alone fail to generate public support during project implementation. It recommends developing a comprehensive public participation framework to facilitate effective engagement between authorities and the public, promoting sustainable transport infrastructure. Future research should assess public participation processes from project conceptualization rather than solely post-implementation evaluation.

Keywords: Public, participation, infrastructure, transport, sustainability

1 INTRODUCTION

The pursuit for urban sustainability necessitates a well understood processes of examining factors which affects public participation during the planning and implementation of transport infrastructure projects. Of late, geographic scholars have been paying attention to how infrastructure impact citizens who live around them and use them (Lesutis 2022:303). Prioritizing the interaction between human activities and transport infrastructure is in crises, and this has become a major challenge for transport planners (Verlinghieri 2020: 364). It is not surprising that public participation discussions are also gaining momentum within Africa, especially in South Africa. Public participation processes within transportation planning are at times managed by consultants not qualified or accredited by any professional body (Grossardt & Bailey 2018:1). Consequently, a gap exists, which necessitates the establishment of a public participation framework for the implementation of effective participation processes. Based on existing participation ideologies, it was established that significant gaps exist between what is perceived as the actual level of public participation and the desired level (Wang & Chan 2020:1). Therefore, in response to bridging the gap to achieve desired levels, this paper complements existing studies by examining the limits of public participation in transport infrastructure planning and implementation. Firstly, the paper explores the conceptual framework and the relevant theories to public participation and the relationships to transport infrastructure planning. It goes on to describe the research methodology which define the data collection and analysis procedures. Lastly, it recommends successful public participation strategies to achieve sustainable transport infrastructure solutions and the conclusion.

2 CONCEPTUAL FRAMEWORK FOR TRANSPORTATION AND PUBLIC PARTICIPATION

Accommodating the public during the development of mega transport infrastructure projects can be a complex exercise. Public participation is necessary for aiding legitimacy to decision making and providing various perspectives when dealing with complex issues (Tippett & How 2020:109).

2.1 *Public participation concept*

The concept of public participation captures the process that recognizes the concerns, interest, views, and values of ordinary people in decision-making (Ho *et al.* 2023:2). Agenda 2030 adopted by the United Nations member states in 2015 emphasised the importance of public participation for the successful implementation of Sustainable Development Goals (SDG). According to the United Nations (2019:25), it is important to include the public in designing of infrastructure, urban spaces, and services as a way of facilitating SDGs.

2.2 *Mega transport infrastructure projects*

The mammoth range of megaprojects make them appealing for gauging the impact of public participation during the conceptualization, planning and implementation. According to Wang, *et al.* (2019:10), public attention has been increasing towards megaprojects because of the social effects of such schemes. While the projects are meant to benefit the public, the schemes maybe less understood by ordinary citizens. The objectives of providing infrastructure are generally to lessen the negative environmental effects and concurrently stabilizing the social-economic sustainability (Mehlawat 2019:8).

3 THEORETICAL FRAMEWORK

In line with the objectives of this paper, it becomes relevant to understand the main ideologies behind public participation. Although there are numerous theories of public participation, the paper only focuses on theories giving the primary background to the core values of collective decision making.

4 SCHUMPETER'S THEORY

The notion of public participation became popular during the late 1960's, yet the advocates of democratic theories have viewed participation practices as minimum and consequently dangerous for democracy (Pateman 1976:1). The 1943 Joseph Schumpeter's theory challenged authorities to allow for open debates without modelling the outcome. To this end, democracy challenges authorities to choose democracy over efficiency by ensuring that decisions are based on effective public participation as opposed to a trade-off that abandon effective participation (Elrahman 2019:464.)

5 SHERRY ARNESTEIN THEORY

The Arnestein ladder is one of the most significant conceptualizations of public participation and the eight categories presented through a ladder, capture the degree of citizen power (Awuh 2022:274). Figure 1 illustrates the levels of citizen participation, ranging for the lower levels of non-participation to the highest level of citizen control. The Arnestein ladder has been influential in the field of urban planning, and it is considered as a model, influencing democratic public participation (Panyavaranant *et al.* 2023:4). This theory focuses on the power of the public and it suggests that, for the public to gain power over decision-making, lower levels of participation must be avoided (Dai *et al.* 2022:2).

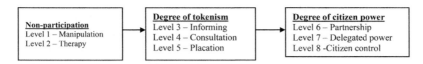

Figure 1. Arnstein ladder of public participation.
Source: adopted from Wang & Chan (2020:2)

6 THE HIGH-PERFORMANCE PUBLIC INVOLVEMENT FRAMEWORK

The purpose of the High-Performance Public Involvement framework was to bridge the Arnestein Gap, by introducing the quality, inclusion, clarity, and efficiency as a system of measurement for public participation (Bailey et al. 2015:45). The gap identified discrepancies between the observed levels of public participation and the quality that the public wants from such participation (Grossardt & Bailey 2018:3). Granted, public participation can be viewed by authorities as successful, yet the public can view the same process as unsatisfactory due to varying issues. Improving on the Arnsetein gap requires strategies that can increase the effectiveness of public process (Nasr-Azadani et al. 2023:3); as proposed through the High-Performance Public Involvement framework.

7 THE INTERNATIONAL ASSOCIATION FOR PUBLIC PARTICIPATION SPECTRUM

Even though the Arnestein Ladder was sought after for decades, it focused more on power and did not offer practical public participation guidelines. The Arnestein Ladder was criticized for focusing on public participation approaches as ascending as opposed to recognizing that diverse situations necessitate diverse public participation (Akerboom & Craig 2021:234). The International Association for Public Participation (IAPP) spectrum has been closely associated with facilitating practical public participation guidelines internationally. The IAPP proposed a five-stage spectrum of public participation, ranging from the authority's commitment to inform, consult, involve, collaborate, and empower the public during decision making (Özden & Velibeyoğlu 2022:35). The highest levels of participation whereby decision-makers collaborate and empower the public are desirable. Collaborations in governance requires an ongoing process and commitment for empowering stakeholders and sharing skills and knowledge (Perera et al. 2023:498).

8 METHODOLOGY

This paper employs a case study phenomenological research design. This design emanates from the philosophy of lived experiences regarding a phenomenon (Creswell & Creswell 2018:13). GFIP is the first electronic tolling (e-toll) project to be implemented within South Africa. This urban road tolling project is implemented within the economic hub of the country hence the strength of this case study. A mixed-method approach was used to explore public participation processes during the planning and implementation of GFIP and thereafter evaluate the impacts of the improved freeways. The mixed method approach always employs mechanisms of qualitative, together with quantitative mechanism due to the collective strengths (Merriam & Grenier 2019:11). The paper relies on both the primary and secondary data collection methods. To gain insights, experts from the national road's agency were purposefully approached for in-depth understanding of GFIP. Secondary data was largely reviewed to acquire insights into public participation processes employed during the planning and implementation of GFIP. Through content analysis, the data was examined. Methodical techniques were used to review and analyse data to identify themes. To improve on the data, journal articles from search engines such as Google Scholar, Science Direct and Elsevier were used. Critical to this paper are the traffic patterns, to determine if improving the transport infrastructure alleviate traffic congestion over extended periods of time. The traffic survey along the improved freeways was collected by Syntell (Pty) Ltd through its subsidiary company Mikros Systems (Pty) Ltd. The traffic data was organized into an EXCEL file and analysed using Statistica 14.

9 RESULTS AND DISCUSSIONS

Although public participation has been the foundation of democracy within South Africa, Fagbadebo & Ruffin (2019:248) recommend that the country should implement strategies to enhance participation by ensuring that it is more accessible. The legislation shortfalls of providing steps to be followed during public participation and it does not indicate how participation should be measured.

10 GFIP PUBLIC PARTICIPATION DURING THE PLANNING PROCESS

The public participation for GFIP was initiated by the agency in 2007 and published in the Government Gazette and other local newspapers, inviting the public to comment within 30 days. For the agency to comply with the statutory requirements, only the location of the toll booths was supposed to be mentioned. The e-toll fees were only published for public comments in 2011 and yet phase 1 upgrades were at final construction completion. Nevertheless, the authorities had complied with in terms of the statutory requirements. The higher levels of IAPP spectrum and Arnestein ladder of participation require that the public be involved in the development of alternatives during decision making. Such a process cannot be achieved through newspaper adverts calling for comments on the position of tolls as was advertised for GFIP. Allowing the public to comment on a small component of a project is non-participations because the authorities lead the projects with the purpose of merely informing and not allowing stakeholders to be part of the process (Panyavaranant *et al.* 2023:19). This is what Arnetein refers to as manipulation. Through a top-down approach, the government had already directed the projects and the role of the public was to follow suite (Mai *et al.* 2022:9).

11 GFIP TRAFFIC FLOW ALONG IMPROVED FREEWAYS

Upgrading the Gauteng freeways was meant to alleviate congestion. Although the overall objective of the study is public participation, the principles that inform successful transport systems must be exhausted, with the public at the center of the proposed solutions. The traffic flow data (volume, directional distribution) were collected in hours per year, and these were regarded as variables for the purpose of this analysis. The scatter plots are presented and the linear regression (Demidova *et al.* 2016:2) lines are fitted and equations given as well as the correlations with years. The coefficient of x (year) is the increasing traffic flow per hour per year. The p-value indicates the relationship between the two variables which is statistically significant, therefore the p-value must be lower than 0.05 (Kabashkin *et al.* 2020:313). The r-value (spearman correlation coefficient) indicates the strength of the relationship between the two variables, and the closer to 1 or -1, the stronger the relationship. The r-value is the regression demonstrating the strength of a relationship between two variables.

11.1 *Ben Schoeman highway (N1) traffic patterns*

The data presented is limited to sites along the N1 connecting the City of Johannesburg and the City of Tshwane.

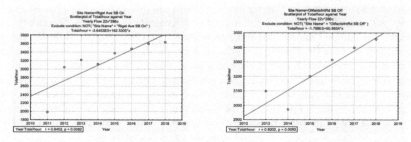

Figure 2. N1 Highway south bound from Ccty of Tshwane towards city of Johannesburg at two sites.

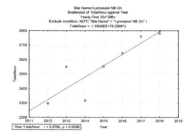 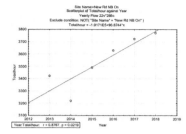

Figure 3. N1 Highway north bound from city of Johannesburg towards city of Tshwane at two sites.

The Figures 2 and 3 indicate that traffic counts increased over the years. The r-value is more than 0.5 across the selected routes, the regression demonstrating the strength of a relationship between the year and the traffic increase. The data indicated that traffic volumes have been increasing over time and therefore the freeway upgrades along the N1 freeway south bound and north bound were necessary.

Due to the public defying the payments, to this end only Phase 1 has been completed. This equals 185km upgrades of the proposed 560km upgrades. Phase 1 was meant to only relief traffic congestion for approximately 3 to 5 years and thereafter Phase 2 and 3 will have to be implemented (Makhura 2014:35). During the data collection, 5 key officials from the national roads' agency confirmed that the freeways will be congested; pushing traffic to where it was before the Phase 1 implementation if Phases 2 and 3 are not implemented. The respondents also confirmed that Phases 2 and 3 are needed immediately.

Based on the responses, it can be deduced that the long-term positive impacts of GFIP will only be realized when all the phases of the project are implemented. This is line with the assumption that authorities spend funds on highway expansions because of traffic congestion and yet the highways will only be able to reduce congestion for a couple of years (Speck 2018:65). The boycotting of the payment system by the public let to unforeseen GFIP implementation disruption, thereby affecting the sustainability of the project. Resorting to protest to air complaints between the public and government is an indication of a disconnect (Biljohn & Lues 2019:232).

B) The agency was requested to respond if they are satisfied with the information to enable the collection of e-toll from most freeway users. A total of 40% are very satisfied, 20% satisfied and 40% are fairly satisfied that information of freeway users is available to facilitate the collection of e-tolls. The agency was subsequently asked if, for the e-toll system to be successful, the agency required the residents to willingly give personal data. The results indicate that the agency does not need residents to willingly share personal data for the e-toll system to be successful, with 40% of the respondents fairly dissatisfied and 40% very dissatisfied. Only 20% of the respondent acknowledged that the success of e-toll depends on residents willingly issuing personal information to the agency. Without the personal information, the outstanding payments cannot be recovered. Subsequent phases were supposed to be implemented from 2016 and yet finance Minister Enoch Godongwana during the Medium-Term Budget Policy Statement on the 26th of October 2022, in parliament announced the scrapping of the GFIP user-pay component. The announcement was immediately embraced by the provincial government as indicated in Figure 4. The oblivious view that the project would succeed without the willingness of the public to pay for e-tolls was proven in 2022. This mega project is one where the public had the great power in changing the project direction because of the lack of public inclusion (Andersen et al. 2021:4).

Citizen corporation is important for the success of public projects, especially where funds must be collected from the citizens. This concurs with Grossardt & Bailey (2018:9), who note that the broad citizens are regarded as the financier of the proposed projects and therefore they are the client sponsoring the public sector agency. The implications of the GFIP protest after the implementation of the project highlight the serious disruptions because of lack of

Figure 4.   Lesufi 2022: GFIP media statement.
Source: https://twitter.com/GautengProvince/status/1585261403140653058/photo/1

public acceptance. Urban infrastructure project's complexities can be exacerbated by possible protests because of possible public opposition and protest (Dai *et al.* 2022:1).

12 TOWARDS SUCCESSFUL PUBLIC PARTICIPATION FOR SUSTAINABLE TRANSPORT INFRASTRUCTURE PROJECTS

The insufficient collection of GFIP tolls affected the sustainability of the project as indicated in previous sections. The dynamic nature of public engagement necessitates the establishment of an appropriate approach which is central to the success of the process as captured in Figure 5. The involvement of the public should be enforced as an everyday concern throughout the project lifecycle (Amadi 2020:545). This means not just satisfying the requirements in terms of the bylaw, but extensively engaging the public on various platforms. The public participation framework should consider evaluating the flexibility of plans

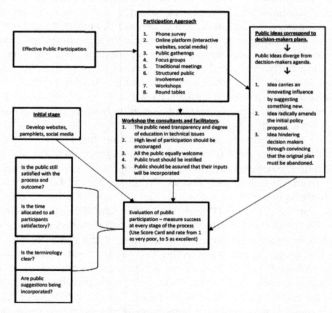

Figure 5.   Guidelines for effective public participation for sustainable transport infrastructure projects.
Source: Own construction (2023) from Bailey et al. (2015:47–50); Sagaris & Ortuzar (2015:21); Wang & Chan (2020:3); Grossardt & Bailey (2018:7–13); Vrydagh (2022:69)

to adapt to diversity of ideas from participants. Ultimately, the role of the public should be viewed as a formal function for providing authorities with suggestions and ideas (Vrydagh 2022:67). The decision-makers must engage genuinely for the public to be on equal power share to promote collaboration and empowerment as proposed through IAPP spectrum. This requires authorities to team up with the public in all the decision making and, in some instances, allow the public to make final decisions (Özden & Velibeyoğlu 2022:35).

13 CONCLUSION

The paper concludes that effective guidelines are key for achieving successful public participation. Within democratic states, all available platforms must be used by authorities to engage the public during the conceptualization, planning and implementation of transport infrastructure projects. The framework for effective public participation highlights the guidelines for effective public participation. Of importance is the debates and agreements for the development of sustainable transport infrastructure and incorporating ideas and suggestion from the public. The paper ends by recommending meaningful engagements between all stakeholders, including ordinary citizens to enhance, user acceptance and desirable socio-economic outcomes and transport infrastructure development. Critical to this paper is authorities incorporating new ideas and changing policies to accommodate new perspectives. To this end, it is not clear how the agency facilitates comprehensive public participation during the initiation of mega projects and the planning and implementation phases. Future research is necessary to evaluate a project from initiation stages as this paper reviewed a project already implemented. The success of the transportation planning entities to mobilise the ordinary public must be evaluated throughout the lifecycle of the project, not just after the implementation stage.

REFERENCES

Akerboom S and Craig RK. 2021. How Law Structures Public Participation in Environmental Decision Making: A Comparative Law Approach. *Environmental Policy and Governance*, 32:232–246. https://onlinelibrary.wiley.com/doi/10.1002/eet.1986

Amadi C, Carrillo P and Tuuli M. 2020. PPP Projects: Improvements in Stakeholder Management. *Engineering, Construction and Architectural Management*, 27(2):544–560. https://doi.org/10.1108/ECAM-07-2018-0289

Andersen PD, Hansen M and Selin C. 2021. Stakeholder Inclusion in Scenario Planning—A Review of European Projects. *Technological Forecasting & Social Change*, 169. Elsevier Inc. https://doi.org/10.1016/j.techfore.2021.120802

Awhu HE. 2022. Geography of Participation: Deepening the Understanding of the Participation Process in Time and Space. *Geography of Participation*, 113(3):273–289. https://onlinelibrary.wiley.com/doi/10.1111/tesg.12514

Bailey, Grossardt T and Ripy J. 2015. High-Performance Public Involvement: Frameworks, Performance Measures, and Data. *Journal of the Transportation Research Board*, 2499: 45–53. https://doi.org/10.3141%2F2499-07

Creswell JW and Creswell JD. 2018. *Research Design: Qualitative, Quantitative, and Mixed Methods Approaches*. SAGE Publication. London

Dai L, Han Q, Vries D and Wang Y. 2022. Exploring Key Determinants of Willingness to Participate in EIA Decision-making on Urban Infrastructure Projects. 76:103400 https://doi.org/10.1016/j.scs.2021.103400

Demidova L, Ivkina M, Zhdankina E, Krylova E, Sofyin E, Reshetova V, Stepanov N and Nikita T. Software Package STATISTICA and Educational Process. *SHS Web of Conferences*, 9:02011. https://doi.org/10.1051/SHSCONF%2F20162902011

Elrahman OA. 2019. Governance of Environmental Health and Transportation Decisions: The Case of New York City. *Case Studies on Transport Policy* 7:463–469. https://doi.org/10.1016/j.cstp.2019.01.006

Fagbadebo O and Ruffin F. 2019. *Perspectives on the Legislature and the Prospects of Accountability in Nigeria and South Africa*. Springer International Publishing. eBook. https://doi.org/10.1007/978-3-319-93509-6

Grossardt T and Bailey K. 2018. *Transportation Planning and Public Participation*. Elsevier Inc. United States of America.
Ho S, Choudhury PR, and Joshi R. 2023. Community Participation for Inclusive Land Administration: A Case Study of the Odisha Urban Slum Formalization Project, *Land Use Policy* 125:106457, https://doi.org/10.1016/j.landusepol.2022.106457
Kabashkin I, Yatskiv I and Prentkovskis O. 2020. *Reliability and Statistics in Transportation and Communication*. Springer Nature. Switzerland.
Lesufi. P 2022: *Gauteng Provincial Government- GFIP Media Statement*. https://twitter.com/GautengProvince/status/1585261403140653058/photo/1
Lesutis G. 2022. Infrastructure as Techno-politics of Differentiation: Socio-political Effects of Mega-infrastructures in Kenya. *Trans Inst Br Geogr*, 47:302–314. https://doi.org/10.1111/tran.12474
Mai NTT, Mai TP, and Linh NH. 2022. Conceptualizing a Model of Community Participation in Sustainable Development in Asian Heritage Context: Research Framework and Agenda. *Journal of Management and Development Studies*, 11 (2): 5–16. https://www.researchgate.net/publication/368883454
Makhura D. 2014. Socio-economic Impact of the Gauteng Freeway Improvement Project and E-tolls Report: Report of the Advisory Panel. http://www.gautengonline.gov.za/
Mehlawat MK, Kannan D, Gupta P and Aggarwal U. 2019. Sustainable Transportation Planning for a Three-stage Fixed Charge Multi-objective Transportation Problem. *Annals of Operations Research*. Springer. https://doi.org/10.1007/s10479-019-03451-4
Merriam SB and Grenier RS. 2019. *Qualitative Research in Practice: Examples for Discussion and Analysis*. Jossey-Bass publishers. United States of America.
Nasr-Azadani E, Wardrop DH and Brooks RP. 2023. Pathways for the Utilization of Visualization Techniques in Designing Participatory Natural Resource Policy and Management. *Journal of Environmental Management* 333:117407 https://doi.org/10.1016/j.jenvman.2023.117407
Özden P and Velibeyoğlu K. 2023. Citizen Science Projects in the Context of Participatory Approaches: The Case of Izmir. *Journal of Design for Resilience in Architecture and Planning*, 4(1): 31–46. https://doi.org/10.47818/DRArch.2023.v4i1081
Panyavaranant P, Nguyen TPL, Santoso DS, Nitivattananon V and Tsusaka TW. 2023. Analyzing Sociodemographic Factors Influencing Citizen Participation: The Case of Infrastructure Planning in Khon Kaen, Thailand. *Social Sciences*, 12:225. https://doi.org/10.3390/socsci12040225
Pateman C. 1976. *Participation and Democratic Theory*. Cambridge University Press. United Kingdom
Percy M and Wanna J. 2018. Road Pricing and Provision: Changes Traffic Conditions Ahead. https://books.google.co.za
Perera ED, Moglia M, Glackin S and Woodcock I. 2023. The Intention-implementation Gap for Community Involvement in Urban Waterways Governance: A Scoping Review. *Local Environment*, 28(4):495–517. https://doi.org/10.1080/13549839.2022.2155941
Sagaris L and Ortuzar J. 2015. Reflection of Citizen-technical Dialogue as Part of Cycling-Inclusive Planning in Santiago, Chile. *Research in Transportation Economics* 53(20–30). http://dx.doi.org/10.1016/j.retrec.2015.10.016
Speck J. 2018. *Walkable City Rules: 101 Steps to Making Better Spaces*. Island Press. United States of America
Tippett J and How F. 2020. *Where to Lean the Ladder of Participation: A Normative Heuristic for Effective Coproduction Processes*. Liverpool University Press. https://doi.org/10.3828/tpr.2020.7
United Nations, 2019. *The Sustainable Development Goals Report* 2019. United Nations Publication
Verlinghieri E. 2020. Learning from the Grassroots: A Resourcefulness-based Worldview for Transport Planning. *Transportation Research Part A* 133: 364–37. https://doi.org/10.1016/j.tra.2019.07.001
Vrydagh J. 2022. Measuring the Impact of Consultative Citizen Participation: Reviewing the Congruency Approaches for Assessing the Uptake of Citizen Ideas. *Policy Sciences*, 55:65–88. https://doi.org/10.1007/s11077-022-09450-w
Wang A and Chan EHW. 2020. The Impact of Power-geometry in Participatory Planning on Urban Greening. *Urban Forestry & Urban Greening* 48:126571. https://doi.org/10.1016/j.ufug.2019.126571
Wang Y, Li H, Zuo J and Wang Z. 2019. Evolution of Online Public Opinions on Social Impact Induced by NIMBY Facility. *Environmental Impact Assessment Review*, 78:106290. https://doi.org/10.1016/j.eiar.2019.106290

Integration of environmental sustainability practices in real estate development in emerging markets

J. Mahachi*
University of Johannesburg, Johannesburg, Gauteng, South Africa

L. Kumalo*
Novare

ABSTRACT: The global real estate industry has a substantial environmental impact, consuming 40% of global energy and contributing 20% of greenhouse gas emissions. To align with the Paris climate conference's 2°C goal, the industry aims to reduce CO2 emissions by 36% by 2030. This paper presents a Zambian case study of a real estate developer's application of green building principles for EDGE certification. The study also assesses challenges faced by property developers in emerging economies when implementing sustainable practices. Findings show the developer achieved EDGE certification by implementing energy-efficient systems, water conservation techniques, and waste reduction practices. These efforts resulted in a 50% energy usage decrease, a 55% water usage decrease, and a 34% embodied energy reduction in materials. The study highlights the importance of green building certifications in promoting sustainable real estate development and suggests regulatory support and public-private partnerships can address challenges in emerging markets.

Keywords: Green building, EDGE Certification, Environmental performance

1 INTRODUCTION

Real estate development is a vital contributor to economic growth and development in emerging markets (Sassen 2001; UN-Habitat 2016). However, the construction and operation of buildings have significant negative impacts on the environment. The United Nations Environment Programme (UNEP) has estimated that buildings account for 40% of global energy consumption and 30% of greenhouse gas emissions (UNEP 2018). These impacts are particularly pronounced in emerging markets, where rapid urbanization and population growth are driving demand for new buildings and infrastructure (UN-Habitat 2016).

Despite these challenges, there is a growing recognition of the importance of sustainability in real estate development. Sustainable development, which emphasizes the integration of economic, social, and environmental considerations, has gained widespread acceptance since the 1980s (Brundtland 1987). Real estate developers are increasingly adopting sustainability practices, such as green building standards, to reduce their environmental footprint and enhance their reputation (Jones Lang LaSalle 2019; Mahachi 2021).

One such green building standard that has gained popularity in both developed and emerging markets is the Excellence in Design for Greater Efficiencies (EDGE) certification system. Developed by the International Finance Corporation (IFC), a member of the World Bank Group, EDGE promotes resource-efficient building design and reduces the environmental

---

*Corresponding Authors: jmahachi@uj.ac.za and lerato@novare.com

impact of buildings (IFC 2019). The system has been implemented in over 160 countries (IFC 2021).

This paper examines the integration of environmental sustainability practices in real estate development in emerging markets, the role of the EDGE certification in advancing sustainability goals, and the challenges property developers face in implementing sustainable practices. The research is based on a case study of a development in Zambia, which applied EDGE certification, and a review of relevant academic articles and reports.

It is anticipated that this study will contribute to the growing body of literature on sustainable real estate development in emerging markets and provide valuable insights for property developers, policymakers, and investors interested in advancing sustainability goals.

The real estate sector in emerging markets is a key driver of economic growth and a significant contributor to environmental degradation. With increasing recognition of the importance of sustainability, real estate developers are adopting various sustainable practices such as green building standards, to reduce the sector's environmental footprint. However, integrating sustainable practices in real estate development in emerging markets still faces various challenges. The EDGE certification system is a green building standard that has gained popularity in emerging markets. However, its effectiveness in promoting sustainable development and the challenges property developers face in implementing sustainable practices are still not fully understood. Therefore, there is a need to assess the integration of environmental sustainability practices in real estate development in emerging markets, with a particular focus on the role of EDGE certification in advancing sustainability goals and the challenges property developers encounter in implementing sustainable practices.

2 LITERATURE REVIEW

Green building standards such as EDGE certification have gained popularity worldwide as a means of promoting sustainable real estate development practices. The EDGE certification system was developed by the International Finance Corporation (IFC), a member of the World Bank Group, to promote resource-efficient building design and reduce the environmental impact of buildings. EDGE certification is based on three criteria: energy efficiency, water conservation, and materials selection.

The EDGE system uses a software tool to assess the resource efficiency of a building design. The software analyses various design parameters, such as building orientation, insulation, and lighting, and calculates the expected resource consumption of the building. The software also provides recommendations for improving the resource efficiency of the building design.

To achieve EDGE certification, a building design must meet a minimum threshold in each of the three criteria. The minimum threshold is based on the average resource consumption of buildings in the relevant market. Buildings that achieve EDGE certification are expected to be at least 20% more resource-efficient than the average building in the market.

EDGE certification offers several benefits to real estate developers. First, it provides a clear and measurable standard for assessing the sustainability of a building design. Second, it offers a practical and cost-effective approach to reducing the environmental impact of buildings. Third, it enhances the reputation of real estate developers by demonstrating a commitment to sustainability.

EDGE certification has gained popularity in emerging markets due to its practicality and affordability. The software tool used to assess building designs is user-friendly and can be used by architects and designers without extensive sustainability expertise. The certification process is also relatively inexpensive, making it accessible to smaller real estate developers.

Several studies have investigated the effectiveness of EDGE certification in promoting sustainable real estate development practices. For example, Alshuwaikhat & Abubakar (2017) conducted a study in Saudi Arabia to evaluate the potential environmental benefits of applying EDGE certification to commercial buildings. The study found that applying EDGE certification could reduce energy consumption and greenhouse gas emissions by up to 35% and 29%, respectively, while also reducing water consumption and solid waste generation. The study concluded that EDGE certification could be an effective tool for promoting sustainable real estate development practices in Saudi Arabia.

Similarly, a study by Nishijima *et al.* (2017) evaluated the effectiveness of EDGE certification in promoting sustainable real estate development practices in Brazil. The study found that applying EDGE certification to residential buildings could reduce energy consumption by up to 27%, water consumption by up to 43%, and greenhouse gas emissions by up to 37%. The study concluded that EDGE certification could be an effective tool for promoting sustainable real estate development practices in Brazil.

Other studies have investigated the effectiveness of EDGE certification in promoting sustainable real estate development practices in other countries. For example, a study by Mansur *et al.* (2021) evaluated the effectiveness of EDGE certification in promoting sustainable real estate development practices in Bangladesh. The study found that applying EDGE certification to commercial buildings could reduce energy consumption by up to 20%, water consumption by up to 42%, and greenhouse gas emissions by up to 38%. The study also concluded that EDGE certification could be an effective tool for promoting sustainable real estate development practices in Bangladesh.

Several case studies have also been conducted on applying EDGE certification in real estate development projects. For example, a case study by Boonlertvanich *et al.* (2019) evaluated the effectiveness of applying EDGE certification to a real estate development project in Thailand. The study found that applying EDGE certification to the project resulted in significant reductions in energy consumption, water consumption, and greenhouse gas emissions. Another case study by Razak *et al.* (2021) evaluated the effectiveness of applying EDGE certification to a real estate development project in Malaysia. The study found that applying EDGE certification to the project resulted in significant reductions in energy consumption, water consumption, and greenhouse gas emissions.

Building certifications are recognized programs that provide guidelines for constructing sustainable buildings. Besides EDGE, there are several other green building certifications available worldwide. LEED, BREEAM, and Green Star are some of the most widely recognized and applied certifications (Pomponi *et al.* 2020).

Leadership in Energy and Environmental Design (LEED) is an American-based certification system that sets guidelines for constructing sustainable buildings. LEED provides a framework for evaluating a building's energy efficiency, indoor environmental quality, water efficiency, and sustainable site development (Bastianoni *et al.* 2018). LEED has been recognized as a gold standard in green building certification, with more than 90,000 registered and certified projects worldwide (USGBC 2022). Building Research Establishment Environmental Assessment Methodology (BREEAM) is a UK-based certification system that evaluates the sustainability of buildings. BREEAM provides guidelines for energy efficiency, water usage, and waste management in a building (Bastianoni *et al.* 2018). BREEAM has been recognized as the most widely used certification system in Europe, with over 500,000 registered projects worldwide (BRE 2022). Green Star is an Australian-based certification system that evaluates the environmental impact of buildings. Green Star provides guidelines for energy efficiency, water usage, materials selection, and indoor environmental quality. Green Star has been recognized as a prominent certification system in the Asia-Pacific region, with over 2,800 certified projects worldwide (Green Building Council of Australia 2022).

Each certification system has its advantages and disadvantages. For example, LEED has been criticized for being too prescriptive and costly, while BREEAM has been criticized for not being challenging enough (Hartman *et al.* 2020). However, all of these certifications have helped to promote sustainable building practices and have made significant contributions to the global efforts to mitigate climate change.

Overall, the literature suggests that green building standards such as EDGE certification can be an effective tool for promoting sustainable real estate development practices in emerging markets. Several studies have shown that applying EDGE certification to real estate development projects can result in significant reductions in energy consumption, water consumption, and greenhouse gas emissions while also reducing solid waste generation. However, the literature also highlights several challenges property developers encounter in implementing sustainable practices, such as the lack of awareness, inadequate funding, and the need for technical expertise (Hailu *et al.* 2019). Addressing these challenges will be crucial to promoting sustainable real estate development practices in emerging markets.

3 RESEARCH METHODOLOGY

The research methodology for this study aimed to assess the integration of environmental sustainability practices in real estate development in emerging markets, using both qualitative and quantitative research methods. Qualitative research methods were employed through interviews and focus groups with stakeholders in the real estate industry, such as property developers, architects, engineers, and government officials. These interviews and focus groups were conducted to gain insights into stakeholders' perspectives and experiences related to sustainable practices in real estate development, including their understanding of EDGE certification, their motivations for adopting sustainable practices, and the challenges they face in implementing these practices.

Quantitative research methods were also utilized, involving the collection and analysis of data on the environmental performance of real estate developments in emerging markets that have implemented EDGE certification or other sustainable practices. This included data on energy consumption, water usage, waste management, and greenhouse gas emissions, among other factors. Comparative analysis was also conducted between real estate developments that have implemented sustainable practices and those that have not, to assess the environmental benefits of sustainable practices.

Furthermore, a case study approach was adopted, focusing on a specific real estate development in an emerging market that had implemented EDGE certification. The case study involved collecting data on the environmental and economic performance of the development, as well as conducting interviews with stakeholders involved in the development process. Specifically, the case study was based on a project undertaken by a real estate developer (Novare), which involved the construction of a head office for a multinational corporation based in Lusaka, Zambia. Novare incorporated EDGE certification for this project as a sustainable design approach that reduces energy consumption, environmental impact, and running costs while also creating a more pleasant working environment and boosting property value for the investor.

Gross Building Area (GBA) refers to the total floor area of a building, including all interior and exterior spaces, but typically excluding any unenclosed areas such as open balconies or patios. It is calculated by measuring the building from the exterior walls, including all levels and areas within its footprint, such as stairwells, hallways, and mechanical rooms. The office building, which was designed to suit the specific operational requirements of the client, has a total GBA of over 24,000m^2. The six-storey building has a Gross Internal Area of 10,880m^2 excluding parking. Broadly speaking, the whole enclosed area of a building within the external walls taking each floor into account and excluding the thickness of the external walls, is defined as the GIA. The ground floor of the building was designed to accommodate retail operations leading onto a street entrance, while the upper floors are predominantly open-plan office spaces. The building also incorporated a building management system to facilitate efficient management and monitoring of utilities and security. Occupancy sensors in various parts of the building also help to manage energy usage. This case study provided valuable insights into the practicalities and benefits of implementing sustainable practices in real estate development in emerging markets.

Overall, the research methodology employed a mixed-methods approach to capture both qualitative and quantitative data, including a case study to provide a comprehensive analysis of the integration of environmental sustainability practices in real estate development in emerging markets.

4 FINDINGS

One of the main findings of the study was the effective execution of measures to attain the EDGE certification for the multinational corporation building. The following is an assessment of the outcomes in accordance with the EDGE certification.

4.1 *Energy*

Energy efficiency is an essential consideration in building design and construction due to the significant impact buildings have on the environment and energy consumption. Energy consumption in buildings is responsible for a significant amount of greenhouse gas emissions and a significant percentage of global energy use. Therefore, it is critical to implement energy-efficient methods and materials during the design and construction phases of a building to reduce energy consumption and minimize its environmental impact.

Several energy-saving methods and materials were used in the building design of the multinational corporation building to achieve energy efficiency. The following summarises the methods and materials include:

- Window-to-wall ratio: A window-to-wall ratio is a measure of the amount of window area there is on a building relative to the total amount of exterior wall area.

  The most efficient window-to-wall ratios are those of approximately 20% window-to-wall area. With a north-east orientation, the building's window-to-wall ratio was calculated at 34.4%
- Insulation: Building insulation is an effective method of reducing energy consumption by minimizing heat transfer between the interior and exterior of a building. Greater thermal efficiency was achieved through the application of an insulating screed to the roof slab. Later on, the addition of solar panels also ultimately provided further shading.
- High-performance windows: Higher thermal performance glass (high-spec double glazing or "low E") was used for curtain walling/glass façade of the building to significantly reduce heat loss during cold seasons and minimize heat gain during hot seasons. This type of glass has advanced glazing and low-emissivity coatings that minimize the amount of heat that passes through them.
- Efficient lighting: Lighting can account for a significant percentage of a building's energy use. Therefore, energy-efficient lighting solutions such as Light-Emitting Diode (LED) bulbs, motion sensors, and daylight harvesting systems were used to reduce energy consumption.
- Efficient heating, ventilation, and air conditioning (HVAC) systems: HVAC systems account for a significant percentage of energy consumption in buildings. The use of energy-efficient HVAC systems, including variable refrigerant flow cooling systems, was used to significantly reduce energy consumption.
- Renewable energy sources: Renewable energy sources such as solar power was used to generate energy for the building, significantly reducing energy consumption from conventional sources.

Thus, energy efficiency is critical in building design and construction, and energy-saving methods and materials should be prioritized to minimize a building's environmental impact and reduce energy consumption. By implementing energy-saving methods and materials, buildings can achieve significant energy savings, reduce their carbon footprint, and contribute to a more sustainable future. The study found that the multinational corporation building not only met but exceeded the minimum requirement of 20% savings in energy, achieving over 50% reduction in energy usage.

4.2 *Water*

Water is a precious resource that is essential for the survival of all living organisms. Unfortunately, the world is currently facing a water crisis, with increasing demand and decreasing supplies of clean water. The real estate sector is a major consumer of water, and as such, must play a significant role in conserving water resources. Water conservation not only helps to ensure the sustainability of water resources but also provides several other benefits such as cost savings, reduced energy consumption, and reduced greenhouse gas emissions.

Several mechanisms can be implemented to reduce water usage and wastage in real estate development. The mechanisms used in the office building were the installation of low-flow fixtures, such as low-flow faucets and water-efficient toilets. These fixtures are designed to use less water while still providing adequate performance. For example, a low-flow faucet can reduce water usage by up to 30%, while a dual flush toilet can reduce water usage by up to 67% compared to a conventional toilet.

Boreholes are also an alternative source of water used to supplement municipal water supply. These mechanisms not only reduce the demand for municipal water but also provide a backup source of water in the event of a water shortage.

Additionally, other simple mechanisms are implemented in the day-to-day running of the complete building to conserve water, such as fixing leaks, repairing faulty fixtures, and educating building occupants on water conservation practices. Implementing these mechanisms can significantly reduce water usage and wastage, thus contributing to a sustainable future.

The study found that water conservation is an important aspect of sustainable real estate development and that the multinational corporation building exceeded the EDGE certification requirement of 20% savings in water, achieving a 55% reduction in water usage.

4.3 *Materials*

Volatile organic compounds (VOCs) are a group of chemicals that contain carbon and easily evaporate at room temperature, and are released into the atmosphere. They can be emitted from various sources, including industrial processes and consumer products such as paints, adhesives, and cleaning products. VOC emissions contribute to the formation of ground-level ozone and particulate matter, which are air pollutants that can have adverse health effects and impact the environment. Therefore, reducing VOC emissions is an important aspect of air pollution control and environmental sustainability.

Medium-weight hollow concrete blocks are precast blocks made of cement, sand, and lightweight aggregates such as expanded clay or shale, which have a relatively lower density than conventional concrete blocks. These blocks offer good thermal insulation properties and can be used for load-bearing walls, non-load-bearing walls, and partition walls. Insulated drywall, also known as thermal drywall or insulated gypsum board, is a type of drywall that is made with an insulating material such as expanded polystyrene (EPS) or extruded polystyrene (XPS) foam. It is used for internal partitioning and offers good thermal insulation properties.

Both medium-weight hollow concrete blocks and insulated drywall are sustainable building materials that can help reduce embodied energy in construction. Embodied energy is the energy consumed in the manufacturing, transport, and installation of building materials. By using materials that require less energy to manufacture and transport, and that offers good thermal insulation properties, the overall embodied energy of a building can be reduced. However, it is important to note that their embodied energy does not solely determine the environmental impact of building materials. Other factors, such as resource depletion, pollution, and end-of-life disposal, should also be considered when evaluating the sustainability of building materials. Additionally, the performance of these materials in terms of structural stability, fire resistance, and sound insulation should also be considered when selecting materials for internal partitioning.

The study on the building found that the building used materials with low VOC emissions, including medium-weight hollow concrete blocks and insulated dry-walling for internal partitioning. The building achieved a 34% reduction in embodied energy in materials, exceeding the EDGE certification requirement of a 20% reduction.

Overall, the study concluded that the successful implementation of these sustainability practices could lead to significant reductions in energy and water usage, as well as embodied energy in materials. These findings can inform real estate developers in emerging markets to incorporate similar sustainable practices in their projects, promoting environmental sustainability and reducing carbon footprints.

5 DISCUSSION

The property development industry is significant to the economy and the environment. In recent years, there has been growing concern about the negative impact of traditional property development practices on the environment. As a result, sustainable property development practices have gained momentum globally, and many certification schemes have been developed to recognize buildings that meet specific sustainability criteria. One of the most recognized certification schemes is the EDGE certification. Despite the increasing importance of sustainable development practices, property developers in emerging economies face several challenges in implementing them. In this section, the challenges expressed by property developers in the implementation of sustainable practices are highlighted. The challenges were consolidated from the interviews held with the practitioners:

- Financial Constraints: One of the most significant challenges property developers face in implementing sustainable practices is financial constraints. Sustainable practices require a significant upfront investment, which can be challenging for developers to finance, especially in emerging economies where access to capital is limited. Additionally, the cost of implementing sustainable practices is often higher than traditional practices, which may deter developers from adopting them. The return on investment for sustainable practices may not be immediately realized, further reducing their attractiveness to developers. A case study of the multinational corporation building revealed that the building required an estimated additional investment of USD 236,348.00 to implement identified interventions to reduce utility costs by an estimated USD 1,726/month while bringing down operational carbon emissions to a projected 317.75 tCO_2/Year.
- Limited Government Intervention: Another challenge faced by property developers in implementing sustainable practices is limited government intervention. Governments in emerging economies often lack the resources and political will to support sustainable development initiatives. This can result from inadequate regulations, weak enforcement, and a lack of incentives for sustainable practices. Although the property market in Zambia has shown much growth in promoting sustainable building practices and a general positive support for greener developments by the government, there are limited government regulations on the construction and operation of buildings to ensure green performance objectives.
- Inadequate Infrastructure: Inadequate infrastructure is another challenge property developers face in implementing sustainable practices. In many emerging economies, the lack of basic infrastructure, such as reliable electricity, water supply, and waste management systems, can make it difficult to implement sustainable practices.

- Limited Awareness and Education: Limited awareness and education is another challenge property developers face in implementing sustainable practices. Many developers and stakeholders in emerging economies may not be aware of the benefits of sustainable development or the negative impacts of traditional practices. This can result in a lack of demand for sustainable practices and a reluctance to invest in them. Additionally, there may be a lack of technical expertise and knowledge in sustainable practices, which can hinder their implementation. However, some developers have attempted to address this challenge by partnering with contractors and suppliers to facilitate skills transfer and promote sustainable business practices.

6 CONCLUSION

In conclusion, the importance of sustainable development practices in the real estate industry cannot be overstated. The EDGE certification scheme is an effective tool to promote sustainability goals in real estate development. However, property developers in emerging economies face several challenges in implementing sustainable practices, which require the collaboration of various stakeholders to overcome. The challenges of financial constraints, limited government intervention, inadequate infrastructure, and limited awareness and education can be addressed by policymakers, real estate developers, and financial institutions working together to promote sustainable development practices in emerging economies. To ensure the widespread adoption of green building standards, such as EDGE certification, policymakers should provide incentives and invest in education and training programs to increase awareness and capacity building. Additionally, real estate developers should view sustainability practices as a long-term investment, as they can result in significant cost savings and improve environmental performance and reputation over the life cycle of a building. Ultimately, integrating sustainability practices in real estate development is crucial for promoting sustainable economic growth in emerging markets and ensuring a more sustainable future for all.

ACKNOWLEDGEMENT

The authors would like to sincerely thank Novare for providing valuable data and information on the environmental sustainability practices incorporated in the design of the multinational corporation building. Their contribution was invaluable to the success of the research project, and we are grateful for their support and collaboration.

REFERENCES

Alshuwaikhat, H. M., & Abubakar, I. I. (2017). An Assessment of the Effectiveness of the EDGE Green Building Certification System in Promoting Sustainable Construction in Saudi Arabia. *Sustainability*, 9(2), 219. https://doi.org/10.3390/su9020219

Balaras, C. A. (2007). *Energy Performance of Buildings: Current Trends and Perspectives*. London, UK: RoutlEDGE.

Bastianoni, S., Pulselli, F. M., & Tiezzi, E. (2018). Sustainability Assessment of Buildings: A Review of Rating Systems. *Sustainability*, 10(11), 3906. https://doi.org/10.3390/su10113906

Boonlertvanich, K., & Suwannarat, K. (2019). The Relationship Between EDGE Certification and Financial Performance of Thai Listed Companies. *Sustainability Accounting, Management and Policy Journal*, 10(4), 543–562. https://doi.org/10.1108/SAMPJ-05-2018-0140

BRE. (2022). *BREEAM*. https://www.breeam.com/about-us/breeam/ [Last accessed 30 April, 2023].

Brundtland, G. H. (1987). *Our Common Future: The World Commission on Environment and Development*. Oxford University Press.

Green Building Council of Australia. (2022). *Green Star*. https://new.gbca.org.au/green-star/ [Last accessed: 30 April, 2023].

Hailu, A. T., Tafesse, B. A., & Yadav, R. K. (2019). Does Environmental Certification Matter? An Empirical Study of Certified and Non-certified Firms in Ethiopia. *International Journal of Environmental Science and Technology*, 16(8), 4295–4306. https://doi.org/10.1007/s13762-019-02335-6

Hartman, J., Adams, R., & Scanlon, A. (2020). Green Building Rating Systems: A Critical Review of Their Evolution, Application and Implications. *Journal of Cleaner Production*, 246, 118993. https://doi.org/10.1016/j.jclepro.2019.118993 https://sustainabledevelopment.un.org/content/documents/21252030%20Agenda%20for%20Sustainable%20Development%20web.pdf [Last accessed 10 April, 2023].

International Finance Corporation. (2019). *EDGE Green Building Certification System: Standard for Emerging Markets*. Washington, DC: World Bank Group.

International Finance Corporation. (2021). *EDGE Overview*. Retrieved from https://EDGEbuildings.com/EDGE-overview/ [Last Accessed: 8 May, 2023].

Jones Lang LaSalle. (2019). *Global Real Estate Transparency Index 2018*. Retrieved from https://www.jll.com.mo/content/dam/jll-com/documents/pdf/research/Global_Real_Estate_Transparency_Index_2018.pdf [Last Accessed: 5 May, 2023].

Kibert, C. J. (2008). *Sustainable Construction: Green Building Design and Delivery*. John Wiley & Sons.

Lamond, B. F., & Li, V. (2019). *Embedding Sustainability in Real Estate Development: An International Perspective*. RoutlEDGE.

Mahachi, J. (2021). Development of a Construction Quality Assessment Tool for Houses in South Africa. *Acta Structilia*. Vol 28, No. 1. DOI: 18820/24150487/as28i1.4.

Mansur, M. A., Islam, M. A., & Hossain, M. S. (2021). The Effect of EDGE Certification on the Financial Performance of Bangladeshi Companies. *International Journal of Economics, Commerce and Management*, 9(2), 1–19. https://doi.org/10.31580/ijecm.2021.9.2.1096

Moreno, M., & Garcia, M. (2019). Energy Efficiency and the Built Environment: Opportunities and Challenges for Emerging Markets. *Energy Research & Social Science*, 54, 83–91.

Nishijima, M., Shibata, T., & Fujii, H. (2017). Environmental, Social and Governance (ESG) Performance and Corporate Value: Empirical Evidence from Japanese Companies. *Journal of Business Ethics*, 145(2), 429–449. https://doi.org/10.1007/s10551-015-2833-0

Pomponi, F., Moncaster, A., & De Wolf, C. (2020). A Review of Sustainable Building Rating Systems: A Focus on Their Evolution and Process. *Journal of Cleaner Production*, 244, 118892. https://doi.org/10.1016/j.jclepro.2019.118892

Razak, D. A., Abdallah, A. E., Abdul-Razak, S., & Abdul-Razak, R. (2021). Exploring the Impact of EDGE Certification on Corporate Social Responsibility Performance: A Case Study of Malaysian Companies. *Sustainability*, 13(2), 603. https://doi.org/10.3390/su13020603

Sassen, S. (2001). *The Global City: New York, London, Tokyo*. Princeton University Press.

UN-Habitat. (2016). *World Cities Report 2016: Urbanization and Development – Emerging Futures*. United Nations Human Settlements Programme.

United Nations Environment Programme. (2018). *Global Status Report for Buildings and Construction 2018*. Nairobi: UNEP.

United Nations. (2015). *Transforming Our World: The 2030 Agenda for Sustainable Development*. Retrieved from:

World Green Building Council. (2018). *Bringing Embodied Carbon Upfront: Coordinated Action for the Building and Construction Sector to Tackle Embodied Carbon*. Retrieved from https://www.worldgbc.org/news-media/bringing-embodied-carbon-upfront-coordinated-action-building-and-construction-sector. [Last accessed: 6 May, 2023]

Exploring development control tools for improved sustainability in land use compatibility: Experiences from Vuwani town, Collins Chabane local municipality

R. Mulokwe & B. Risimati
Department of Urban and Regional Planning, University of Venda, Thohoyandou, Limpopo Province, South Africa

ABSTRACT: At present, in many rural small towns of developing countries, effective and efficient land use planning and development control are not well established. This is particularly so in small towns, characterized by the patent manifestation of the chaotic state of land use activities. Thus, this aimed at exploring development control tools for improved sustainability of land use compatibility in a growth point area of Vuwani Town. The research employed a phenom-enological case study design comprising of mixed method approach. A interviews were held with officials from Collins Chabane Local Municipality. Questionnaires were distributed to the developers and informal traders in Vumani town. The study results reveal that policy documents are key development control tools associated with land use compatibility. Drawings insights from the findings, this study recommends various initiatives to ensure the implementation of development control tools to curb land use incompatibility in small rural towns.

Keywords: Development control, land use, sustainability, town, tools

1 INTRODUCTION

In many rural towns of developing countries, effective and efficient land use planning and development control are not well established. Generally, for any system to work as expected there is always the need for control and, balance which is a form of regulation for necessary operation (Bryson & George 2020). Ratcliffe, Stubbs & Keeping (2021) state that in the built environment, development control regulations provide for these controls and balance, and that development control regulations (development guidelines and specifications) are provided by the various development control tools ranging from development plans to development legislations. These development control tools are intended to provide the strategic framework and policy context for all local planning decisions. In Africa, it is estimated that be-tween 2000 and 2030, the urban population will increase from 294 million to 742 million (UN-Habitat 2012). This phenomenon comes along with distressing problems especially with regard to the management of space, particularly in developing countries satellite towns. This is corroborated by Makato (2016) who maintains that in developing countries like Kenya, urban development has been taking place in a haphazard way especially in Nairobi's satellite towns such as Kitengela.

Amongst a wide range of development plans and development legislations, less attention is given on which of these tools positively affect the conformity of physical developments. This is corroborated by Abugtane (2015) who stated that most of the studies conducted in relation to the development control tools focused on their application, effectiveness as well as their compliance on physical development. The study seeks to assess how effective development control tools are in improving land use compatibility in a commercial district under the case study of Vuwani Town, Collins Chabani Municipality, South Africa.

2 RELATED WORK

Contemporary literature reveals several challenges with effective developments control. Most of the maps used to prepare planning schemes from the planning authorities are outdated (Lara. & Gómez-Urrutia 2022). It is claimed that the Planning Department does not visit the fields anymore but uses old maps to prepare land use plans. These plans are seen sometimes depict a different situation from that on the ground. The implication is that the prepared development plans will not be effectively implemented because they will be addressing situations which have long changed in the area.

The second challenge of effective development control traced is issue of public participation. A survey that has been conducted in Ghana about the level of public participation in planning activities have resulted in partial participation as corroborated by Lara & Gómez-Urrutia (2022). His results being that participation in planning activities is not encouraging as citizens complained that they do not see or hear what the planning authorities do in their communities. He further asserted that this has made some citizens to pass comments like —they are useless, and —I don't even know if we exist, among others. Moreover, even those who do participate like chiefs are unimpressed about the level of participation, as it is only limited to information giving. It is important for ordinary citizens to participate in planning process. However, certain technical details require technical persons and therefore might exclude them. The extent to which people are aware of the existence of planning activities and development control mechanisms is important because it partly determines the extent to which people will comply with these control mechanisms. Lack of public participation and awareness of urban development plans and planning legislations lowers the chance of successful implementation of development control tools and the degree of compliance with the required regulations (Peter & Yang 2019).

Spatial Planning and Land Use Management Act 16 of 2013 stipulates that any documents and plans that are provided to the local authority while applying for development permission have to be prepared by a qualified, registered and professional individual or entity to ensure that development proposals are prepared by the relevant professionals with adequate knowledge on the requirements of the law concerning development proposal (South Africa 2013). However, the professional fees charged by various professionals such as Architects, Planners and Civil Engineers can be identified as a serious hindrance to development within the legal frameworks of South Africa. These excessive fees are sometimes unaf-fordable thus discouraging developers to hire for services of these professionals, therefore leaving room for private developers to ruin the build environment (Lara & Gómez-Urrutia 2022).

Inadequate awareness of the development control tools has the negative impact on both the developers, local planning authority as the first instance for development control, as well as the society at large. These negative impact include inadequacy of provision for day-to-day physical shopping structures such as retail, lack of aesthetics, and development of human activities in unhealthy environment. The extent to which the public is aware of the existence of the development control tools is vital because it partly determines the extent to which people will comply with the standard/regulations of development control. A large portion of people are not aware of the development standards (Simmie 2020). The results being that there will be no- compliance to development control regulations negatively affecting the implementation of development plan, land use plans or scheme and policies.

3 METHODOLOGY

The research approach adopted in this study is qualitative research approach. As this study aims at unearthing development control mechanisms/tools that are known and offices involved in development control practices are al-so known, the study will therefore utilize purposive sampling technique, and snowball sampling techniques respectively. Purposive sampling technique was effective since the research targeted officials or key institutions based on their roles that are relevant to physical development control and enforcement of development control tools in relation to land use compatibility. These institutions included members of the development control section who are the implementers of various development control mechanisms, and these are the department of spatial planning officials, local economic development unit, and the municipal building inspectorate department. The heads of these departments were selected purposively because they had an in-depth knowledge on development control tools, and the enforcement of development control regulations, which can foster land use compatibility from the study area.

For the purpose of this study, the researcher have used an interview guide and questions that are structure and unstructured to solicit information from several municipal officials. Figure 1, is provided to explain the main players on development control mechanisms of Collins Chabane local municipality.

Figure 1. Hardware and liquor restaurant in Vuwani town. Source: Author's field data (2022).

4 RESULTS AND DISCUSSION

The study further sought to examine to what extent the relevant development control tools will fulfil its intended purpose. From the study analysis, it is evident that development control in Vuwani area is at low level against a backdrop of myriad of factors inter-alia lack of adequate human resource at the municipality, political interference, and poor communications. However, these issues ought to be addressed and the possible solutions be implemented for development control tools to fulfil its intended purpose for the matter in question from the study area.

The study revealed that the planning and development section in Collins Chabane municipality is facing the challenge of inadequate human resource. This simply indicates that development control tools are not fully given the necessary attention it deserves, and this greatly affect the discharge of duties as the limited officials are over burdened with duties where they concentrate on matters they deem as important. Abugtane (2015) corroborate that a country can only be sure of orderly development in urban areas if there is effective and qualified human resources to combat and address contemporary planning problems as cited in Nyadombo (2018). Furthermore, the respondents interviewed on the challenges also pointed out that on legal backings, the planning and development control section have just one personnel that attend to legal backings of which if the person is on leave, everything stops. Alluko (2011) posited that institutional capacity with respect to human resource has greater influence in enforcing development control effectively (Ibid). Therefore, lack of adequate staff to implement development control tools compromises and results to the sprouting of non-conforming land use developments as it prevails in Vuwani study area.

Thus, in order for development control tools that are associated with land use compatibility such as land use scheme as per the study findings not to remain a mirage, more workforce should be employed and deployed to Vuwani study area, and they should also be educated enough on the concepts of such development control tools. Following are pictures conducted through author's field-work, depicting different types of land use activities in Vuwani study area adjacent to each other.

Figure 1 above illustrates two different land use activities in Vuwani study area that are located in an adjacent location where one that is on the left is a hardware (building material retail) and one on the right is a liquor restaurant operating at the same building. The exercise of each these land use activity are questionable if one land use activity conform with the adjacent land use and whether are there any development control protocols that have been followed, hence the study in question. However, the study prevails that there are no development control implemented for the functioning of these activities. Following are the other pictures depicting other land use activities that are also adjacent to each other taken from the study area.

Figure 2. A funeral pallor and food retail store in Vuwani town. Source: Author's field data (2022).

Figure 2 is comprised of three sub-plates (A, B & C). These sub-plates indicates other dominant questionable land use activities operating at an adjacent location. A illustrates a funeral pallor whilst B indicates a food and accessories retails store. C illustrates how the two land use activities are adjacent to one another on a back-to-back situation, thus one is facing to the right and the other facing to the left. Apart from having development activities that are adjacent to each other as shown above, and which are questionable enough on the protocols that they undergo for their operations, there are visible informal structures throughout the study area.

Political interference have been identified as one of the major challenge revealed by the study that imped against the implementation of development control tools that are associated with land use compatibility from the study area. One of the critical motives behind the burning of shops and other properties were the influence of political leaders who thought they were to lose power since they no longer fall under Makhado municipality where they had power and while joining a newly established municipality of Collins Chabane local municipality. Hence, the very same limited human resource/municipal officials who deals with development control find it hard to do site visits to areas such as Vuwani study area since they are afraid of the threats from the community members under the influence of political leadership. Thus, the findings here imply that politics is also a factor for improving land use compatibility that needs to be addressed, not until this matter is resolved, the development control tools associated with compatible land use will remain a mirage. Following are the pictures depicting chaos that occurred in Vuwani during protests against their incorporation into Collins Chabane local municipality.

Figure 3. Vuwani residents protesting.

Part (A) the community members together with politicians are wearing red and blue t-shits written '100% Makhado', a local municipality in which protests wanted to remain in for service delivery. Furthermore, (B) depicts community members looting and burning infrastructures around Vuwani area. For the purpose of this study, the key informant interviewee (KII) have acknowledge that municipal officials in the Planning and Development department of Collins Chabane municipality are not practising myriad of their primary duties in Vuwani study area. This include an assessment of land development applications, the facilitation for the removal of illegal structures, conducting awareness and workshops with regards to policies and making sure that everything is as approved in terms of quality control and management etc.

The study further reveals that there are communication challenges between municipal officials and the society of Vuwani town at large, wherein the community members are not well-informed about the processes to follow during their developments (it can either be a street vendor or any other business) and all the by-laws available from the municipality regarding land use. This militates against the implementation of development control initiatives by the municipal officials as it makes execution of duties and enforcements to be in contrary with what wishes the community since there is no clear communication channel. Nduthu (2010) posited that successful development control services require clear management direction and commitment whereby the authority provide adequate publicity in the area planned to be incorporated in their plan, give those interested an adequate opportunity to make representation and make them aware of this opportunity as cited in Fourie (2019).

However, the key interview respondents pointed out that poor communication was partly because of the threats that comes out of the society of Vuwani town due to their incorporation to the newly established municipality while they still intend to remain to their previous municipality. Thus, the findings here imply that in order for development control tools associated with land use compatibility to fulfil its intended purpose, the challenge of poor communication should first be addressed. From the study analysis, it is evident that in order to address the issue of communication, there should be an effective community participation in the preparation of development plans where community members are given an adequate opportunity to make representations while informing them on the processes to follow during their developments and all the by-laws available.

Furthermore, one of the key informant from the department of planning and development when responding to the question of 'what do you think are the main causes of conflicting land use activities from the study area?' specifically said that

' ...*my personal understanding is that political issues that took place in the year/s 2015/2016 when the demarcation board took the decision to establish Collins Chabane local municipality incorporating parts Makhado local municipality and including Vuwani town, is one of the motives Vuwani is experiencing such burning issues because there is lack of development control. The reason why development control is poor is that many officials are not comfortable to visit the study area because they are afraid of threats from the general public who are influenced by politicians'.*

The responses from the key informant interviews were many but pointed out to the same meaning thus the study specifically chosen this one that incorporated the issues of political influences. Hence, one can note that political influence play a vital role in development control and is also a factor to curb or otherwise conflicting land use activities, thus development control tools can either be implemented or otherwise through the influence of politicians for land use compatibility to be possible from the study area. The key informants further provides that though there are such development control tools such as the land use scheme which are associated with land use compatibility, their implementation remains difficult from the study area due to several complications which includes political interference and others. One of the key informant had this to say:

'... though the municipality have some forms of development control initiatives in case of such conflicting land use development, the implementation remains a mirage since only few people submit applications and there are also threats from the public, hence few workforce are deployed to Vuwani area'.

Another key informant interviewee stated that:

'... Without the application of the development control tools land use compatibility cannot be possible because in order to obtain land use rights you need approval for all the tools'

and that

'... We are short of man power in terms of inspectors on site and legal law enforcers, due to the fact that there is no budget for them yet'

The findings here implies that there are various forms of development control initiatives as provided by the key informants themselves. However, there are also various challenges to the implementation of those development control tools. This concurs to what Abukari & Dinye (2011) who says that the inability (specifically short in man power) of development control departments to ensure effective implementation causes unauthorized developments and encroachment on public open spaces and government land (Nyadombo 2018), hence the study area is experiencing such non-conforming and conflicting land use developments.

5 CONCLUSIONS

The study focused at determining the conceptual issues that links development control tools with land use compatibility on a municipal growth point area (Vuwani town). The study identified myriad of factors that links development control tools with land use compatibility inter-alia awareness into development control tools associated with land use compatibility, staff establishment into planning and development control unit, the infiltration of political influence into land management and development matters through demarcations, as well as communications amongst development control stakeholders including the general society. The study found out that development control in Vuwani town is at a low level against a backdrop of 68 % of the questionnaire respondents who posits they are not aware of what development control tools entails. This shows that there is partial knowledge of what development control tools such as LUS entails, and who are the affected parties.

Furthermore, 71% of the key informants corroborate there is poor staff establishment in the planning and development section of Collins Chabane municipality. Moreover, political interference was also revealed by the study that it imped against the implementation of development control tools where politicians are against taking advices to decide on the best way to utilize land and buildings from officials in Collins Chabane municipality. These findings dented the land use conformity and harmonious land use locations in Vuwani town. In addition, the study also analysed various instruments of development control associated with land use compatibility for the purpose of this study. The study found out that there are various development control tools/instruments associated with land use compatibility inter-alia spatial development framework (SDF), Spatial Planning and Land Use Management Act (SPLUMA by-laws), Land Use Scheme (historically known as Town Planning Scheme in South Africa), Rezoning, and Consent use to name few. More emphasis should be placed on the results part whereby they concentrate on inputs from different stakeholders so as to effectively and efficiently use the limited resources. Therefore, future studies are encouraged to focus more on the appropriateness of the development control tools as well as the technical capacity of the development control unit.

REFERENCES

Abdulrashid, I.A.I., Mohamad, D.B. and Badrulzaman, N., 2022. The Growth And Transformation of Katsina Metropolis between 1987 and 2017: The Physical Planning Implications, *Fudma International Journal of Social Sciences*, 3(1), pp.74–83.

Abugtane, A. 2015. *Assessing the Effectiveness of Physical Development Planning and Control Mechanisms in Ghana: The Experience of Wa Municipality*. Kumasi: Kwame Nkrumah University of Science and Technology.

Bibri, S.E., Krogstie, J. and Kärrholm, M., 2020. Compact City Planning and Development: Emerging Practices and Strategies for Achieving the Goals of Sustainability. *Developments in the Built Environment*, 4, p.100021.

Biel, W., Ariola, M., Bolshakova, I., Brunner, K.J., Cecconello, M., Duran, I., Franke, T., Giacomelli, L., Giannone, L., Janky, F. and Krimmer, A., 2022. Development of a Concept and basis for the DEMO Diagnostic and Control System. *Fusion Engineering and Design*, 179, p.113122.

Boamah, N., Gyimah, C., and Nelson, J. (2012). Challenges to the Enforcement of Development Controls in the Wa Municipality. *Habitat International*(36), 136–142.

Daniel, G. 2000. *"Development Control in Introduction to Planning Practice"*. "West Sussex, John Wiley and Sons Limited". London.

Davies H.W. E, Steeley, G. C., John, F. and Roger, W. Suddards. 1980. Policy Forum: The Relevance of Development Control: *The Town Planning Review*. Vol. 51, pp. 5The Town Planning Review24.

Dawaba, J. 2016. *Evaluation of the Effectiveness of Development Control in Minna Metropolis, Nigeria*. Department of Geography Ahmadu Bello University Zaria, Nigeria.

Dawson, V.P., 2019. Protection from Undesirable Neighbors: The Use of Deed Restrictions in Shaker Heights, Ohio. *Journal of Planning History*, 18(2), pp.116The Town Planning Review136.

Keeble, L. 1972. *"Town Planning Made Plain"*. "London and New York, Construction Press".

Khanzode V.V., (1995), *Research Methodology: Technique & Trends*, New Delhi: APH Publishing Corporation.

Kjærås, K., 2021. Towards a Relational Conception of the Compact City. *Urban Studies*, 58(6), pp.1176–1192.

Koroso, N.H., Zevenbergen, J.A. and Lengoiboni, M., 2020. Urban Land Use Efficiency in Ethiopia: An Assessment of Urban Land use Sustainability in Addis Ababa. *Land Use Policy*, 99, p.105081.

Long, Y., Han, H., Lai, S.K., Jia, Z., Li, W. and Hsu, W., 2020. Evaluation of Urban Planning Implementation from Spatial Dimension: An Analytical Framework for Chinese Cities and Case Study of Beijing. *Habitat International*, 101, p.102197.

Lu, P. and Zhang, W., 2022. Exploring Practical Village Planning in the Context of Territorial Spatial Planning-An Example of Baozi Village in Baihe Town. *Academic Journal of Science and Technology*, 2(1), pp.94–102.

Lubida, A., Veysipanah, M., Pilesjo, P. and Mansourian, A., 2019. Land-use Planning for Sustainable Urban Development in Africa: A Spatial and Multi-objective Optimization Approach. *Geodesy and Cartography*, 45(1), pp.1–15.

Makato, W. 2016. *Analysis of Development Control Regulations Compliance in Kitingela Town, Kajiado County, Kenya*. University of Nairobi.

Ntlhe, D., 2023. An Evaluation of the Relationship Between Spatial Form and Transport Mode in Elim, South Africa. *International Journal of Environmental, Sustainability, and Social Science*, 4(1), pp.73–80.

Peter, L.L. and Yang, Y., 2019. Urban Planning Historical Review of Master Plans and the Way Towards a Sustainable City: Dar es Salaam, Tanzania. *Frontiers of Architectural Research*, 8(3), pp.359–377.

Phosho, M.H. and Gumbo, T., *South Africa's Pursuit of Sustainable Urban Development: A Reality or Rhetoric?*.

Ratcliffe, J., Stubbs, M. and Keeping, M., 2021. *Urban Planning and Real Estate Development*. Routledge.

Ren, C., Yu, C.W. and Cao, S.J., 2023. Development of Urban Air Environmental Control Policies and Measures. *Indoor and Built Environment*, 32(2), pp.299–304.

Sachs, J.D., Schmidt-Traub, G., Mazzucato, M., Messner, D., Nakicenovic, N. and Rockström, J., 2019. Six Transformations to Achieve the Sustainable Development Goals. *Nature Sustainability*, 2(9), pp.805–814.

Schiavina, M., Melchiorri, M., Freire, S., Florio, P., Ehrlich, D., Tommasi, P., Pesaresi, M. and Kemper, T., 2022. Land Use Efficiency of Functional Urban Areas: Global Pattern and Evolution of Development Trajectories. *Habitat International*, 123, p.102543.

Varpio, L., Paradis, E., Uijtdehaage, S. and Young, M., 2020. The Distinctions Between Theory, Theoretical Framework, and Conceptual Framework. *Academic Medicine*, 95(7), pp.989–994.

Zhang, J., Li, S., Lin, N., Lin, Y., Yuan, S., Zhang, L., Zhu, J., Wang, K., Gan, M. and Zhu, C., 2022. Spatial Identification and Trade-off Analysis of Land Use Functions Improve Spatial Zoning Management in Rapid Urbanized Areas, China. *Land Use Policy*, 116, p.106058.

Zhou, Y., Chen, M., Tang, Z. and Mei, Z., 2021. Urbanization, Land Use Change, and Carbon Emissions: Quantitative Assessments for City-level Carbon Emissions in Beijing-Tianjin-Hebei region. *Sustainable Cities and Society*, 66, p.102701.

Zhou, Y., Wu, T. and Wang, Y., 2022. Urban Expansion Simulation and Development-oriented Zoning of Rapidly Urbanising Areas: A Case Study of Hangzhou. *Science of the Total Environment*, 807, p.150813.

Zhu, Q., Lei, Y., Sun, X., Guan, Q., Zhong, Y., Zhang, L. and Li, D., 2022. Knowledge-guided Land Pattern Depiction for Urban Land Use Mapping: A Case Study of Chinese Cities. *Remote Sensing of Environment*, 272, p.112916

Development of South African structural performance criteria for innovative building systems

N. Bhila & J. Mahachi*
University of Johannesburg, Johannesburg, Gauteng, South Africa

ABSTRACT: The utilization of innovative building systems to address shortage of low-cost housing in South Africa gained interest in recent years. However, low adoption rate of these systems is partly due to concerns surrounding their structural performance compared to conventional building systems. Determining reliable structural performance criteria is crucial to address this issue. This study aims to address this gap in literature by investigating the development of structural performance criteria for innovative building systems. The study employs both qualitative and quantitative research methods, including numerical simulations. Current building codes, regulations and unique challenges of various climatic conditions are considered. The results provide a critical evaluation of current knowledge on structural performance criteria for innovative building systems and identifies areas for improvement. The findings will contribute to developing updated criteria for assessing these systems, resulting in quality low-cost housing solutions and greater adoption rate.

Keywords: Innovative building systems, Performance-based, Structural reliability

1 INTRODUCTION

In the past decade, South Africa has been faced with a shortage of housing due to fast-growing population. In most cases, conventional building materials and methods are used to address this problem. However, it cannot be entirely addressed using conventional methods due to many households not affording, due to the increase in the building materials and services. To lower the costs, the government attempted to find solutions through large-scale production, economies of scale, and subsidies have resulted. This has resulted in the switch from conventional methods to innovative building systems to overcome the constraints (Wienecke, 2010). Innovative building systems (IBS) comprise new structural designs, construction methods and materials which are used solely or in conjunction with conventional building materials (Agrément SA, 2021).

SANS 10400 is currently used as the South African performance building code to assess these systems. SANS 10400-B provides structural performance criteria to be used in the assessment of these systems and a set of deemed-to-satisfy provisions, in which when building systems are built according to the provision, they are deemed to have satisfied the requirements of the criteria. Assessment and certification by Agrément SA are mandatory for all IBS prior to construction (SABS standard division, 2019). The evaluation is done by the assessment of the building system against performance-based criteria (Agrément SA, 2021).

However, there have been concerns about their structural performance following several failures of these systems under extreme climate conditions. Studies have been conducted to investigate the structural performance of innovative building systems assessed and

*Corresponding Author: jmahachi@uj.ac.za

constructed according to SANS 10400. Even though there are other concerns about poor material quality and inadequate construction monitoring, most of the studies warrant the review of the technical standards used to assess these systems (Mahachi et al., 2018). In a study assessing the performance of deemed-to-satisfy SANS 10400 masonry wall provisions, de Villiers et al (2021) concluded that the wall provisions require reconsideration to align with current South African loading conditions. In another study assessing the performance of deemed-to-satisfy SANS 10400 roof anchor systems, Van Der Merwe & Mahachi (2021) suggested the revision of the deemed-to-satisfy provisions in accordance with the criteria that is suitable for the current climate conditions.

This research aims to review the South African performance-based National Standard SANS 10400-B, and supplementary standards used by Agrément South Africa and other testing bodies to assess the structural performance of innovative building systems. The study aims to further look at conducting a numerical analysis that will contribute to the development of structural performance criteria for innovative building systems.

2 LITERATURE REVIEW

2.1 *Performance-based assessment criteria for innovative building systems*

Literature indicates that there has been an extraordinary development and evolution of innovative building systems considered receptive to the urban poor over the past years. However, "there are no appropriate frameworks or methodologies that can be used to assess the response of these innovative building systems holistically" (Odhiambo & Wekesa, 2010). Although there is no standard way of assessing innovative building systems, there is a collection of performance criteria used to assess the building features. Performance-based building assessment criteria comprise performance requirements for structural safety, fire resistance, serviceability, energy proficiency, and environmental effect (Coglianese et al., 2003). Structural performance must be evaluated mainly from three viewpoints, specifically safety, serviceability, and stability. Standard assessments can be agreed upon to verify the fitness for purpose of specific materials for example hard body and soft body impact tests, tests for fittings, resistance to door banging, storm resistance, and so forth (Watermeyer & Pham, 2011).

The performance-based building (PBB) concept offers a technically non-prescriptive and flexible framework for structural design and construction (Jasuja, 2005). While construction regulations and design codes are generally associated with specific construction materials and design solutions, performance criteria are derived from client needs and are independent of specific innovative solutions. The successful implementation of performance criteria relies upon the feasibility of determining the criteria. Performance criteria define the necessary attributes of a building in terms of the functions the structure must perform. These qualities are well-defined in terms of the user's needs for safety, activity support, maintenance expenses, comfort, and visual adequacy (Yokel, 1972). This building approach is mainly concerned with the result of the required outcome and not prescribing how the outcomes must be achieved, 'the means' (Foliente, 2005).

The performance of the building components and materials determines the total performance of a building. However, the performance of the components and the material must be assessed to ensure that the building meets the required performance level. Performance features of components, materials and arrangements cannot adequately and completely describe the performance of the structure (Trinius et al., 2005). The quality of the building depends on the interactions of various elements. While no standard assessment methods are available to evaluate the structural performance for innovative building systems. The only way of assessing is by simulation of the building's response to certain conditions (Foliente, 2000).

The development of innovative solutions by regulating outcomes in contrast to prescriptive methods is the fundamental principle of performance-based code. This is facilitated through a progression of mandatory qualitative statements. These qualitative statements define the building occupant's requirements instead of setting a quantitative measurement for compliance. However, updating the building code allows for making corrections, including omissions, introducing new concepts or methods for smooth implementation, responding to changes from research findings, gaining experience, and meeting up with the expectation of society (Nwadike *et al.*, 2019).

Many countries such as Australia, the UK, New Zealand, the USA, Japan, etc. use three major methods to demonstrate compliance with building regulations in line with the performance-based approach. The first approach is the verification method which uses the testing method (laboratory testing), experiments (tests-in-situ), and engineering analysis like calculations which when followed are deemed fit for compliance. The second method is the acceptable solution which prescribes how the structure should be built to demonstrate compliance, which is mainly adopted in simple residential buildings. Lastly, is the alternative solution where the primary innovation and uniqueness are rooted in performance-based regulation because it empowers the designers to introduce a new solution (Nwadike, 2021).

2.2 Adoption of performance-based building in Africa and abroad

There are ongoing studies on performance-based building in Africa. However, there are very few countries that have adopted this method into their regulation. The adoption of the performance-based concept at the regulatory level has been reviewed for the following countries since they have a more developed framework in terms of performance-based building:

Australia – Australia uses the Australian National Construction Code (NCC) as a performance-based and plumbing code. This code provides the minimum requirements for the construction of new buildings and new construction work within an existing building. The Australian code gives performance requirements and compliance solutions. These compliance requirements can either be a performance solution or a deemed-to-satisfy (DTS) solution or both (Australian Building Codes Board, 2022). The list of actions that the structure must withstand are provided in the NCC. The five-percentile method is used to determine the resistance of materials. When using the performance provision verification method, the structural reliability method is used for components with a resistance coefficient between 10% and 40% (Australian Building Codes Board, 2019, Kemp et al., 1998).

To demonstrate compliance, a comparative study between the performance solution and the DTS provisions is carried out. These methods will determine whether the performance solutions have equivalent or better performance as compared to the DTS provisions, if so then the performance solution is deemed to satisfy the NCC performance requirements. Other verification methods such as tests, calculations, inspections, or a combination of these which assess performance are used (Australian Building Codes Board, 2022).

Europe – Technical assessments are issued for construction products in accordance with the guidelines for European technical approval established under Directive 89/106/EEC. Specific assessment methods for a particular building system must be established to suit its properties. Criteria for performance assessment of conventional building systems are specified in the harmonized standards. The manufacturer must request the European Technical Assessment for products that are not covered or not fully covered in the harmonized standards then the European Technical Document will be drafted. The structural performance criteria and method for assessing the performance of the product in relation to its properties are contained in the European Assessment Document. Where the performance parts of the product can be assessed per the criteria and method already established in the harmonized standard such existing methods and criteria is incorporated in the European Assessment Document.

United States of America – Most parts of the International Building Code (IBC) which is used in the USA for the assessment of innovative buildings is performance-based. However, prescriptive standards are referenced in the code to assist in the application of the code.

Chapter 16 of the IBC provides the minimum load requirements. Loading conditions, combinations, and criteria for load combination are also stated. The design requirements are provided in terms of general requirements, strength, and serviceability requirements. Chapter 17 provides testing details for innovative building systems with alternative materials or construction methods that cannot be designed by conventional engineering analysis and are not covered in referenced standards (International Building Code, 2012).

2.3 *South African performance-based regulation framework*

The South African performance code was established in 1990 and only revised in 2008 due to the change in the apartheid rules, the fast-growing population, and the changes in regulations. Figure 1 shows the framework used for the performance-based assessment of systems in South Africa including the innovative building systems. The following levels are used along with the regulations/standards stipulated in each level.

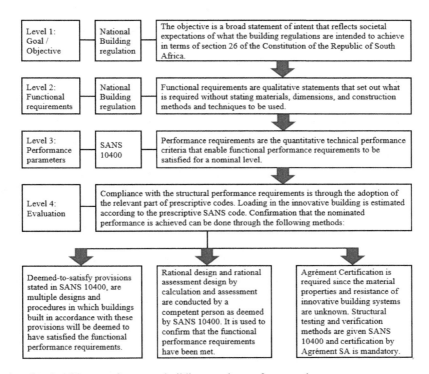

Figure 1. South African performance-building regulatory framework.

3 METHODOLOGY

Both qualitative and quantitative research methods were used. The qualitative approach is used for exploring the current structural performance of IBS in South Africa. Also, to understand the adoption of performance-based regulation for assessing the performance of IBS in SA and abroad. Literature is typically collected from building regulations/standards and past studies. The quantitative approach is used to determine the quantitative performance requirements (structural performance criteria). Numerical analysis has been used to calculate load parameters for innovative building systems including wind, dead and seismic loading. The latest South African prescriptive loading code SANS 10160 was used for the calculations. Load parameters calculated in the study are analysed and compared with the parameters provided in SAN 10400-B. Methods for determining each loading parameter are explained below.

3.1 Wind loading

SANS 10160-3 was used for the calculation of peak wind loads. The South African wind loading map provided in SANS 10160-3 provides fundamental basic wind speeds ($v_{b,0}$) to be considered when calculating wind loading for different regions in South Africa. This study considered all four regions with fundamental basic speeds of 44m/s, 40m/s, 36m/s, and 32m/s. Table 1 gives wind load parameters based on conditions more likely to be true for innovative building systems used in low-cost housing in South Africa.

Table 1. Wind load parameters to SANS 10160-3, 2019.

Parameter	Symbol	Value	Clause
Fundamental basic wind speed	vb,o	44, 40, 36 and 32m/s	Figure 1
Terrain Category	–	C	Table B.1
Probability of exceedance	P = 1/T	0.02	Cl. 7.2.3
Terrain roughness/Height factor	cr(z)	0.73	Table 3
Topography (Orography) factor	co(z)	1.00	Cl. 7.3.1.1
Air density	Qp(z)	1.2 kg/m3	Table 4

The South African National Builders Manual requires the design working life of a home in respect to the structural system to be more than 30 years (NHBRC, 2014). Therefore, the mean return period is taken as 50 years as per SANS 10160-1 which classifies residential housing in design category 3. The terrain category is selected as the one more likely to have low-cost residential buildings such as villages and suburban terrain. The terrain topography is taken as flat. Site altitude is taken approximately at sea level. The wind velocity is considered for structures less than 4m above the ground.

The peak wind pressure for each region has been calculated as per this procedure using the parameters provided in Table 1

$$\text{Step 1 – Basic wind speed } v_b = c_{prob} \times v_{b,0} \tag{1}$$

$$\text{Where } C_{Prob} = \left[\frac{1 - k \times \ln\{-\ln(1-p)\}}{1 - k \times \ln\{-\ln 0.98\}} \right]^n \tag{2}$$

$$\text{Step 2 – Peak wind speed } v_{p(z)} = C_r(z) \times C_o(z) \times v_{b,peak} \tag{3}$$

$$\text{Where } v_{b,peak} = 1,0 \times v_b \tag{4}$$

$$\text{Step 3 – Peak wind speed pressure } q_p(z) = 1/2 \times p \times v_p^2(z) \tag{5}$$

3.2 Seismic loading

SANS 10160-4 was used to derive the guidelines for the configuration of innovative building systems in seismic loading zones. Most parts of South Africa will not experience excessive seismic loading. However, some areas will experience seismic loading which is categorized into two zones according to SANS 10160-4. Consideration of seismic loading in a building system shall depend on the seismic zone of the area where the building will be constructed and the importance class. Table 2 gives seismic loading parameters to be used in the calculation of seismic loading for innovative building systems used in low-cost housing in South Africa where the recommended configuration cannot be achieved.

Table 2. Parameters for seismic loading per SAN10160-4.

Parameter	Symbol	Value	Clause
Peak ground acceleration (a_g)	a_g	0.15g	–
Ground type	–	4	Table 1
Building importance factor	Y_1	2	Table 3
Behaviour factor	q	1.0	Cl 5.3.4
Factor for Fundamental period of vibration	C_T	0.05	Cl 8.5.2.1
Soil factor	S	1.35	–

Based on the available seismic hazard zones, the highest peak ground acceleration is provided as 0,15g. The ground type with the worst bearing capacity has been selected. The importance factor is selected for the residential type building. Since innovative building systems comprise various material compositions, a behaviour factor of 1 is selected. The factor for the fundamental period is determined based on the type of building.

3.3 Imposed load and self-weight

SANS 10160-2 has been used for determining the criteria for the imposed load. Literature has been used to determine the criteria for self-weight.

4 FINDINGS

The following loading criteria which can form part of the revision of the current structural performance criteria for IBS have been determined.

4.1 Wind loading

The peak wind pressures are calculated for all the fundamental wind speeds in South Africa namely 44m/s, 40m/s, 36m/s, and 32m/s. Figure 2 shows the calculated unfactored peak wind pressure based on SANS 10160-3. Figure 3 shows the calculated factored peak wind pressure based on the partial wind factor of 1.6 as required by SANS 10160-1.

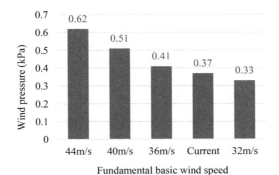

Figure 2. Unfactored peak wind pressure.

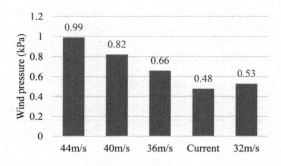

Figure 3. Factored peak wind pressure.

Based on the results shown in Figure 2, the current service peak wind pressure as recommended by SANS 10400-B is 40% lower than the highest peak wind pressure which is in zones with 44m/s fundamental wind speed. However, areas with a basic wind speed of 32m/s still have a service wind speed slightly below the one started in the current SANS 10400-B. Based on Figure 3 the current ultimate peak wind is 52% lower than the highest required peak wind pressure. None of the areas meet the current code requirements for factored peak wind pressure.

4.2 *Seismic loading criteria*

The first approach to address seismic loading will be the configuration of the building. The geometry, openings, and roof of innovative building systems in seismic hazard zones shall comply with the following:

Geometry – A one-storey building shall be designed to ensure that there is a good distribution of bracing walls. Plans that provide symmetrical resistance in both directions are preferred.

Openings – The total area of openings shall be less than one-third of the wall area. The positioning of openings should ensure uniform distribution of strength. Large openings in external walls should be avoided and placing openings near the corners is not preferred. In cases where this condition cannot be avoided, lateral stability must be ensured by providing adequate detailing.

Roofs – Heavy roof structures are not preferred, particularly on lightweight wall constructions. Where heavy roofs are used, lateral stability must be ensured by providing adequate detailing for heavy roof structures on the lightweight wall.

Gables and parapet walls – Lateral stability shall be ensured but providing proper reinforcement or support with transverse walls.

In areas of high seismicity, hipped roofs are preferred as compared to gables.

In cases where the configuration of the building cannot be proven adequate to resist seismic loading, the seismic loading resistance shall be assessed. Load due to seismic action shall be calculated according to SANS 10160-4. The building systems are suitable for the use of the lateral force method, therefore Equation 6 and Equation 7 shall be used

$$\text{Design base shear force } V_n = S_d(T) \times W_n \qquad (6)$$

$$\text{Sustained vertical load } W_n = G_n + \psi_i \times Q_{ni} \qquad (7)$$

Sustained vertical load is calculated using Equation 7 as recommended by SANS 10160-4 since the nominal weight of the building and sustained portions of imposed vertical loads are unknown. The correction factor of 0,85 shall be used where proper reinforcement and detailing is provided.

4.3 Imposed loading criteria

Table 3 shows the criteria for imposed loading as derived from SANS 10160-2. It is selected based on the focus of the study which is low-cost residential housing. These criteria include uniformly distributed loads to be applied on roofs as well as floors.

Table 3. Imposed loading for a residential dwelling.

Building component	Stages of loading	Uniformly distributed load (kN/m²)
Inaccessible roof	Construction	0.75 for area $\leq$ 3 m² and 0.25 for area $>$ 15 m²
	Maintenance	0.50 for area $\leq$ 3 m² and 0.25 for area $>$ 15 m²
Accessible roof	–	2.0 or 2kN over an area of 0.1m x 0.1m
All floors	–	1.5 or 1.5 kN

4.4 Permanent loading criteria

The self-weight of the building system and fixtures shall be taken into consideration in the design of building systems as per SANS 10160-1 (SABS Standards Division, 2018). The self-weight for building systems that consists of known material shall be calculated based on the density of the material constituents as provided in SANS 10160-2 Table A.1 to Table A.6. The self-weight for a building material that consists of unknown material, the density shall be experimentally determined and used for the calculation of self-weight (de Villers et al., 2021). Equation 8 shall be used for the determination of the self-weight load effect.

$$E = \int_V I\gamma \mathrm{d}V \tag{8}$$

4.5 Combination effect for structural resistance

Using the partial factors for the STR and STR-P combination case as given in SANS 10160-1. The following load combinations shall be used.
Load combination for transient design situation (STR)

$$E_{d,STR} = 1,2G_k + 1,6Q_{k,1} + \sum_{i>1} \psi 1,6Q_{k,1} \tag{9}$$

If the combination of the permanent action and appropriate leading variable action leads to the most unfavourable action, then Equation 10 is used. This equation is only used when the permanent action is larger than the variable action, if not, Equation 9 is used.
Load combination for permanent design situation (STR-P)

$$E_{d,STR-P} = 1,35G_k + 1,0Q_k \tag{10}$$

5 DISCUSSION AND PRACTICAL IMPLICATIONS

The results of the study show that there has been a significant change in the South African wind loading conditions since the last revision of SANS 10400. The highest calculated peak wind pressure of 0.62 kPa is more than the 0.37 kPa specified in SANS 10400-B (SABS Standards Division, 2020), only buildings in areas with the fundamental wind speed of 32m/s meet the currently used criteria for service wind speed and none for the ultimate peak wind pressure. Therefore, building systems are more likely to pass the wind loading assessments but fail under the actual site conditions. The study gave the criteria for imposed, permanent loading and seismic loading which are not adequately stated in SANS 10400. The load combinations have been changed due to the change in partial factor for wind load from 1.3 to 1.6 in the current SANS 10160 code.

The study focused on the development of structural performance criteria at a broader scope. Therefore, for some innovative building systems, further calculations will be necessary to suit the material properties and configuration of the structure. The results will offer a technical guide for calculating the exact loading to be applied and contribute to the body of knowledge in performance-based building regulation formation. Numerical findings can be used to improve the current structural performance criteria. Application of loading criteria that is in line with the current climate conditions as per the results will improve the performance of these systems under actual site conditions. Changes in partial load factors and load combinations will increase the performance reliability and thus increase the level of adoption of IBS in low-cost housing.

6 CONCLUSION

South Africa has a well-developed performance-based regulation framework for the assessment of innovative building systems. However, revision of the technical criteria is necessary to keep up with climate change. Alternative building systems are strongly gaining force in construction to tackle socio-economic crisis faced in South Africa. Ongoing revisions and development of codes to cater for these systems is recommended. Socio-economic crisis faced in South Africa are greatly faced in other African countries. Adoption of the PBB concept can be the first step for the introduction of innovative building systems in other African countries.

The results of the study also focused on the development of technical structural performance criteria. The constant change in wind conditions has a major impact on the structural performance criteria. Frequent review of the wind loading criteria for innovative building systems may be necessary. Seismic loading has also become a concern in some South African regions, the inclusion of such actions in the criteria is recommended. The study suggests further research on the material resistances of innovative building systems. Deemed-to-satisfy solutions are designed to meet the requirements of the structural criteria. Therefore, revision of these provisions will be mandatory upon the revision of the criteria.

REFERENCES

Agrément South Africa 2021. *Performance Criteria: Building and Walling Systems – Structural Strength and Stability*. Pretoria: Agrément South Africa.

Australian Building Codes Board 2019. *National Construction Code Series, Volume 1—Class 2 to Class 9 Buildings, 2019 edition, vols. 1, 3*. Canberra: Australian Building Codes Board.

Australian Building Codes Board 2022. *Understanding the NCC Assessment Methods*, Canberra.

Coglianese, C., Nash, J. & Olmstead, T. 2003. Prospects and Limitations in Health, Safety, and Environmental Protection. *Performance-based Regulation* 55: 705.

de Villiers, W.I., van Zijl, G. & Boshoff, W.P. 2021. Review of Compatibility Between SANS 10400 Deemed-to-satisfy Masonry Wall Provisions and Loading Code. *Journal of the South African Institution of Civil Engineering* 63(1): 45–60.

Foliente, G., Boxhall, P. and Pham, L. 2005. Facilitating Innovation & Enhancing Trade–The Performance-Based Building Networks in Australia & Asia. *CIB Symposium Combining Forces*: 55–59.

Foliente, G.C. 2000. Developments in Performance-based Building Codes and Standards. *Forest Products Journal* 50(7/8): 12.

International Code Council 2021. *International Building Code*. Falls Church: International Code Council.

Jasuja, M. 2005. The Performance Based Building Network: Impacts and Perspectives. *CIB Symposium Combining Forces*: 1–13).

Kemp, A.R, Mahachi, J. and Milford, R.V. 1998. Comparisons of International Loading Codes and Options for South Africa. *South African National Conference on Loading – Towards the Development of a Unified Approach to Design Loading on Civil and Industrial Structures for South Africa.*

Mahachi, J., Bradley, R. and Goliger, A. 2018. Windstorm Damage to Houses: Planning and Design Considerations. *Proceedings, Out-of-the-Box Human Settlements Conference, CSIR, Pretoria.*

National Home Builders Regulation Council 2014. *Home Builders Manual*. Johannesburg: National Home Builder's Regulation Council.

Nwadike, A., Wilkinson, S. and Clifton, C. 2019. Comparative Insight on Building Code Paradigm shift Practice and Updates: International Perspectives. In *The 4th International Conference on Civil, Structural and Transportation Engineering (ICCSTE'19)*. Ottawa, Canada.

Nwadike, A.N. 2021. *The Impacts of Building Code Amendments in New Zealand. A thesis submitted in partial fulfilment of the requirements for the degree of Doctor of Philosophy in Construction*. Auckland, New Zealand: Massey University.

Odhiambo, J. & Wekesa, B. 2010. A Framework for Assessing Building Technologies for Marginalised Communities. *Human Settlements Review*: 59–85.

SABS Standards Division 2011. *SANS 10160-2, Basis of Structural Design and Actions for Buildings and Industrial Structures – Part 2: Self-weight and Imposed Loads*. Pretoria: SABS Standards Division.

SABS Standards Division 2017. *SANS 10160-4, Basis of Structural Design and Actions for Buildings and Industrial Structures Part 4: Seismic Actions and General Requirements*. Pretoria: SABS Standards Division.

SABS Standards Division 2018. *SANS 10160-1, Basis of Structural Design and Actions for Buildings and Industrial Structures – Part 1: Basis of Structural Design*. Pretoria: SABS Standards Division.

SABS Standards Division 2019. *SANS 10160-3, Basis of Structural Design and Actions for Buildings and Industrial Structures –Part 3: Wind Actions*. Pretoria: SABS Standards Division.

SABS Standards Division 2022. *SANS 10400-A, The Application of the National Building Regulations – Part A: General principles and Requirements*. Pretoria: SABS Standards Division.

SABS Standards Division 2022. *SANS 10400-B, The Application of the National Building Regulations – Part B: Structural Design*. Pretoria: SABS Standards Division.

Trinius, W., Sjöström, C., Chevalier, J.L. and Hans, J. 2005. Life Performance and Innovation on Construction Materials and Components. *11th Joint CIB International Symposium Combining Forces-Advancing Facilities Management and Construction through Innovation*: 14–24.

Van Der Merwe, R. & Mahachi, J. 2021. An Investigation of South African Low-income Housing Roof Anchor Systems. *Journal of the South African Institution of Civil Engineering* 63(4): 24–34.

Watermeyer, R. & Pham, L. 2011. A Framework for the Assessment of the Structural Performance of 21st Century Buildings. *Structural Engineer* 89(1): 19–25.

Wienecke, M.A. 2010. Promoting Alternative Technologies: Experiences of the Habitat Research and Development Centre (HRDC). *Human Settlements Review* 1(1): 12–33.

Yokel, F.Y. 1972. *Structural Performance Evaluation of Innovative Building Systems. National Bureau of Standards*. Washington, DC: US Government Printing Office.

Smart infrastructure and cities, digital innovation in construction & digital transition

Municipal free Wi-Fi, governance and service delivery in the city of Tshwane

T.P. Mathane & T. Gumbo
University of Johannesburg, Johannesburg, South Africa

ABSTRACT: After the year 2010, big cities in South Africa started introducing free data programmes. The interventions are commendable because they are in sync with the Sustainable Development Goal (SDG) 11, i.e. of building sustainable cities/communities. The main question tackled by the study was to establish the extent to which the Tshwane free municipal Wi-Fi is assisting citizens to realize improvements on the right to the city, good governance, and enhanced service delivery. The researchers used a case study research design, qualitative in nature. Primary and secondary data sources were used. Fifty residents responded to a randomly administered online survey. The survey results were analyzed by categorical aggregation and content analysis. The study finds that the residents feel that the Tshwane free Wi-Fi is making some positive impact on their lives. The authors recommend that the city explore innovative funding/revenue monetization models to make free Wi-Fi financially sustainable.

Keywords: Sustainable Development, Smart city, Free Wi-Fi. Tshwane, Digitisation, Digitalization

1 INTRODUCTION

In the past two decades, there have been different academic contributions on the role of smart technologies in urban sustainability. One contribution is that smart city technologies can marginalize the poor in a city (Willis 2019) and produce inequalities (Caragliu & Del Bo 2022). Yet another contribution is that the smart city technologies can make cities more sustainable (Hamza 2021), thus bringing positive impacts on communities (Alavi *et al.* 2018). However, despite all these contributions, there are still some knowledge gaps. One of the gaps is that the majority of the studies are not based on empirical data. Another gap is that the majority of the studies are done in developed economies, and very few in the African context.

2 CONCEPTUAL FRAMEWORK

The concept of sustainable development encapsulates different pillars, e.g. economic, social, environmental, and institutional/governance (Cecchin *et al.* 2021). However, in the context of a digital world, some scholars aptly posit that the technological pillar is very relevant (Mondini 2019). Sustainable development is about raising the living standards of the next generations (Kopnina 2020; Schopp *et al.* 2020). Many other scholars opine that smart city digital technologies successfully assist cities to address urban management challenges (Allen *et al.* 2020; Onyango *et al.* 2021).

In the context of the 4IR, the creation of sustainable cities is one of the most pressing challenges (Yang & Taufen 2022), especially due to increasing technological advances (Steputat *et al.* 2020). There is a need to build sustainable cities/communities, as espoused by

Sustainable Development Goal 11 (Mondini 2019; Zielinska-Dabkowska & Bobkowska 2022). Sustainable smart cities deploy digital technologies to improve the quality of life for residents (Zvolska *et al.* 2019). This brings the concept of the right to the city to the fore. Influenced by the thoughts of Henri Lefebvre's work The Right to the City, Willis (2019) asks a question: *"Whose Right to the Smart City?"* Ben-Lulu (2021) asks yet another important question: who has the right to the city? Urban policies should promote the notion of citizenship for all (Borja 2019), because the right to the city is an agenda for social change (Vergara-Perucich & Arias-Loyola 2019), especially in smart cities (Van der Graaf 2020). The deployment of smart cities technologies can benefit citizens in various ways (Timeus *et al.* 2020), including efficiency in the delivery of public services (Yang *et al.* 2019). Most citizens use e-government platforms to access basic services (Allen *et al.* 2020). However, Aurigi & Odendaal (2022) caution the use of smart digital technologies should be context and place-based, to enhance their sustainability. E-platforms can encourage more citizen participation (Gade 2019), and thus improve service delivery (Gil *et al.* 2019). In addition, digital platforms can strengthen governance systems (Rodríguez Bolívar 2019), and enhance accountability (Basu 2019).

3 METHODOLOGY

This research investigated the views of the residents in the Pretoria CBD and surrounding suburbs in the City of Tshwane (CoT), South Africa, as far as the impact of the Tshwane free Wi-Fi on the citizen's right to the city, good governance, and service delivery. The study used an online survey tool, which was administered to fifty (50) residents in the Pretoria CBD covering the following areas: Hatfield (48%), Sunnyside (24%), and the main Pretoria CBD (12%). The profile of respondents were mixed and dynamic: 54% of the respondents were students; 30% were employed, and 10% and 6% were unemployed graduates and self employed respectively. Approximatly 90% were between the ages of 18 and 34, followed by 8% between 34 and 45. No less than 42% had a 3-year post-matric qualification, 44% with an honors degree and above, and 14% with matric. The data from these residents were collected through an oline survey tool, captured, synthesized, and analyzed. In addition, secondary sources (reports, academic journals, books, etc) were used. So, the research is qualitative in nature and was guided by the qualitative approach.

4 FINDINGS

The findings as follows are based on the results the residents who responded to a randomly administered online survey. This section begins with a discussion about the SWOT analysis of free Wi-Fi in Tshwane. It then focuses on the views of the respondents, focusing on the impact of free Wi-Fi on the right to the city, service delivery, and good governance.

4.1 *SWOT analysis of Tshwane free Wi-Fi*

Ramokgopa's (2018:226) provides insightful thoughts about the SWOT analysis of free Wi-Fi in Tshwane. From the perspective of strengths, strong executive commitment, coupled with administrative and technical internal competency in ICTs is key. From a Business Continuity Management perspective, it is also commendable that when the Democratic Alliance (DA) took over governance in the CoT, the free Wi-Fi programme was continued. From the perspective of weaknesses, the fact that the city does not own bandwidth is one of the strategic risks to be managed judiciously. It can also be added that the fact that the CoT has not yet found a sustainable solution for theft and vandalism constitutes a major threat. Another major weakness relates to the lack of clarity on the funding model.

Table 1. The SWOT analysis of Tshwane free Wi-Fi.

STRENGTHS	WEAKNESSES
• Executive Leadership is committed to the programme. • Internal CoT ICT technical team. • Partnership model with competent service providers.	• The funding model is not yet clarified • Lack of CoT ownership of bandwidth • Lack of specialized and experienced legal team for contract management

OPPORTUNITIES	THREATS
• Increased smartphone usage in CoT. • Universal access to the internet • Increasing education levels of CoT residents • CoT enjoyed first-mover advantages • Online learning opportunities in CoT. • Revenue generation opportunities • Internet-based businesses can grow. • More private-sector investments • Increases engagements between city management and communities • Real-time communication	• The perception that the internet is a luxury can result in unfavourable funding decisions. • CoT can face possible litigation by some private sector internet providers. • Likely to commit mistakes as first movers • Negating the potential scale benefits. • Financial unsustainability. • Cyber security risks • Delayed legislative clarity because municipalities are not designated as internet service providers. • Impersonal community interactions

From the perspective of opportunities, there are many positives – increasing internet access, accelerating usage of smartphones, internet-based and/or online learning for teachers and learners, as well as opportunities for local online businesses Ramokgopa (2018). However, there are also missed opportunities. Currently, opportunities for the CoT to partner with CBOs, NGOs, and the residents themselves are not fully optimised. In addition, opportunities for crowd souring funding need to be explored. The City is not optimally pursuing monetizing opportunities through the free Wi-Fi platform. From the perspective of a threat, some of the key challenges include cyber security risks and financial unsustainability. Significantly, lack of legislative clarity considering that municipalities are not designated as internet service providers; including the fact that free data is not yet gazetted as a basic function of municipalities in South Africa is one of the strategic threats as well.

4.2 Tshwane free Wi-Fi and the right to the city

Because the right to the city is a multilayered concept, different aspects were dealt with in this study. Overall, generally, 32% indicated that the free Wi-Fi is making a lot of contribution, whereas 45% belive that it is making modest contribution, followed by 14% who

Figure 1. Tshwane free Wi-Fi and the right to the city, social justice.

belive that it is making low contribution. Only 8% of the respondensts indicate that the free Wi-Fi is making no contribution in terms of improving chances to the right to the city.

One aspect relates to the right to the city is about informational rights. In this regard, the respondents were asked their opinion regarding the Tshwane free Wi-Fi's contribution to enhancing the capacity of residents to exercise their informational rights to the city of Tshwane. The results are shows below:

Figure 2. Tshwane free Wi-Fi and the contribution to informational rights of citizens.

A majority of respondents (42%) felt that the free Wi-Fi is making a modest contribution, followed by 30% who believe that it is making a lot of contribution. About 22% felt that the contribution is modest. Only 6% of the respondentss belive that the free Wi-Fi is making no impact on the informational rights to the city.

Another aspect related to the right to the city is about social justice. The respondents were asked their opinion regarding the Tshwane free Wi-Fi's contribution to enhancing social justice in Tshwane. The results are as follows:

Figure 3. Tshwane free Wi-Fi and contribution to social justice.

A majority of respondents (56%) agree that the Tshwane free Wi-Fi's is making positive contribution to enhancing social justice in Tshwane, followed by 20% who strongly agree. Another 20% is not sure; meanwhile those who disagree make less than 4% of the respondnets.

4.3 *Tshwane free Wi-Fi and service delivery and governance*

The respondents were asked their opinion regarding the Tshwane free Wi-Fi's contribution to enhancing service delivery in Tshwane. The results are as follows:

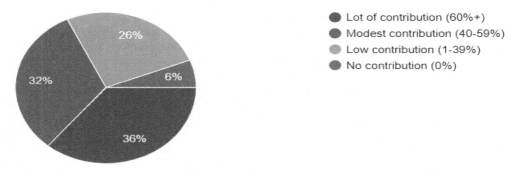

Figure 4. Tshwane free Wi-Fi and contribution to service delivery.

More residents (36%) feel that the Tshwane fre Wi-Fi is making a lot of contribution to enhance access to services; whilst 32% believe that the contribution is modest. About 26% of the respondents are not sure, whilst only 6% belive that the free Wi-Fi is not making any contribution.

Another critical question relates to the use of free Wifi by the residents to engage the City on service delivery matters. Interestting results are depicted below:

Figure 5. Tshwane free Wi-Fi and contribution to acces to services.

Approximately 44% of the respondents believe that many residents in Tshwane rely on the Tshwane Free Wi-Fi to engage the city on service delivery matters. About 34% are not sure; whilst 22% believe that generally, Tshwane residents prefer traditional (non-digital) ways of engaging the city.

Respondents were also asked if they themselves use the Tshwane free Wi-Fi to engage the city of service delivery matters. The results are depicted below:

Figure 6. Tshwane free Wi-Fi and contribution to engagements on services.

About 22% of the respondents indicated that they are part of the Tshwane residents who use Tshwane Free Wi-Fi to engage the city on service delivery matters. Interestingly, 42% indicate that they use Tshwane free data for service delivery engagements only when their private data is finished. The rest (22%) indicate that they prefer residents prefer traditional (non-digital) ways of engaging the city.

Respondents were also their views about the use of tshwane free Wi-fi to do online access of statements, bills, payment of bills, etc. The results are depicted below:

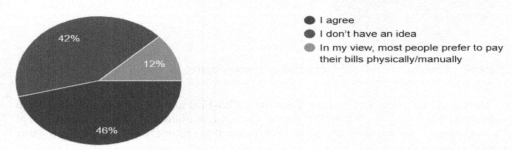

Figure 7. Tshwane free Wi-Fi and its use towards online access of statements, bills, payment of bills.

Nearly half of the respondents (46%) believe that many Tshwane residents use the Free Wi-Fi to pay their bills online, check statements online, etc. And, 12% believe that residents still prefer the traditional (non-digital) ways to pay their utility bills. Approximately 42% were not sure.

Finally, the respondents were asked to share their opinions regarding whether the Tshwane Free Wi-Fi is contributing to allowing residents to participate in governance processes in the City of Tshwane.

Figure 8. Tshwane free Wi-Fi and its use to empower residents to participate in governance processes.

Some 36% of the respondents feel that the Tshwane free Wi-Fi is making a lot of contribution in enhancing local participatory democracy. However, another 36% believe that even though the contribution does exist, is it just modest. 20% of the respondents feel that free Wi-Fi is making low contribution in this regard. Finally, only 8% feel that the free Wi-Fi program is not making any contribution whatsoever in this regard.

5 CONCLUSIONS AND RECOMMENDATIONS

A key aspect of the Sustainable Development Goals (SDG) 11 is about building sustainable cities/communities. The study finds that the residents belive that the free Wi-Fi is making positive contribution to SDG 11 in at least three ways/areas. First, as far as service delivery access and governance, more than 57% of the residents feel that the Tshwane free Wi-Fi is making useful contribution. Another area is in relation to the to the right to the city (informational rights). Even on this one, a more than 70% of the users feel that the free Wi-Fi is making psitive contibution. The third area is on the role of the free Wi-Fi in enhancing local participatory democracy. Even here, more than 56% of the users believe that the Tshwane free Wi-Fi is making useful contribution. In conclusion, the Tshwane free Wi-Fi seem to be enhancing the possibilities of making the Pretoria CBD more liveable, sustainable; supporting principles of social justice, fairness and equity. Importantly, the city need to explore innovative funding and revenue monetization models so that the free Wi-Fi is financially sustainable.

REFERENCES

Allen, B., Tamindael, L. E., Bickerton, S. H., & Cho, W. 2020. Does Citizen Coproduction Lead to Better Urban Services in Smart Cities Projects? An Empirical Study on E-participation in a Mobile Big Data Platform. *Government Information Quarterly*, *37*(1), 101412.

Aurigi, A., & Odendaal, N. 2022. From "Smart in the Box" to "Smart in the City": Rethinking the Socially Sustainable Smart City in Context. In *Sustainable Smart City Transitions* (pp. 53–68). Routledge.

Basu, I. 2019. Elite Discourse Coalitions and the Governance of 'Smart Spaces': Politics, Power, and Privilege in India's Smart Cities Mission. *Political Geography*, *68*, 77–85.

Borja, J. 2019. The Right to the City: From the Street to Globalisation. *Monografias CIDOB*, *76*, 33–42.

Cecchin, A., Salomone, R., Deutz, P., Raggi, A., & Cutaia, L. 2021. What is in a Name? The Rising Star of The Circular Economy as a Resource-related Concept for Sustainable Development. *Circular Economy and Sustainability*, *1*(1), 83–97.

Gade, D. 2019. Technology Trends and Digital Solutions for Smart Cities Development. *International Journal of Advance and Innovative Research*, *6*(1), 29–37.

Gil, O., Cortés-Cediel, M. E., & Cantador, I. 2019. Citizen Participation and the Rise of Digital Media Platforms in Smart Governance and Smart Cities. *International Journal of E-Planning Research (IJEPR)*, *8*(1), 19–34.

Kopnina, H. 2020. Transitioning to Quality Education: Examining Education for Sustainable Development Goals, Its Limitations, and Alternatives. *Quality Education*, 1.

Mondini, G. 2019. Sustainability Assessment: From Brundtland Report to Sustainable Development Goals. *Valori e Valutazioni*, (23).

Onyango, J., Kiano, E., & Saina, E. 2021. Environmental Damage Theory Applicable to Kenya. *Asian Journal of Business Environment*, *11*(1), 39–50.

Rodríguez Bolívar, M. P. 2019. In the Search for the 'Smart' Source of the Perception of Quality of Life in European Smart Cities.

Schopp, K., Bornemann, M., & Potthast, T. 2020. The Whole-institution Approach at the University of Tübingen: Sustainable Development Set in Practice. *Sustainability*, *12*(3), 861.

Steputat, C.C., Ural, D. and Nanni, A. 2020. Sustainable Cities and Communities Through GFRP Secant-pile Seawall Innovation, Sustainability, Fortification and Hurricane Storm Surge Protection. In *IOP Conference Series: Earth and Environmental Science*, 588 (4), IOP Publishing.

Timeus, K., Vinaixa, J., and Pardo-Bosch, F. 2020. Creating Business Models for Smart Cities: A Practical Framework. *Public Management Review*, *22*(5), 726–745.

Van der Graaf, S. 2020. The Right to the City in the Platform Age: Child-friendly City and Smart City Premises in Contention. *Information*, *11*(6), 285.

Vergara-Perucich, J. F., & Arias-Loyola, M. 2019. Bread for Advancing the Right to the City: Academia, Grassroots Groups and the First Cooperative Bakery in a Chilean Informal Settlement. *Environment and Urbanization*, *31*(2), 533–551.

Willis, K. S. 2019. Whose Right to the Smart City?. In *The Right to the Smart City*. Emerald Publishing Limited.

Yang, L., Elisa, N., & Eliot, N. 2019. Privacy and Security Aspects of E-government in Smart Cities. In *Smart Cities Cybersecurity and Privacy* (pp. 89–102). Elsevier.

Zielinska-Dabkowska, K. M., & Bobkowska, K. 2022. Rethinking Sustainable Cities at Night: Paradigm Shifts in Urban Design and City Lighting. *Sustainability*, *14*(10), 6062.

Zvolska, L., Lehner, M., Voytenko Palgan, Y., Mont, O., & Plepys, A. 2019. Urban Sharing in Smart Cities: The Cases of Berlin and London. *Local Environment*, *24*(7), 628–645.

Municipal free Wi-Fi intervention: Mamelodi and Soshanguve comparative analysis

T.P. Mathane & T. Gumbo
University of Johannesburg, Johannesburg, South Africa

ABSTRACT: Access to data is no more a luxury; it is a basic need for daily livelihood in the context of the fourth industrial revolution. Circa 2013, the City of Tshwane introduced a free Wi-Fi programme, with at least three strategic objectives: (a) social justice, (b) economic enablement, and (c) digitalising service delivery access and public engagements. A case study research design was used. Primary and secondary data sources were used. One hundred (100) free Wi-Fi users from Mamelodi (50) and Soshanguve (50) responded to a randomly administered online survey. The study articulates the lived experiences of the users of the Tshwane free Wi-Fi. The findings reveal that the intensity and scale of use of the free Wi-Fi is not the same for different communities. This study suggests some recommendations on improving the possibilities of the realisation of the three objectives stated.

Keywords: Fourth Industrial Revolution, Smart City, City of Tshwane, Tshwane Free Municipal Wi-Fi

1 INTRODUCTION

In the past two decades, there have been different academic contributions on the role of smart technologies in urban sustainability. One contribution is that smart city technologies can marginalize the poor in a city (Willis 2019) and produce inequalities (Caragliu & Del Bo 2022). Yet another contribution is that the smart city technologies can make cities more sustainable (Hamza 2021), thus bringing positive impacts on communities (Alavi et al. 2018). However, despite all these contributions, there are still some knowledge gaps. One of the gaps is that the majority of the studies are not based on empirical data. Another gap is that the majority of the studies are done in developed economies, and very few in the African context.

2 METHODOLOGY

This research investigated the views of the Tshwane free Wi-Fi users (residents) in Mamelodi and Soshanguve townships in the City of Tshwane (CoT), South Africa. The focus of the study was on the extent to which the Tshwane free Wi-Fi is meeting its stated strategic objectives of (a) social justice, (b) economic enablement, and (c) digitalising public consultations. The study relied on an online survey tool to gather the data. Using a case study research design, a total of one hundred (100) users in these two townships (50 each) responded to the online survey. Secondary data sources were also used in this study; and the results were analysed by categorical aggregation and content analysis.

3 CONCEPTUAL FRAMEWORK

The concept of sustainable development encapsulates different pillars, e.g. economic, social, environmental, and institutional/governance (Cecchin *et al.* 2021). However, in the context of the fast-changing digital world, some scholars aptly posit that the technological pillar is very relevant (Mondini 2019). It is believed that sustainable development is about raising the living standards of the next generations (Kopnina 2020; Schopp *et al.* 2020). Many other scholars opine that smart city digital technologies successfully assist cities to address urban management challenges (Allen *et al.* 2020; Onyango *et al.* 2021).

4 STUDY SETTING

4.1 *Profiling the townships of Soshanguve and Mamelodi*

The Soshanguve Township (affectionately called *Sosh*) is situated about 30 km north of Pretoria. It is geographically located in Region 1 of the City of Tshwane. Established in 1974, it is one of the oldest Black-dominated townships in the City of Tshwane; and one of the top 10 biggest townships in South Africa. The name Soshanguve is an acronym for Sotho, Shangaan, Nguni and Venda. This makes Soshanguve a multi-ethnic community. Currently, the population of Soshanguve estimated at 879 000; and is estimated to reach 1 million in 2025 (City of Tshwane 2023). The Soshanguve based free Wi-Fi users who participated in the survey came from the following areas: Block L, DD, BB, VV, P, S, XX, R, EXT, EXT 2, EXT 4, EXT 7, EXT 1, M EXT, Ext 4, and Ext 2. Geographically located in Region 6 of the City of Tshwane, the Mamelodi Township is situated about 20km east of the City of Tshwane. Mamelodi is one of the oldest townships in Tshwane. This township was established in June 1953, then called Vlakfontein, according to the Group Areas Act. Mamelodi (affectionately called *Mams*) is one of the oldest Black-dominated townships in the City of Tshwane; and one of the top 10 biggest townships in South Africa. The name Mamelodi was given by the then South African State President, Paul Kruger because he thought Africans were able to whistle like birds (melody). The population of Mamelodi is approximately 334 557 and nearly all residents are African (98.8%), and 61% reside in formal dwellings (City of Tshwane 2023). In Mamelodi, the respondents came from the following townships: Mamelodi West, Lusaka, East, Ext 7, Ext 8, Ext 4, Nellmapius, Ext 2, Mamelodi Gardens, Ext 5, Industrial View, Mahube Valley, C1, Gardens, Moretele View, Section D, and Sun Valley.

5 DISCUSSION AND FINDINGS

5.1 *Tshwane free Wi-Fi as an enabler for social justice*

5.1.1 *Contribution of the Tshwane free Wi-Fi on social justice*

When asked about the extent to which the Tshwane Free Wi-Fi is contributing to enhancing social justice in Tshwane, Table 1 below shows that a majority, i.e. 78% of respondents in Mamelodi agreed or strongly agreed that the Tshwane Free Wi-Fi is contributing to enhancing social justice in Tshwane. In the case of Soshanguve, 73% of the respondents similarly agreed or strongly agreed with the strategic objective of ensuring that the Tshwane Free Wi-Fi is positively contributing towards enhancing social justice in the city. Specifically, over 51% of respondents in Mamelodi agreed, while 27% *strongly agreed*. In the case of Soshanguve, specifically 44% agreed, and 29% strongly agreed. Those that disagree make 13% and 17% in Mamelodi and Soshanguve respectively.

Table 1. Respondent's views about the contribution of the Tshwane free Wi-Fi on social justice.

Mamelodi	Responses	Soshanguve
50.9%	I agree	44.1%
27.3%	I strongly agree	28.8%
12.8%	I disagree	16.9
9.1%	I am not sure	10.2%
Total: 100%		**Total: 100%**

5.2 *Tshwane free Wi-Fi as an enabler for participation on social media platforms*

When asked about the extent to which the Tshwane Free Wi-Fi is enabling participation on social media platforms, as shown in Tshwane, Table 2 below, a majority, i.e. 84% of respondents in Mamelodi agreed or strongly agreed that the Tshwane Free Wi-Fi is enabling participation on social media platforms. In the case of Soshanguve, 88% of the respondents similarly agreed or strongly agreed that the Tshwane Free Wi-Fi is immensely enabling participation on social media platforms. Specifically, over 51% of respondents in Mamelodi agreed, while 31% strongly agreed. In the case of Soshanguve, specifically 42% agreed, and 46% strongly agreed. Those that disagree make 11% and 7% in Mamelodi and Soshanguve respectively.

Table 2. Tshwane free Wi-Fi as an enabler for participation on social media platforms.

Mamelodi	Responses	Soshanguve
52.7%	I agree	42.4%
30.9%	I strongly agree	45.8%
10.9%	I disagree	6.8%
5.5%	I am not sure	5.1%
Total: 100%		**Total: 100%**

When the respondents were asked about whether they themselves are part of the Tshwane residents who use Tshwane Free Wi-Fi to participate on social media platforms, Table 3 below shows that 35% of respondents in Mamelodi agreed, compared to 44% in the case of Soshanguve. About 46% of the respondents in Mamelodi indicated that they their private data for social media, compared to 27% in Soshanguve.

Table 3. Respondent's view on their usage of Free Wi-Fi to participate on social media platforms.

Mamelodi	Responses	Soshanguve
34.5%	Yes, I always use Tshwane free Wi-Fi data for social media	44.1%
45.5%	No, I use my private data for social media	27.1%
20.0%	I only use Tshwane free Wi-Fi data for social media when my private data is finished	28.8%
Total: 100%		**Total: 100%**

Those who only use Tshwane free Wi-Fi data for social media when their private data is finished account for 30% in Mamelodi, and 29% in Soshanguve. On average, the results show that more respondents (40%) always use Tshwane Free Wi-Fi to participate on social media platforms, compared to about 37% of respondents who use their private data for social media purposes. Approximately 25% of the respondents indicate that they only use the free Wi-Fi when their private data is finished.

5.3 Tshwane free Wi-Fi for economic enablement

5.3.1 Tshwane free Wi-Fi as an enabler for enhancing economic justice

When asked about the extent to which the Tshwane Free Wi-Fi is enabling economic justice, as shown in Tshwane, Table 4 below, 40% of respondents in Mamelodi are of the view that the free Wi-Fi is making a lot of contribution, whilst 36% believe that it is making modest contribution. In the case of Soshanguve, 41% that the free Wi-Fi is making a lot of contribution, whilst 39% believe that it is making modest contribution. Specifically, 13% Mamelodi are of the view that the free Wi-Fi is making low contribution, whilst 11% believe that it is making no contribution at all. Thus, is comparable to Mamelodi, where 14% are of the view that the free Wi-Fi is making low contribution; and 7% believe that it is making no contribution at all. What is key to observe is that less than half of the respondents in both townships believe that the free Wi-Fi is making 'a lot of contribution' to enhance economic justice in the city.

Table 4. Respondent's views about the Tshwane free Wi-Fi as an enabler for enhancing economic justice.

Mamelodi	Responses	Soshanguve
40.0%	Lot of contribution (60%+)	40.7%
36.4%	Modest contribution (40%–59%)	39.0%
12.7%	Low contribution (1%–39%)	13.6%
10.9%	No contribution (0%)	6.8%
Total: 100%		**Total: 100%**

5.4 Tshwane free Wi-Fi as an enabler for job opportunities and business transactions

When asked about their views regarding whether are many Tshwane residents who rely on the Tshwane Free Wi-Fi for economic opportunities such as job application and business transactions, as shown in Tshwane, Table 5 below, a majority, i.e. 83% of respondents in Mamelodi agreed or strongly agreed that the Tshwane Free Wi-Fi comes handy for residents

Table 5. Respondent's views about the use of the Tshwane free Wi-Fi to enhance economic justice.

Mamelodi	Responses	Soshanguve
50.9%	I agree	39.9%
32.7%	I strongly agree	37.3%
7.3%	Neutral	10.2%
3.6%	I disagree	8.5%
5.5%	I strongly disagree	5.1%
Total: 100%		**Total: 100%**

to pursue economic opportunities such as job application and business transactions. In the case of Soshanguve, 77% of the respondents similarly agreed or strongly agreed that the Tshwane Free Wi-Fi comes handy for residents to pursue economic opportunities such as job application and business transactions. Specifically, over 51% of respondents in Mamelodi agreed, while 333% strongly agreed. In the case of Soshanguve, specifically 40% agreed, and 37% strongly agreed. Those that disagree make 10% and 14% in Mamelodi and Soshanguve respectively.

When the respondents were asked about whether they themselves are part of the Tshwane residents who use Tshwane Free Wi-Fi to participate in economic opportunities such as job application and business transactions, Table 6 below shows that 45% of respondents in Mamelodi agreed, compared to 49% in the case of Soshanguve. About 40% of the respondents in Mamelodi use their private data for as job application and business transactions, compared to 25% in Soshanguve.

Table 6. Respondent's views about their usage of Tshwane Free Wi-Fi to participate in economic opportunities.

Mamelodi	Responses	Soshanguve
45.5%	Yes, I always use Tshwane free Wi-Fi data for economic or business opportunity purposes	49.2%
40.0%	No, I use my private data for economic or business opportunity purposes	25.4%
14.5%	I only use Tshwane free Wi-Fi data for economic or business opportunity purposes when my private data is finished	25.4%
Total: 100%		**Total: 100%**

Those who only use Tshwane free Wi-Fi data for to participate in economic opportunities such as job application and business transactions account for 15% in Mamelodi, and 25% in Soshanguve. On average, the results show that more respondents (46%) always use Tshwane Free Wi-Fi to participate in the economic transactions above, compared to about 20% of respondents who only use the free Wi-Fi for such purposes when their private data is finished.

5.5 *Tshwane free Wi-Fi as an enabler of digitalising service delivery access and online public engagements*

5.5.1 *Tshwane free Wi-Fi as an enabler for online/digital payments*

When asked about whether they believe that there are many residents in Tshwane who rely on the Tshwane Free Wi-Fi to pay their bills, check statements, etc. Table 7 below shows that a majority, i.e. 46% of respondents in Mamelodi agreed that many residents in Tshwane do rely on the free Wi-Fi to pay their bills, check statements. In the case of Soshanguve, even

Table 7. Respondent's views about the Tshwane free Wi-Fi as an enabler for online/digital payments.

Mamelodi	Responses	Soshanguve
45.5%	I agree	52.5%
34.5%	I don't have an idea	28.8%
20.0%	In my view, most people prefer to engage the CoT physical/manually	18.6%
Total: 100%		**Total: 100%**

more respondents (53%) agreed. Specifically, 35% of respondents in Mamelodi 'don't have an idea', compared to 29% in Soshanguve. Those that believe that most people prefer to engage the city physically for purposes of checking bills, statements, etc make 20% and 19% in Mamelodi and Soshanguve respectively.

When the respondents were asked about whether they themselves are part of the Tshwane residents who use Tshwane Free Wi-Fi to participate in economic opportunities such as job application and business transactions, Table 8 below shows that 42% of respondents always use their private data, 25% in the case of Soshanguve. Those who prefer to engage the city physically account for 27% in Mamelodi, and 28% in Soshanguve.

Table 8. Respondent's responses on their usage of Tshwane Free Wi-Fi to participate in economic opportunities.

Mamelodi	Responses	Soshanguve
41.8%	Yes, I always use my data all the time	25.4%
30.9%	Sometimes: when my private data is finished	45.8%
27.3%	I prefer to engage the City of Tshwane physical/manually	28.8%
Total: 100%		**Total: 100%**

5.6 *Tshwane free Wi-Fi as an enabler for digitalising service delivery access*

When asked about the extent to which the Tshwane Free Wi-Fi is enabling digitalising service delivery access, as shown in Tshwane, Table 9 below, 36% of respondents in Mamelodi are of the view that the free Wi-Fi is making a lot of contribution, whilst 27% believe that it is making modest contribution. In the case of Soshanguve, 37% that the free Wi-Fi is making a lot of contribution, whilst 41% believe that it is making modest contribution. Specifically, 27% Mamelodi are of the view that the free Wi-Fi is making low contribution, whilst 9% believe that it is making no contribution at all. In Mamelodi, 14% are of the view that the free Wi-Fi is making low contribution; and 9% believe that it is making no contribution at all. What is key to observe is that less than half of the respondents in both townships believe that the free Wi-Fi is making 'a lot of contribution' to enabling digitalising service delivery access in the city.

Table 9. Respondent's views about the Tshwane free Wi-Fi as an enabler for digitalising service delivery access.

Mamelodi	Responses	Soshanguve
36.4%	Lot of contribution (60%+)	37.3%
27.3%	Modest contribution (40%–59%)	40.7%
27.37%	Low contribution (1%–39%)	13.6%
9.1%	No contribution (0%)	8.5%
Total: 100%		**Total: 100%**

When asked about whether they believe that there are many residents in Tshwane who rely on the Tshwane Free Wi-Fi to engage the city on service delivery matters, etc. Table 10 below shows 47% of respondents in both Mamelodi and Soshanguve believe that many residents in Tshwane use the Tshwane Free Wi-Fi to engage the city on service delivery matters. Specifically, 27% of respondents in Mamelodi 'don't have an idea', compared to 37% in Soshanguve. Those that believe that most people prefer to engage the city physically

Table 10. Respondent's views about the Tshwane free Wi-Fi as an enabler for engagements on service delivery matters.

Mamelodi	Responses	Soshanguve
47.3%	I agree	47.5%
27.3%	I don't have an idea	37.3%
25.5%	In my view, most people prefer to engage the City of Tshwane physical/ manually	15.3%
Total: 100%		**Total: 100%**

for to engage the city on service delivery matters make 26% and 15% in Mamelodi and Soshanguve respectively.

When the respondents were asked about whether they themselves are part of the Tshwane residents who use always use their data to engage the city digitally on service delivery matters, Table 11 below shows that 51% of respondents in Mamelodi always use their data to engage the city digitally on service delivery matters, compared to 25% in the case of Soshanguve. About 31% of the respondents in Mamelodi indicated that they use Tshwane free Wi-Fi data sometimes, i.e. when their private data is finished, compared to 42% in Soshanguve. Those who prefer to engage the city physically on service delivery matters account for 18% in Mamelodi, and 32% in Soshanguve. The findings show that there is more appetite for Mamelodi respondents to use their own data to engage the city digitally on service delivery matters compared to Soshanguve respondents. However, the findings also show that that there is more appetite for Soshanguve respondents to engage the city physically on service delivery matters compared to Soshanguve respondents.

Table 11. Respondent's responses about their use of the Tshwane free Wi-Fi to engage the city digitally on service delivery matters.

Mamelodi	Responses	Soshanguve
50.9%	Yes, I always use my data all the time	25.4%
30.9%	Sometimes: when my private data is finished	42.4%
18.2%	I prefer to engage the City of Tshwane physical/manually	32.2%
Total: 100%		**Total: 100%**

5.7 Tshwane free Wi-Fi as an enabler for digitalising citizen participation in good governance systems and processes

When asked about their views about the extent to which the Tshwane Free Wi-Fi is contributing to allowing residents to participate in governance processes in the City of Tshwane, as shown in Tshwane, Table 12 below, 38% of respondents in Mamelodi are of the view that the free Wi-Fi is making a lot of contribution, whilst 23% believe that it is making modest contribution. In the case of Soshanguve, 41% that the free Wi-Fi is making a lot of contribution, whilst 31% believe that it is making modest contribution. Specifically, 29% Mamelodi are of the view that the free Wi-Fi is making low contribution, whilst 9% believe that it is making no contribution at all. In Mamelodi, 14% are of the view that the free Wi-Fi is making low contribution; and 15% believe that it is making no contribution at all. What is key to observe is that less than half of the respondents in both townships believe that the free Wi-Fi is making 'a lot of contribution' to allowing residents to participate in governance processes in the city.

Table 12. Respondent's views about the contribution of the Tshwane free Wi-Fi in enabling residents to participate in governance processes in the City of Tshwane.

Mamelodi	Responses	Soshanguve
38.2%	Lot of contribution (60%+)	40.7%
23.6%	Modest contribution (40%–59%)	30.5%
29.1%	Low contribution (1%–39%)	13.6%
9.1%	No contribution (0%)	15.3%
Total: 100%		**Total: 100%**

6 CONCLUSSIONS AND RECOMMENDATIONS

More than 70% of the free Wi-Fi users in Mamelodi and Soshanguve townships feel positive about the contribution of the free Wi-Fi on social justice and enabling participation on social media platforms, respectively. However, when it comes to economic justice, the picture is different: less than half (40%) of the respondents believe that the free Wi-Fi is making 'a lot of contribution'. Also, as far as free Wi-Fi as an enabler of economic justice, less than half of the respondents believe that the free Wi-Fi is making 'a lot of contribution'. However, specifically on job application and business transactions, on average, 80% of respondents feel positive about free Wi-Fi. With regard to the Wi-Fi as an enabler of online/digital payment services, on average, 86% of respondents feel positive. When it comes to digitalising service delivery access in the city, less than 40% of the respondents believe that the Wi-Fi is making 'a lot of contribution'. Similarly, only 47% of respondents believe that residents in Tshwane use the free Wi-Fi to engage the city on broad service delivery matters. In addition, less than half of the respondents believe that the free Wi-Fi is making 'a lot of contribution' in allowing e-governance in the city.

In conclusion, this study shows that generally, communities in the two townships seem to be (potentially) happy with the use the free Wi-Fi in terms of objectives of social justice, increasing social media participation, and job applications. However, the study also shows that there could be less use of the free Wi-Fi to engage the city on governance and service delivery matters. Significantly, the study seem to suggest that the intensity and scale of use of the free Wi-Fi in different communities may not be exactly the same. For instance, respondents seem to imply that Mamelodi residents have more appetite to use data to engage the city digitally, compared to Soshanguve residents, who seem to have more appetite to engage the city physically. Thus, it is recommended that going forward, the city should do regular extensive reviews per region/area, with a view to understand the specific needs and utility trends analysis; and action area specific empirically facts-based interventions rather than a universal/one size fits all approach of implementation.

REFERENCES

Cecchin, A., Salomone, R., Deutz, P., Raggi, A., & Cutaia, L. 2021. What is in a Name? The Rising Star of the Circular Economy as a Resource-related Concept for Sustainable Development. *Circular Economy and Sustainability*, *1*(1), 83–97.

Mondini, G. 2019. Sustainability Assessment: From Brundtland Report to Sustainable Development Goals. *Valori e Valutazioni*, (23).

Onyango, J., Kiano, E., & Saina, E. 2021. Environmental Damage Theory Applicable to Kenya. *Asian Journal of Business Environment*, *11*(1), 39–50.

Schopp, K., Bornemann, M., & Potthast, T. 2020. The Whole-institution Approach at the University of Tübingen: Sustainable Development Set in Practice. *Sustainability*, *12*(3), 861.

Steputat, C. C., Ural, D., & Nanni, A. 2020. Sustainable Cities and Communities Through GFRP Secant-pile Seawall Innovation, Sustainability, Fortification and Hurricane Storm Surge Protection. In *IOP Conference Series: Earth and Environmental Science* (Vol. 588, No. 4, p. 042063). IOP Publishing.

Yang, Y., & Taufen, A. 2022. Sustainable Cities and Landscapes: Cultivating Infrastructures of Health. In *The Routledge Handbook of Sustainable Cities and Landscapes in the Pacific Rim*. Taylor & Francis.

Advancements in E-mobility: A bibliometric literature review on battery technology, charging infrastructure, and energy management

T. Moyo*, A.O. Onososen, I. Musonda & H. Muzioreva
Centre for Applied Research + Innovation in the Built Environment (CARINBE), Faculty of Engineering and the Built Environment, University of Johannesburg, South Africa

ABSTRACT: Contemporary innovations in the public transportation are at the heart of transformation and increasing opportunities globally. This has led to an increased investment in alternative energy sources such as e-mobility. Despite the immense benefit of the adoption and implementation of e-mobility are low. The study utilizes web of science to conduct bibliometric literature review on electric vehicles. 578 articles were included in the analysis. The findings reveal China, the European Union and the United States have the largest number of publications. The review of literature reveals several key findings and implications to inform deployment of e-mobility in Africa namely the development of a robust charging infrastructure is a critical factor in facilitating widespread electrical vehicle adoption. The study recommends that the government should develop policies and incentives to facilitate the transition to e-mobility.

Keywords: e-mobility, electrical vehicle, Africa, energy, public transportation

1 INTRODUCTION

Contemporary innovations in public transportation are at the heart of transformation and increasing opportunities globally. In many communities, e-mobility transitioning is fast becoming a key instrument in driving innovative, inclusive, and sustainable growth, improved processes, and contributing to sustainable development (Esteban & Otero 2023). Despite the immense benefit of the adoption and implementation of e-mobility are low (Kirpes *et al.* 2019). The challenge is more evident in developing counties, where for many countries, access to basic public transportation and infrastructure is still a major challenge. Literature reveals tech-enabled initiatives in African cities are critical drivers for adopting and exploring the benefits of emerging e-mobility technologies as African cities push to leverage tools and technology enabled by the fourth industrial revolution (Buresh *et al.* 2020; Collett *et al.* 2021). Specifically, the need to improve efficiency, effectiveness, and sustainability of development in the wake of constrained resources has meant that new technology-based solutions are needed for sustainable development.

Given the high fuel prices constantly increasing, Africa has the lowest use of motorized transport globally and contributes only 5% of total global transport CO2 emissions (calculated in 2019). In many developing countries informal public transportation modes are the most common such as the mini-bus taxis. A significant demand exists to develop an inclusive and safe, sustainable, and low-cost mass-transportation system suitable for developing countries (Abraham *et al.* 2021; Booysen *et al.* 2022; Onososen & Musonda 2022). Therefore, an alternate mode of transport which provides good ventilation, does not require sharing space in an enclosed environment with multiple other people, is low-cost and uses renewable energy.

---

*Corresponding Author: tmoyo@uj.ac.za

Globally there has been an increase in investment electrical vehicle in response to Climate change concerns to reduce emissions from public transportation sector (Pietrzak & Pietrzak 2020). Although these technological advancements are expected to contribute to the transportation sector decarbonization, the great turning point relies on vehicle electrification (Rodrigues & Seixas 2022). The direct and indirect electrification technologies such as hybrid electric vehicles (HEV), battery electric vehicles (BEV) and fuel cell vehicles (FCV) have the potential to reduce the transportation sector emission levels considerably (Husain *et al.* 2021; Taiebat & Xu 2019). One of the primary motivations for investing in electric vehicles is their potential to reduce greenhouse gas emissions and mitigate climate change.

Literature consistently demonstrates that electric vehicles produce lower or zero tailpipe emissions, thereby contributing to improved air quality and reduced local pollution (Liu *et al.* 2021). Moreover, the adoption of electric vehicles can facilitate the integration of renewable energy sources, such as solar and wind, into the power grid, further reducing carbon emissions. In addition to helping to diminish the dependency on fossil fuels and reduce emissions if powered with renewable energy sources (RES), the deployment of electric vehicles (EV) in an urban context contributes to improving the air quality, reducing noise pollution and increasing energy efficiency (Colmenar-Santos *et al.* 2019; Li *et al.* 2022; Teng *et al.* 2020).

To inform the transition to e-mobility and promote African sustainability, it is essential to recognize undertake a review on emerging literature on e-mobility to inform its transition in Africa. Therefore, this study aims to provide a comprehensive bibliometric analysis on scientific literature of electric vehicles. The following section presents the methodology adopted in the study. This is followed by a presentation of the findings and discussion on the implications for e-mobility transition in Africa. The last section provides a conclusion.

2 METHODOLOGY

Academic databases, present a knowledge reservoir to assist researchers, policymakers, and communities to access scientific databases. This literature review opted to use Clarivate Web of Science (WoS) to acquire research publications on electric vehicles focusing on public transportation, as this database credited for being a significant scientific database providing high quality academic literature (Birkle *et al.* 2020). Data were searched and retrieved on 20 June 2022 from WoS via the University of Johannesburg Library database website. The following query string was used in the Boolean search "electric bus*" AND ("hybrid" OR "fuel cell" OR "battery") to obtain literature on public transportation electric vehicles which yielded 613 results. The selection criteria included articles written in English published between 2010 to 2023. Figure 1 present the selection approach. The VOS Viewer software was then selected as the supporting tool to identify different key terms and research cluster of the 578 selected articles.

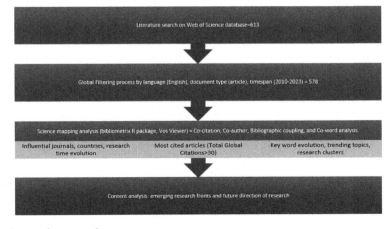

Figure 1. Research approach.

A content literature was then undertaken on key articles in the field of electric vehicles. Firstly, the most cited articles were selected using a citation method. This step was undertaken to systematically obtain a list of influential articles. Given that papers published recently still require more time to receive more citations, the second step was to select, analyse and include recent papers to enlarge the analysis. The last step was to observe further the research approach, findings and implications from the selected literature.63 papers were identified for the content analysis.

3 FINDINGS AND DISCUSSIONS

3.1 *Distribution of publications*

Research on the electric bus is not new. The literature collected in this analysis were from 2010 to 2023. Before 2010 there was minimal scientific literature on electric buses (Tan *et al.* 2019). Figure 2 outlines contemporary there has been an increase in research on the electric bus. From 2010 to 2013 there was minimal production of articles, most of these articles focused on the early development of sustainable alternative technologies for vehicles. Post 2014 there was an increase in publication of literature on electric vehicles, this was in response to global calls to action to address climate change by the United Nations (Creutzig *et al.* 2015). This period led to increased investment in alternative energy and development of partnerships to address emission rates. Given the exponential growth in publications from 2017 to 2022, this trend in the number of publications is expected to increase in future research.

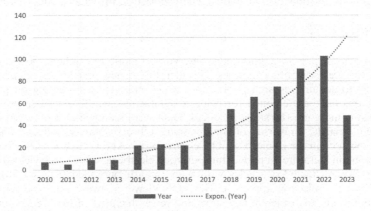

Figure 2. Scientific production in electric bus articles from 2010–2023.

3.2 *Publications per country*

The results reveal the global trend of countries investing more in the development of electric vehicles and development of policies to inform deployment of alternative energy technologies (see Figure 3). Globally China, the Europe Union (EU) and the United States of America (USA) present the largest market of electrical vehicles. Given these countries account for the highest emissions levels globally, the adoption of electric vehicles has been in response to climate change mitigation strategies. China has the largest number of publications on electric buses. Du *et al.* (2019) have outlined China has approximately 98% of the global market of operational electrical buses. The USA, China and the EU have developed several policies to govern the deployment of electrical vehicle. These policies relate to regulations, incentives, subsidies for production, subsidies for electric vehicles operations, tax incentives and charging infrastructure regulations. In Africa, given the roll out of electrical vehicles is still new, there have been few publications on electrical vehicles namely South Africa, Ghana and Rwanda.

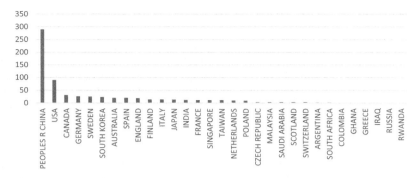

Figure 3. The evolution of the cumulative publication of the leading journals in the field by year.

3.3 Co-occurrence of keywords

An analysis of research streams from 2010 to 2023 reveals thematic clusters of key words. Figure 4 outlines 8 thematic clusters of electric vehicles literature. These clusters were grouped in 'Vehicle Technology', 'Safety', 'Battery Technology', 'Operational Strategies', 'Energy Management', 'Charging Infrastructure', 'Optimisation', and 'Sustainability'. The circle size represents the number of times a selected keyword occurred, while the distance among them defines the relationship between each keyword in the cluster – the closer the distance is, the more related the words are. A further analysis of literature trends reveals between 2010 to 2013 there was more investment in the 'Vehicle Technology' cluster. While from 2014 to 2018 there was an increase in investment in 'Sustainability' cluster. Post 2019 the focus has been on increasing investment in 'Safety', 'Battery Technology', 'Operational Strategies', 'Energy Management', 'Charging Infrastructure', and 'Optimisation' clusters.

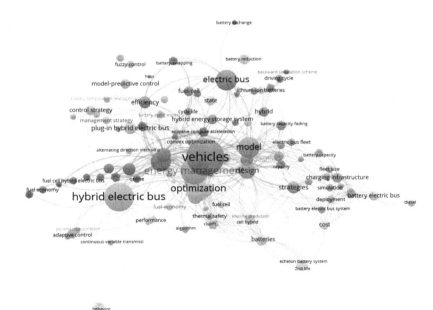

Figure 4. Network analysis mapping of co-occurring keywords in electrical vehicle literature.

3.4 Content analysis

This review unpacks the deeper content review on selected literature which had a high citation. The content analysis sough to unpack the key thematic areas of electric vehicle. The growing concern on the sustainability of electric vehicles and the need to reduce greenhouse gas emissions has led to a surge in the popularity of electric vehicles (Colmenar-Santos *et al.* 2019; Li *et al.* 2022). As electric vehicles continue to gain traction in the automotive market, there is a need to critically review the existing literature on electric vehicles, particularly focusing on the advancements and challenges in vehicle technology. Figure 5 outlines the research areas in relation to investment in electric vehicles deployment.

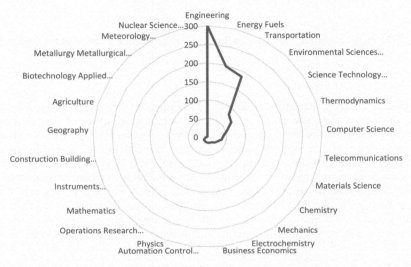

Figure 5. Research areas in electric vehicles deployment.

The battery technology utilised in electrical vehicles is a crucial aspect in literature, and numerous studies have focused on improving energy storage systems (Carrilero *et al.* 2018; Mahmoud *et al.* 2016; Rodrigues & Seixas 2022). The literature reveals significant progress in battery chemistry, such as lithium-ion batteries (Li-ion), which have become the standard in electrical vehicles. Researchers have explored strategies to enhance energy density, increase charge/discharge rates, and improve overall battery performance (An *et al.* 2020; Deliali *et al.* 2021; Du *et al.* 2019). However, from the analysis of literature challenges such as limited driving range, long charging times, and the environmental impact of battery production and disposal remain areas of concern and have been identified as critical areas to be addressed.

The availability and accessibility of a robust charging infrastructure play a crucial role in the widespread adoption of EVs ((Herrenkind *et al.* 2019; Mohamed *et al.* 2017; Peters *et al.* 2017; Wu *et al.* 2019). Literature in this area highlights the need for an extensive network of charging stations, including fast-charging options (Liu *et al.* 2021). Researchers have examined charging protocols, standardization efforts, and innovative charging technologies to address issues like charging time, compatibility, and interoperability (Guschinsky *et al.* 2021; Stumpe *et al.* 2021). Therefore, substantial research has focused on the integration of renewable energy sources and smart grid technologies to facilitate sustainable charging solutions (Dimitriadou *et al.* 2023; Gönül *et al.* 2021; Majhi *et al.* 2021).

Energy management plays a crucial role in the effective utilization and optimization of energy resources in electric vehicles. Efficient energy consumption modelling and optimization techniques are crucial for maximizing the driving range and minimizing energy losses in electrical vehicles. Literature reveals the development of models to predict energy consumption under different driving conditions, considering factors like speed, terrain, and traffic patterns (Deng et al. 2019; Tan et al. 2019; Wu et al. 2020). Researchers have explored optimization algorithms, including dynamic programming, genetic algorithms, and model predictive control, to optimize energy management strategies (Guo et al. 2019; Wang et al. 2020). Optimization options have emerged as a better option to design energy management strategies, these focusing on real-time implementation, driver behaviour prediction, and uncertainty in driving conditions as solutions for energy consumption modelling and optimization.

4 LESSONS LEARNT AND IMPLICATIONS FOR AFRICA

Undertaking transitioning to e-mobility, demands capability in terms of competence of the people involved, resources available and processes or management systems in place that can be deployed, and the requisite knowledge held or shared by key stakeholders. In addition, success depends on effective linkages between researchers and local communities participating in the transition. In Africa Transit oriented developments (TOD) have been informed by a strategy that seeks to improve movement through creating a hierarchy of access to reinforced public transport routes, while also promoting pedestrian walkability and movement. Mixed developments that include different housing typologies and densities are another focus area incorporating hostels and informal settlements through redevelopment and upgrading respectively.

The review of global literature reveals several key findings and implications to inform deployment of e-mobility in Africa. The development of a robust charging infrastructure is a critical factor in facilitating widespread electrical vehicle adoption. The study highlights the expansion of charging networks, the emergence of fast-charging technologies, and the integration of renewable energy sources into charging infrastructure. There is still need for studies to unpack interoperability with existing systems and standardization electrical vehicle infrastructure.

Globally government policies and incentives have played a significant role in promoting electrical vehicles deployment. The literature emphasizes the importance of supportive measures such as financial incentives and stricter emissions regulations. These policies encourage the transition to e-mobility. Lastly the literature emphasizes the need for collaboration among key stakeholders as partnerships and cooperation can help address challenges, leverage expertise, and accelerate electrical vehicles deployment.

5 CONCLUSION

This study conducted a bibliometric literature review on electrical vehicles focusing on electrical bus research through an in-depth analysis of 578 articles identified in the WoS database for the period 2010–2023. In the transition towards e-mobility the study reveal that electric vehicles can be used to reduce emissions from public transport sector. Given the global partnerships on e-mobility have led to the development and streamline of electric vehicle infrastructure plans on land use and ecological management to reduce the impacts of development on climate change. Emerging trends reveal a focus on 8 thematic clusters namely 'Vehicle Technology', 'Safety', 'Battery Technology', 'Operational Strategies', 'Energy Management', 'Charging Infrastructure', 'Optimisation', and 'Sustainability'. Key findings reveal battery technology is at the heart of our understanding of emerging technologies in e-mobility. Studies of energy management show the importance of optimization of energy resources in electric vehicles. Given the are few studies from Africa, future studies should address challenges of development of electric vehicles policies and planning for electric vehicles infrastructure in Africa. The limitation of the study is the use of one scientific database, future studies should include other databases to enhance the robustness of the analysis.

ACKNOWLEDGEMENT

The authors would like to appreciate the South African National Energy Development Institute (SANEDI) and the Centre for Applied Research + Innovation in the Built Environment (CARINBE), Faculty of Engineering and the Built Environment, University of Johannesburg, for financial support.

REFERENCES

Abraham, C.J., Rix, A.J., Ndibatya, I. and Booysen, M.J., 2021. Ray of Hope for Sub-Saharan Africa's Paratransit: Solar Charging of Urban Electric Minibus Taxis in South Africa. *Energy for Sustainable Development*, 64, pp.118–127.

An, K., Jing, W. and Kim, I., 2020. Battery-swapping Facility Planning for Electric Buses with Local Charging Systems. *International Journal of Sustainable Transportation*, 14(7), pp.489–502.

Birkle, C., Pendlebury, D.A., Schnell, J. and Adams, J., 2020. Web of Science as a Data Source for Research on Scientific and Scholarly Activity. *Quantitative Science Studies*, 1(1), pp.363–376.

Booysen, M.J., Abraham, C.J., Rix, A.J. and Ndibatya, I., 2022. Walking on Sunshine: Pairing electric Vehicles with Solar Energy for Sustainable Informal Public Transport in Uganda. *Energy Research & Social Science*, 85, p.102403.

Buresh, K.M., Apperley, M.D. and Booysen, M.J., 2020. Three Shades of Green: Perspectives on At-work Charging of Electric Vehicles Using Photovoltaic Carports. *Energy for Sustainable Development*, 57, pp.132–140.

Carrilero, I., González, M., Anseán, D., Viera, J.C., Chacón, J. and Pereirinha, P.G., 2018. Redesigning European Public Transport: Impact of New Battery Technologies in the Design of Electric Bus Fleets. *Transportation Research Procedia*, 33, pp.195–202.

Collett, K.A., Hirmer, S.A., Dalkmann, H., Crozier, C., Mulugetta, Y. and McCulloch, M.D., 2021. Can Electric Vehicles be Good for Sub-Saharan Africa?. *Energy Strategy Reviews*, 38, p.100722.

Colmenar-Santos, A., Muñoz-Gómez, A.M., Rosales-Asensio, E. and López-Rey, Á., 2019. Electric Vehicle Charging Strategy to Support Renewable Energy Sources in Europe 2050 Low-carbon Scenario. *Energy*, 183, pp.61–74.

Creutzig, F., Jochem, P., Edelenbosch, O.Y., Mattauch, L., Vuuren, D.P.V., McCollum, D. and Minx, J., 2015. Transport: A Roadblock to Climate Change Mitigation?. *Science*, 350(6263), pp.911–912.

Deliali, A., Chhan, D., Oliver, J., Sayess, R., Godri Pollitt, K.J. and Christofa, E., 2021. Transitioning to Zero-emission Bus Fleets: State of Practice of Implementations in the United States. *Transport Reviews*, 41 (2), pp.164–191.

Deng, R., Liu, Y., Chen, W. and Liang, H., 2019. A Survey on Electric Buses—Energy Storage, Power Management, and Charging Scheduling. *IEEE Transactions on Intelligent Transportation Systems*, 22(1), pp.9–22.

Dimitriadou, K., Rigogiannis, N., Fountoukidis, S., Kotarela, F., Kyritsis, A. and Papanikolaou, N., 2023. Current Trends in Electric Vehicle Charging Infrastructure; Opportunities and Challenges in Wireless Charging Integration. *Energies*, 16(4), p.2057.

Du, J., Li, F., Li, J., Wu, X., Song, Z., Zou, Y. and Ouyang, M., 2019. Evaluating the Technological Evolution of Battery Electric Buses: China as a Case. *Energy*, 176, pp.309–319.

Esteban, H. and Otero, B.G., 2023. Is the Transition to E-mobility the Silver Bullet to Achieve Climate-neutral Transport? An Interdisciplinary Review in the Search for Consistency and Collateral Effects. *European Business Law Review*, 34(1).

Gönül, Ö., Duman, A.C. and Güler, Ö., 2021. Electric Vehicles and Charging Infrastructure in Turkey: An Overview. *Renewable and Sustainable Energy Reviews*, 143, p.110913.

Guo, H., Wang, X. and Li, L., 2019. State-of-charge-constraint-based Energy Management Strategy of Plug-in Hybrid Electric Vehicle with Bus Route. *Energy Conversion and Management*, 199, p.111972.

Guschinsky, N., Kovalyov, M.Y., Rozin, B. and Brauner, N., 2021. Fleet and Charging Infrastructure Decisions for Fast-charging City Electric Bus Service. *Computers & Operations Research*, 135, p.105449.

Herrenkind, B., Brendel, A.B., Nastjuk, I., Greve, M. and Kolbe, L.M., 2019. Investigating end-user Acceptance of Autonomous Electric Buses to Accelerate Diffusion. *Transportation Research Part D: Transport and Environment*, 74, pp.255–276.

Husain, I., Ozpineci, B., Islam, M.S., Gurpinar, E., Su, G.J., Yu, W., Chowdhury, S., Xue, L., Rahman, D. and Sahu, R., 2021. Electric Drive Technology Trends, Challenges, and Opportunities for Future Electric Vehicles. *Proceedings of the IEEE*, 109(6), pp.1039–1059.

Kirpes, B., Danner, P., Basmadjian, R., Meer, H.D. and Becker, C., 2019. E-mobility Systems Architecture: A Model-based Framework for Managing Complexity and Interoperability. *Energy Informatics*, 2, pp.1–31.

Li, C., Zhang, L., Ou, Z., Wang, Q., Zhou, D. and Ma, J., 2022. Robust Model of Electric vehicle Charging Station Location Considering Renewable Energy and Storage Equipment. *Energy*, 238, p.121713.

Liu, X., Qu, X. and Ma, X., 2021. Optimizing Electric Bus Charging Infrastructure Considering Power Matching and Seasonality. *Transportation Research Part D: Transport and Environment*, 100, p.103057.

Mahmoud, M., Garnett, R., Ferguson, M. and Kanaroglou, P., 2016. Electric Buses: A Review of Alternative Powertrains. *Renewable and Sustainable Energy Reviews*, 62, pp.673–684.

Majhi, R.C., Ranjitkar, P., Sheng, M., Covic, G.A. and Wilson, D.J., 2021. A Systematic Review of Charging Infrastructure Location Problem for Electric Vehicles. *Transport Reviews*, 41(4), pp.432–455.

Mohamed, M., Farag, H., El-Taweel, N. and Ferguson, M., 2017. Simulation of Electric Buses on a Full Transit Network: Operational Feasibility and Grid Impact Analysis. *Electric Power Systems Research*, 142, pp.163–175.

Onososen, A. and Musonda, I., 2022. Barriers to BIM-based Life Cycle Sustainability Assessment for Buildings: An Interpretive Structural Modelling Approach. *Buildings*, 12(3), p.324.

Pagliaro, M. and Meneguzzo, F., 2019. Electric Bus: A Critical Overview on the Dawn of Its Widespread Uptake. *Advanced Sustainable Systems*, 3(6), p.1800151.

Peters, J.F., Baumann, M., Zimmermann, B., Braun, J. and Weil, M., 2017. The Environmental Impact of Li-Ion Batteries and the Role of Key Parameters–A Review. *Renewable and Sustainable Energy Reviews*, 67, pp.491–506.

Pietrzak, K. and Pietrzak, O., 2020. Environmental Effects of Electromobility in a Sustainable Urban Public Transport. *Sustainability*, 12(3), p.1052.

Rodrigues, A.L. and Seixas, S.R., 2022. Battery-electric Buses and Their Implementation Barriers: Analysis and Prospects for Sustainability. *Sustainable Energy Technologies and Assessments*, 51, p.101896.

Stumpe, M., Rößler, D., Schryen, G. and Kliewer, N., 2021. Study on Sensitivity of Electric bus Systems Under Simultaneous Optimization of Charging Infrastructure and Vehicle Schedules. *Euro Journal on Transportation and Logistics*, 10, p.100049.

Taiebat, M. and Xu, M., 2019. Synergies of Four Emerging Technologies for Accelerated Adoption of Electric Vehicles: Shared Mobility, Wireless Charging, Vehicle-to-grid, and Vehicle Automation. *Journal of Cleaner Production*, 230, pp.794–797.

Tan, H., Zhang, H., Peng, J., Jiang, Z. and Wu, Y., 2019. Energy Management of Hybrid Electric Bus Based on Deep Reinforcement Learning in Continuous State and Action Space. *Energy Conversion and Management*, 195, pp.548–560.

Teng, F., Ding, Z., Hu, Z. and Sarikprueck, P., 2020. Technical Review on Advanced Approaches for Electric Vehicle Charging Demand Management, Part I: Applications in Electric Power Market and Renewable Energy Integration. *IEEE Transactions on Industry Applications*, 56(5), pp.5684–5694.

Wang, J., Kang, L. and Liu, Y., 2020. Optimal Scheduling for Electric Bus Fleets Based on Dynamic Programming Approach by Considering Battery Capacity Fade. *Renewable and Sustainable Energy Reviews*, 130, p.109978.

Wu, J., Wei, Z., Li, W., Wang, Y., Li, Y. and Sauer, D.U., 2020. Battery Thermal-and Health-constrained Energy Management for Hybrid Electric Bus Based on Soft actor-critic DRL Algorithm. *IEEE Transactions on Industrial Informatics*, 17(6), pp.3751–3761.

Wu, Z., Guo, F., Polak, J. and Strbac, G., 2019. Evaluating Grid-interactive Electric Bus Operation and Demand Response with Load Management Tariff. *Applied Energy*, 255, p.113798.

Flood simulation with GeoBIM

K. Kangwa & B. Mwiya
Department of Civil and Environmental Engineering, School of Engineering, University of Zambia, Lusaka, Zambia

ABSTRACT: The integration of BIM and GIS has gained popularity in the AEC and Geospatial Industry, enhancing collaboration and visualization in urban planning, construction processes, and disaster management, among other faculties. This research integrates BIM and GIS in a flood scenario, using a case study of an unplanned settlement in Kanyama Area. The research aimed to establish factors and input data necessary to simulate a flash flood, conduct a flood simulation, and develop flood hazard maps. The research identified influential flood factors such as elevation, slope, topographic wetness index, precipitation, soil type, land use land cover, and proximity to streams. The flood simulation showed that 39% (147 ha) of the case study area is flood-prone, with 2880 buildings at high risk. The research recommends that the local authority should establish a drainage system in the area that is more prone to flooding or restrict building rules to mitigate future floods.

Keywords: Building Information Modelling (BIM), Geographic Information System (GIS), GeoBIM, and Flood Simulation

1 INTRODUCTION

Floods are the most prevalent and destructive natural disasters, posing a constant threat to life and property (Kabir *et al.* 2013). Floods may wreak havoc on communities, causing death and damage to personal property as well as important public health infrastructure, Floods impacted approximately 2 billion people globally between 1998 and 2017 (WHO 2022). According to the United Nations Office for Disaster Risk Reduction (UNDRR), Africa has the highest number of people exposed to flood risk of any continent, with approximately 200 million people exposed to flood risk each year. Zambia endured its worst floods in 40 years in the early months of 2009, injuring over 20,000 people and destroying agricultural areas, roads, and houses. Ever since, rainstorms have left densely inhabited regions of Lusaka flooded in ankle-deep water for extended periods during the rainy season (Start 2017). Floods are generally classified into flash, river, and coastal floods caused by heavy rain, river capacity exceeded, and storm surges from cyclones, respectively (WHO 2022).

The common flood sources that bring devastating environmental events in some areas of Zambia are water rising due to rainfall (known as flash floods). Flash floods occur in small catchments where the drainage basin's response time is short. (Doswell 2003). To reduce floods in cities, experts can conduct a flood risk assessment by analysing floods using new technologies and data that are customized to the local environment and resources. (Suthakaran *et al.* 2018). Flood simulation is a critical tool for predicting the behaviour of floodwaters, assessing their impact, and developing effective mitigation measures. GeoBIM improves flood simulation accuracy and efficiency by integrating geospatial data and BIM, providing a detailed representation of the built environment and enabling better decision-making (Thiecken *et al.* 2005).

This research aims at simulating a flash flood, by integrating Geospatial and Building information with inundation data. Specifically, the research objectives are to establish factors and input data necessary to simulate a flash flood, To Conduct a flood simulation using a case study, and to develop flood hazard maps.

Flood risk assessment requires integrating building information, which can be represented by BIM, with flood information typically handled by GIS (Amirebrahimi *et al.* 2015). GeoBIM flood simulation foresees the effects of flooding, assisting authorities in understanding risks and potential infrastructure damage. In order to manage the risks, it influences risk mitigation strategies and emergency response plans. (Daniela *et al.* 2018). GeoBIM technologies provide detailed and current data on the built environment, improving flood simulation accuracy and effectiveness (Díez-Herrero & Garrote 2020). This helps authorities and planners better understand potential flooding impacts and take appropriate actions to protect people and property, by creating more realistic and accurate flood simulations. (Junxiang 2020).

Interest is growing in integrating GIS and BIM for potential benefits in the Architectural, Engineering & Construction (AEC), and Geospatial industries. (Junxiang 2020). The Geo-BIM (i.e., Geographic Information Systems and The Building Information Modelling (BIM) environment with GIS data can analyze and simulate flood models for better decisions to reduce the danger of natural disasters. Integrating BIM and GIS data into hydrodynamic models provides more realistic data and a comprehensive approach to the problem. (Andre & Stephen 2004). Forecasting the size of flood events is an important part of flood risk management and one of the approaches to take is mapping flood hazards to determine the flooded regions based on predicted discharges for various return periods Flood hazard mapping is used to determine the size and depth of a flood (Daniela *et al.* 2018).

To achieve accurate and comprehensive flood simulation analysis, it's essential to include both building and geospatial information. Integrated models enable stakeholders to make informed decisions regarding restructuring infrastructure, optimizing drainage system design, and implementing diversion procedures to prevent disruption of natural water flow problem statement. Flooding is an environmental challenge that requires serious interventions. Inundating is one of the major causes of the environmental crisis (Magami *et al.* 2014). Individuals and communities are affected by floods, which have a wide range of negative and positive repercussions, depending on the location and degree of the flooding, as well as the sensitivity and value of the natural and built ecosystems they influence (Geoff 2011). This study mainly focuses on the inundation occurrences faced by Kanyama Residents in Lusaka, Zambia where flash floods have significantly impacted the area causing escalating life and property-threatening incidences during the rainy seasons. The floods are largely attributed to heavy rains, low terrain, and highly permeable limestone resting on impermeable bedrock that becomes saturated. Other contributing factors include blocked/silted drains which are aggravated by random waste dumping, poor drainage design, and inadequate solid waste management. Flooding is further intensified by uncoordinated growth and construction on streams and marshlands (Nchito *et al.* 2018).

This research details the essential input data factors for a flood simulation, the role of BIM and GIS in flood simulations and How carrying out a flood simulation assist in mitigating floods. The study covers a 328Ha section, Eastward of Kanyama area of Lusaka, Zambia. The settlement is located approximately 7 km westward of the Central Business District (CBD), bordered by Los Angeles Road on the western side and Mumbwa road on the eastern side. Kanyama falls under unplanned settlements, the Constituency has a population of 364,655 people, whereas the Lusaka district has a population of 1,747,152 people, according to the 2010 census.

2 METHODOLOGY

The research objectives were to simulate a flash flood, conduct flood simulation using a case study, and develop flood hazard maps. The flood simulation using GeoBIM involved selecting a case study, integrating a digital elevation model with BIM and GIS data, and using various geoprocessing tools in ArcGIS to analyze flood influential factors such as slope, elevation, TWI, proximity to streams, precipitation, soil type, and land cover. The flood risk was classified into four categories using the "Reclassify" tool in ArcGIS. Finally, the layers were combined, and a weight was assigned to each layer based on its relative importance in influencing flooding. The resulting flood hazard map can be used to inform flood management and planning decisions

Through the use of an AHP scale, a pair-wise comparison was used to establish the weights for the flood influential factors.

The flood simulation model was validated by comparing it to actual flood data, and the results were analyzed to identify potential flood-prone areas. These areas were presented and evaluated to answer the research questions.

The study covers a portion of 375 ha of the boundary of Kanyama Area, which is an unplanned settlement in Lusaka, Zambia. The settlement is located approximately 7 km westward of the Central Business District (CBD) and lies within two main highways of Los Angeles Road on the western side and Mumbwa road on the eastern side. The Estimated. The Central Statistics Office (CSO) estimates the entire Kanyama population density in 2010 was around 5,636 persons per km^2 (Tembo et al. 2019). The community of Kanyama is one of the settlements in Lusaka, Zambia that experiences persistent floods in the rainy season. Figure 1 shows a detailed map of the study area.

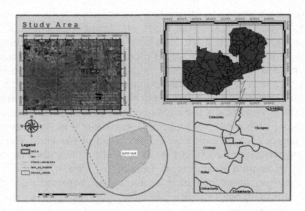

Figure 1. Map of the study area.

2.1 Methods of data collection

GIS data was obtained from Medici governance and DMMU as part of the data collection process. Subsequently, 2D outlines of buildings and roads were extracted from OSM and converted into 3D features using ArcGIS Pro. The GIS data obtained from Medici governance comprised attribute tables for powerline poles, road networks, and digitized land parcels. Additionally, the GIS data from DMMU included the boundary information of the wards of Zambia. These data sets were integrated into ArcGIS and utilized to perform a flood simulation.

2.2 Data collection instruments, procedure, and time line

The raw aerial images collected from Medici governance were primarily imported to Pix4D, where individual images were stitched, then Pix4D was used to process a dense point cloud, DSM, Mesh, and an ortho-mosaic. Subsequently, the DTM was imported into Global

Mapper for further processing. Image spatial rectification and assignment of a projection system were done in Global Mapper.

The Rectified DTM was exported to ArcGIS Pro where the rest of the data processing was accomplished. Several geo-processing processes were run to establish flood-prone areas. Spatial queries were executed to extract the number of buildings affected by the developed classified zones of the flood hazard map.

3 DATA PRESENTATION AND ANALYSIS

GeoBIM interoperability plays a crucial role in flood simulations by enabling the integration of geospatial data and building information models to create a more accurate and comprehensive representation of the physical environment. With interoperable systems, detailed topographic, hydrologic, and hydraulic data can be combined with building information models to simulate the impact of flood events on buildings and infrastructure. This helps engineers and planners to identify potential flood hazards and assess the risk to communities and assets, which in turn supports informed decision-making and flood management planning. Interoperable GeoBIM tools also facilitate collaboration between different stakeholders, including engineers, planners, and policy-makers, to better understand the potential consequences of floods and develop effective response strategies.

3.1 *Flood influential factors analysis and mapping*

Elevation (DEM/DSM) analysis: Figure 6 is a digital surface model (DSM) generated in Global Mapper, showing a smooth gradient of colors from blueish to reddish, indicating deviations from lowest to highest areas. This surface is used as the base ground for flood simulation in the researcher's work. A DSM is preferred over a digital terrain model (DTM) because it incorporates all features on the surface, including man-made and natural barriers, providing a more realistic surface for simulations.

Slope analysis: Slope is an important factor for flood mapping as it affects the flow of water during a flood. Steep slopes are prone to flash flooding, while gentler slopes experience slower flooding. Slope data is collected using various techniques and combined with elevation data to create digital elevation models (DEMs) for predicting floodwaters' flow. Flood maps using both elevation and slope data provide more accurate information. Figure 5 shows the study area's slope, which is generally flat to slightly steeper, meaning that rainwater will flow slowly, leading to overall flooding.

Land use/land cover analysis: Land use and cover are important factors in flood mapping, as they affect water flow and storage during floods. Data is collected through satellite imagery and used to create land use/cover maps that predict flooding impact and identify vulnerable areas. Soil, man-made structures, and watershed shape and size also affect flood mapping. Figure 2 shows the land cover in the study area, with buildings being the most prone to flooding, followed by roads, bare land, and vegetation as the least prone.

Precipitation analysis: Precipitation is an important factor in flood mapping, as high amounts of precipitation, especially heavy rainfall or snowfall, can lead to flooding. Precipitation data is collected using weather stations and instruments to create maps that predict flooding impact and identify vulnerable areas. Figure 9 shows equal amounts of rainfall that impacted the study area for the past two years.

Proximity to streams/channels analysis: Proximity to streams and channels is an important factor in flood mapping, as areas near water bodies are more likely to be affected by flooding. Geographic information system (GIS) data and mapping tools are used to determine the proximity to streams and create flood hazard maps. The Euclidean distance tool is used to analyze spatial relationships and patterns, as shown in Figure 4, where the yellow color represents the most prone and blue/purple represents the least prone to floods.

Topographic wetness index (TWI) analysis: TWI is used in flood mapping to identify areas more prone to flooding due to their hydrologic properties. The topographic wetness index (TWI) is commonly used to express the topographic sizes and locations of water-saturated areas.

Soil Type: Soil type is also an important factor that affects the way water flows during a flood event. Sandy soils have higher infiltration rates, while clay soils have lower infiltration rates. The study area in the research had generally clay/loam soil, which has a slower infiltration rate. The figures below depict the results of the 7 influential factors used in flood mapping simulations.

The following figures illustrate the outcomes of the seven influential flood simulation factors employed in this research, namely land cover, slope, precipitation, soil type, topographic wetness index (TWI), proximity to streams, and topographical variation (digital elevation model or DEM).

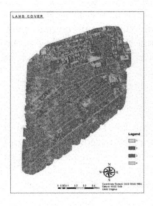

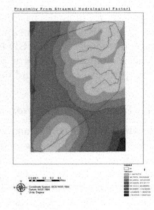

Figure 2. Land cover classification.

Figure 3. Topographic wetness index map.

Figure 4. Proximity to natural stream.

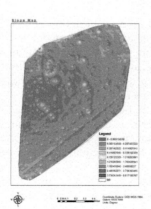
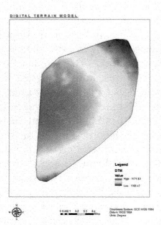
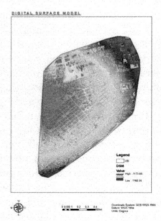

Figure 5. Slope factor map.

Figure 6. Digital Elevation Model (DEM).

Figure 7. Digital Surface Model (DSM).

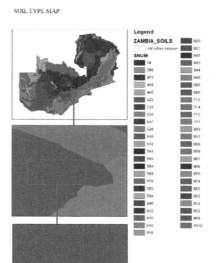

Figure 8. Soil type map.

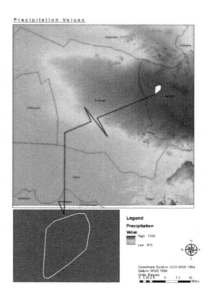

Figure 9. Average precipitation map.

The map in Figure 10 presented illustrates the weighted average of various influential factors in flood analysis such as slope, elevation, TWI, proximity to streams, precipitation, soil type, and land cover. The weighted average technique is commonly employed in GIS to synthesize data from multiple sources, assigning greater importance to more reliable or relevant sources. The flood hazard map was generated by merging the seven influential factors, with each raster image classified into four levels of flood risk ranging from 1 (least prone) to 4 (most prone) to inundation. Figure 10 displays the final flood hazard map, which is based on the weighted average of seven influential factors.

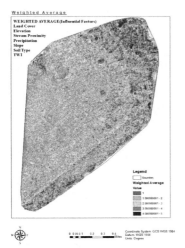

Figure 10. Weighted average of the 7 flood influential factors.

Table 1. The weighted average values obtained from the AHP.

Factors	Weight
Slope	0.139
Elevation	0.205
TWI	0.067
Proximity of Streams	0.122
Precipitation	0.234
Soil Type	0.122
Land Cover	0.112

 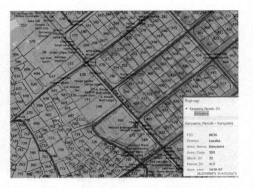

Figure 11. (BIM data with LOD-1 overlaid with GIS data part 1).

Figure 12. Illustration of integrated GIS and BIM data on a flood hazard surface.

Figure 11 shows BIM data at LOD 1 is integrated with GIS shapefiles, where white blocks represent Building Models that are extruded to arbitrary heights. Figure 12 exhibits an overlay of various GIS data, including Local Parcels-Polygons, Roads-Lines, and some building models at LOD-0, on a flood hazard surface. This integrated data is an invaluable tool for stakeholders to readily identify affected plot numbers, buildings, roads, and other GIS data. The varying shades of blue underlying the parcel numbers/buildings illustrate the extent of their susceptibility to flooding. By clicking on view attributes, additional parcel details can be accessed.

The bar chart in Figure 14 illustrates the distribution of the four levels of flood severities across the study area. The findings indicate that the least flood-prone land, approximately 83 hectares in size, is the largest in the study area. The second least prone to flooding covers 144 hectares, followed by the second most flood-prone area, which covers 113 hectares. Finally, the most flood-prone land, represented by the fourth level, spans 34 hectares.

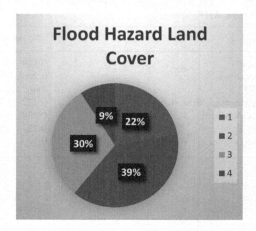 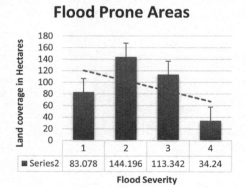

Figure 13. The chart shows the percentages of land covers of the 4 flood susceptibilities.

Figure 14. Land cover vs flood severity.

4 CONCLUSIONS

Flood simulation using geospatial building information modelling (Geo-BIM) is a useful tool for predicting and mitigating flood risks. Incorporating Building Information Models and Geospatial information provides a profound foundation for more realistic flood simulation. The input influential flood factors used in this research were elevation, slope, TWI, land cover, proximity to streams and precipitation, and soil type. In this research the following objectives were addressed;

1. **Factors and input data necessary used in flood simulation:** The input factors necessary to simulate a flood include data pertaining to precipitation, land cover/land use, slope aspect, Topographic wetness index, soil type, elevation and proximity to streams.
2. **Conducting a flood simulation using a case study:** A case study of approximately 375 ha was selected within Kanyama, by initially acquiring an orthmosaic and a DEM. Raster files of input factors were acquired from various, developed, reclassified into 4 levels of severity, and merged using a weighted average geoprocessing tool. Level 1 depicted areas that are the least likely to flood and 4 represented areas that are the most prone to be inundated. A simulation of the BIM data with LOD-1 was run on the merged raster file (tiff).
3. **Develop flood hazard maps**: Based on the results obtained from the flood model and field study, a comprehensive methodology is required to present the data analysis. The flood model generated a detailed flood hazard map, which provided crucial information on the severity of flooding across the study area. The map showed that flood severity was highest in area 2, with a land cover of 144 ha (39%), followed by area 3 with 113 ha (30%), area 1 with 83 ha (22%), and finally, area 4 with 34 ha (9%).

To further analyze the impact of flooding on buildings in the study area, spatial queries were executed in ArcGIS. The analysis showed that out of a total of 8451 buildings, 1856 were situated in areas with the least susceptibility to flooding, 2019 buildings were located in the second most susceptible areas, while 861 buildings were situated in the most vulnerable locations to inundation. To validate the accuracy of the flood model, field interviews were conducted, and the findings revealed that a significant proportion of individuals living in flood-prone areas confirmed the flood severities depicted in the model. Specifically, 87% of respondents in area 1 confirmed the flood severities as modeled, while 84% and 81% confirmed the same for areas 2 and 3, respectively. In area 4, 91% of the respondents validated the flood intensity as depicted in the flood model.

5 RECOMMENDATIONS

The statistics show that around 2880 buildings are at high risk of flooding, and it is recommended to adopt several procedures to mitigate flood hazards, including floodplain management, strict building regulations, emergency preparedness, flood insurance, flood proofing, land use planning, flood warning systems, and demolition and relocation. The use of GeoBIM for flood simulation is also suggested as a powerful tool to reduce the risk of damage to buildings and communities.

Future research in flood simulation with GeoBIM should center on integrating real-time sensor data, investigating social and behavioral aspects, urban drainage system modeling, machine learning and AI applications, cross-disciplinary collaboration and visualization and virtual reality. These additional fields of study can advance our knowledge of Geobim and its use in flood simulation, resulting in more efficient flood risk management and resilient urban design.

REFERENCES

Amirebrahimi, S., Rajabifard, A., Priyan, M. & Ngo, T., 2015. *A Data Model for Integrating GIS and BIM for Assessment and 3D Visualisation of Flood Damage to Building.* Volume I, pp. 1–13.

Andre, Z. & Stephen, W., 2004. *Beyond Modelling: Linking Models with GIS for Flood Risk Management.* Volume I, pp. 191–208.

Daniela, R., Usman, T. K. & Costas, A., 2018. *Flood Risk Mapping Using GIS and Multi-Criteria Analysis: A Greater Toronto Area Case Study.* Volume II, pp. 1–2.

Díez-Herrero, A. & Garrote, J., 2020. *Flood Risk Analysis and Assessment, Applications and Uncertainties: A Bibliometric Review.* Volume I, pp. 1–24.

Doswell, C. A., 2003. *Flooding.* Volume III, pp. 1–8.

Junxiang, Z., 2020. *Automatically Processing IFC Clipping Representation for BIM and GIS Integration at the Process Level.* Volume I, pp. 1–2.

Kabir, U., Deo, G. R. & Amarnath, G. B., 2013. *Application of Remote Sensing and GIS for Flood Hazard Management: A Case Study from Sindh Province, Pakistan.* Volume I, pp. 1–5.

Magami, I. M., Yahaya, S. & Mohammed, K., 2014. Causes and Consequences of Flooding in Nigeria: A Review. *Biological and Environmental Sciences Journal for the Tropics 11*(2), I(0794 – 9057), p. 2.

Nchito, W. et al., 2018. *Policy Brief | Lusaka: Preparing for Increased Flooding.* Lusaka: Future Resilience for african Cities and Lands.

Start, 2017. *Start.* [Online] Available at: https://start.org/highlights/pioneering-new-approaches-to-tackle-urban-flooding-in-lusaka/ [Accessed 19 May 2022].

Suthakaran, S., Withanage, A., Gunawardhane, M. & Gunatilak, J., 2018. Flood Risk Assessment Based on OpenStreetMap Application: A Case Study in Manmunai North Divisional Secretariat of Batticaloa, Sri Lanka. *FOSS4G Asia 2018 Conference*, Volume I, pp. 1–12.

Tembo, J. M. et al., 2019. Pit Latrine Faecal Sludge Solid Waste Quantification and Characterization to Inform the Design of Treatment Facilities in Peri-urban Areas: A Case Study of Kanyama. *African Journal of Environmental Science and Technology*, I(1996-0786), pp. 1–13.

Thiecken, A., Muler, M., Kreibich, H. & Merz, B., 2005. Flood Damage and Influencing Factors: New Insights from the August 2002 Flood in Germany. *Journal of Water Resources Research*, Volume I, pp. 40–42.

WHO, 2022. *World Health Organization.* [Online] Available at: https://www.who.int/health-topics/floods#tab=tab_1 [Accessed 28 March 2022].

Framework for integrated inner-city construction site layout planning

F.G. Tsegay
Department of Civil and Environmental Engineering, School of Engineering, University of Zambia, Lusaka, Zambia
School of Civil Engineering, EIT-M, Mekelle University, Mekelle, Ethiopia

E. Mwanaumo
Department of Civil and Environmental Engineering, School of Engineering, University of Zambia, Lusaka, Zambia
Department of Civil Engineering, School of Engineering, UNISA, Pretoria, South Africa

B. Mwiya
Department of Civil and Environmental Engineering, School of Engineering, University of Zambia, Lusaka, Zambia

ABSTRACT: Inner-city building construction site space is inadequate: needs wisely planned and used professionally to optimize productivity and safety. Thus, mathematical optimization methods have been applied in construction site layout planning (CSLP) to create safe and short moves among the temporary facilities (TFs) in site. This paper is aimed to develop a framework to integrate the space requirements and achieve the optimization goals by reviewing optimization methods and processes of CSLP. The literature on CSLP indicates the consistent and increasing interest in using mathematical optimization methods having a multi-objective function with frequent goals: crew productivity and safety. Existing methods optimize CSLP by minimizing the travel distance among TFs ignoring the working area to TFs distance and the project data was used in fragmented ways. Therefore, this paper has presented a framework for an integrated CSLP process with lean construction workflow principles using Dynamo-based building information modeling to optimize productivity and safety.

Keywords: Mathematical optimization, inner-city, construction site layout planning, and building information modeling

1 INTRODUCTION

Construction work crew spends most of their time within construction sites and require easy, quick, and short moves from the working area to temporary facilities (TFs) to increase productivity and safety. Besides, space in inner-city building construction is the most important but limited resource that needs effective and efficient use. Thus, optimization problems of a construction project such as cost, productivity, and safety often rely on the optimal planning of the construction space by having construction site layout planning (CSLP) that evolved from expert judgment to modern scientific planning of an optimum plan(Elbeltagi & Hegazy 2001; Zolfagarian & Irizarry 2014). However, the construction industry is still suffering from less productivity performance and frequent happening of accidents. Improper planning of space accounts for up to 65% loss in actual efficiency (Huang *et al.* 2010), and the construction work crew spends only 15-20% of the time on direct work; whereas, the major time is spent on indirect work that is directly connected with crew movement and waste due to waiting, disruptions, redoing errors, etc.(Björnfot & Jongeling 2007).

The manufacturing industry productivity has radically changed, due to the transition from craft production via mass production to lean production(Jastia & Kodali 2015). Facility layout planning has played a significant role in the efficiency of material handling by using different mathematical optimization methods and tools(Hosseini-nasab et al. 2017; Izadinia & Eshghi 2016). Thus, the manufacturing industry has been used as the main source of methods, tools, and theories for the construction industry for so long. Lean construction is an example that emerged as a response to the industry challenges: low productivity, errors, delays, cost overruns, and safety. The current status of the construction industry (low productivity) is not taken only as a challenge; rather, lean construction believes that there is potential for improvement in the industry(Howell 1999; Jastia & Kodali 2015). In lean construction, the key to improving the construction process is improving the flow of work in the execution of construction work, i.e., the movement of the work crew in the construction site(Howell 1999; Rybkowski et al. 2020; Vicente et al. 2023).

The crew movement across the site is the main approach while formulating and solving the site space utilization problems(Chavada et al. 2012). There are a lot of optimization methods introduced to solve CSLP problems, which are mainly derived from the manufacturing industry or formulated based on the workflow of the manufacturing process. In the manufacturing process layout, the materials and products move from one workstation to another sequentially without any backtracking; as a result, the optimization methods for facility layout planning focus to optimize the materials and products movement and assure facilities and machines of a similar type are arranged together in one place(Jastia & Kodali 2015). In contrast, building components of a construction project are fixed; that needs establishing a workstation near the permanent location of the components; and the workflow includes the crew travel cycle for resource supply from workstations to TFs(Jongeling 2006).

2 METHODOLOGY

This study relied on a systematic review of the scientific literature presented in electronic databases on mathematical optimization for solving construction site layout planning problems according to the Preferred Reporting Items for Systematic Reviews and Meta-Analyses (PRISMA) guidelines(Page et al. 2021). Mathematical optimization methods and CSLP were used as a generalized search engine for the literature search. Articles were included in the current study if they were published in peer-reviewed journals in English and if the articles are mathematical optimization methods used to solve the construction site layout planning problems.

Different electronic databases were searched for this purpose: Google Scholar and semantic scholar were the main search engines extended to include: Science Direct, Web of Science, IEEE Xplore, Wiley Online, Emerald Insight, and SpringerLink. Repetitive and comparative articles were excluded and initial screening was done to put together all potential studies according to their titles and abstracts. These studies were then filtered after a full review of the article text and prepared to analyze their objective function(s) and optimization methods efficiency for the inner-city site requirements as shown in the next section.

3 LITERATURE REVIEW

In many studies, the CSLP mainly focuses on how to allocate the TFs that range from simple laydown areas to warehouses, fabrication shops, maintenance shops, batch plants, and residence facilities in the available site space(Ghanim 2020; Moradi & Shadrokh 2019). Thus, the CSLP process needs to involve identifying problems and opportunities, developing solutions, and (optimally) positioning TFs in the unoccupied areas on the specified duration within the site boundaries. (Moradi & Shadrokh 2019; Ning et al. 2011; Oral et al. 2018; Thomas & Ellis 2017). The CSLP is the main area in construction planning that applies mathematical optimization methods and the historical development and search approaches evolve through time and consists of three broad classes: classical methods, evolutionary algorithms (EAs), and more recently hybrid methods(Keller 2018; Moraes et al. 2022).

3.1 Mathematical optimization methods for construction site layout

Construction site layout planning problems involve objective functions that are not continuous and/or differentiable. Thus, the classical optimization techniques have limited scope in practical applications compared to (meta)heuristic methods. But still (meta)heuristic methods usually fail to provide a global optimum solution because they are predominantly greedy and do not operate exhaustively on all available data. Thus, to overcome those computational challenges, recently there is an approach of using hybrid methods for the CSLP. The hybrid methods use both exact and heuristic approaches in a combined model (Prayogo et al. 2020; Rao 2020; Wong et al. 2010). The use of mathematical optimization for CSLP and the geographical coverage of research works, the top three countries that published related works are China with 30%, Canada with 14%, and Iran with 8%; but note that no single article was found from sub-Saharan countries as shown figure below (Fikadu et al. 2023).

The evolution of using mathematical methods for CSLP has continuously increased and reached from simple mathematical formulas and calculations to nature-inspired optimization methods and a hybrid of these. So, the CSLP has used different optimization techniques to address the site space requirements for different site conditions. These are grouped into different categories based on the type of modeling classes and computing techniques with continuous increases and introduce new methods in the last three decades as shown in Figure 2 below.

Generally, a CSLP for the limited space of an inner-city site is critically important to have an effective and efficient construction operation by optimizing the overarching goals that are commonly defined in CSLP studies: minimizing cost, increasing productivity, safety, security, and minimizing environmental impacts represented by directly and embedded objective

Figure 1. Optimization methods of CSLP-related research work geographical coverage (after Tsegay et al. (2023)).

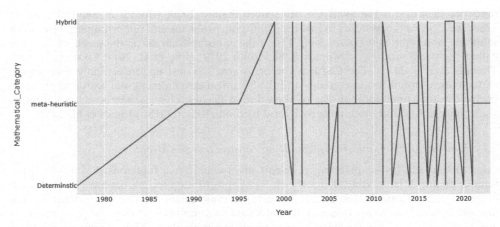

Figure 2. Trends for applying mathematical optimization classes for CSLP.

function(s) to be solved by different methods(Hammad et al. 2021; Song et al. 2018). According to the systematic review of Tsegay et al. (2023) the practice of using mathematical optimization dominantly focuses on cost minimization goal and a productivity optimization goal. Similarly, safety issues are also addressed by some of the reviewed papers but it is embedded in the objective functions of many of the articles included in the review(Fikadu et al. 2023). Thus, the CSLP is responsible for effective and efficient project performance: increasing productivity and safety that is often considered as a multi-objective function in searching for an optimum layout with embedded objectives such as closeness weight preferences among the facilities to keep minimum distance requirements for safety and other performance requirements.

The background goals of objective functions for CSLP increasing productivity and safety of a building construction project are highly dependent on the crew movement from the working area to other space requirements (facilities) and from facility to facility. But most of the CSLP methods were developed to allocate only the TFs and their objective function(s) were formulated by considering the work crew movement from TFs to TFs to minimize travel distance, travel time, or/and travel cost. This approach is based on the facility layout planning of the manufacturing industry that emphasizes the physical prearrangement of temporary facilities at the construction site(Jongeling 2006; Maghfiroh et al. 2023). So, in the systematic review of Tsegay et al. (2023), the optimization objective function(s) of the existing CSLP models tried to minimize travel distance between facilities without considering the crew movement from the working area to the facilities.

3.2 *Commonly used optimization methods for construction site layout*

The evolution of mathematical optimization methods: classic/deterministic, evolutionary algorithms (EAs), and more recently the hybrid methods classes have been applied to solve CSLP problems. The selection of optimization methods depends on the defined and formulated objective function(s) of the construction site layout problem and the preferred search approach by the planner. Thus, researchers have used three search approaches based on how and in what order the space is allocated to the site space requirements: construction approach, improvement approach, and concurrent approach(Sadeghpour & Andayesh 2015).

The decision of selecting a search approach for a construction site layout model is based on what level of accuracy and sophistication in optimization is required and expected to be achieved by the model. So, different optimization techniques have been used to solve the optimization problems with either exact solutions by using techniques from both the deterministic and hybrid optimization classes or not-exact solution(s) by using the (meta)heuristic cluster of techniques.

The deterministic and hybrid approaches don't scale efficiently to large and/or complex problems like CSLP. Hence, different (meta)heuristic techniques have been deployed to solve such problems to find near-optimum solutions. From the review, the Genetic Algorithm (GA) is a method that is dominantly used for CSLP as shown in Figure 3 below.

Genetic Algorithms (GA) are a class of optimization algorithms inspired by the principles of natural selection and genetics that operate on a population of potential solutions, which are represented as individuals or chromosomes. Each chromosome consists of genes that encode specific parameters or variables of the problem and the population evolves through generations(Haupt & Haupt 2004). The fittest individuals have a higher probability of being selected for reproduction during the key components of a genetic algorithm process: selection, crossover, and mutation. Selection favors individuals with higher fitness, allowing them to pass their genetic information to the next generation. Crossover involves combining genetic material from two parents to create offspring with a mixture of their traits. Mutation introduces small random changes in the genes to maintain diversity and explore new regions of the search space(Chambers 2000). GA explores the solution space through repeated iterations by gradually converging towards optimal or near-optimal solutions. This character made GA well-suited for problems with complex, nonlinear, and multi-modal search spaces, where other optimization methods struggle.

GA methods are widely used in various fields to solve complex optimization problems CSLP is not an exception. From the systematic review from Tsegay et al. (2023), GA has been applied to solve both single objectives and the multi objective problems of the CSLP (Fikadu *et al.* 2023). Therefore, the multi-objective genetic algorithm method is proposed for this framework due to the commonness of the method in the CSLP research areas and the availability of programming tools like Python and possibility to integrate with building information modeling software like Revit with the help of dynamo add-ins.

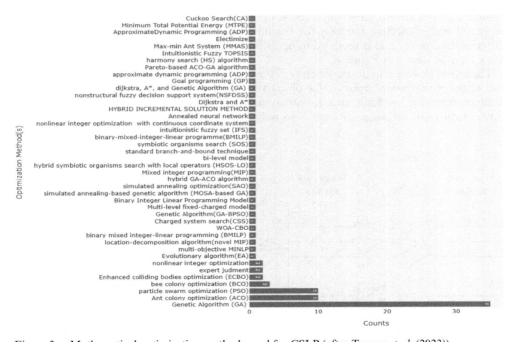

Figure 3. Mathematical optimization methods used for CSLP (after Tsegay *et al.* (2023)).

4 INTEGRATED CONSTRUCTION SITE LAYOUT PLANNING FRAMEWORK IN INNER-CITY

Construction work crews spend most of their time within construction sites and it's needed to have easy and quick moves to save time and increase productivity as well as safety. However, space in inner-city building construction is an inadequate resource that needs to be planned wisely and professionally used. So, mathematical optimization methods have been applied to solve the project optimization problems: cost/productivity and safety that often rely on the optimal planning of the construction space. The CSLP problems have a multi-objective function (s) either directly stated or a single objective with embedded objectives mainly for increasing crew productivity and safety goals. However, the developed methods focus on the optimization problems of CSLP for TFs and established their objective function(s) by considering the crew movement between TFs. So, this study concluded that most of the optimization methods used for CSLP were directly adopted from the facility layout planning approach for manufacturing production, where products and resources move from facility to facility.

4.1 *Objective function for integrated construction site layout planning*

In the construction site, products are stationary that need to establish a working area around them and a supply of resources from the TFs. Few developed methods have considered the building block(s) as a fixed member of Facilities and optimized by minimizing travel distance among all facilities based on the construction workflow. However, the working area for each crew was not clearly stated. So, the existing optimization methods and their objective function(s) luck the lean construction workflow optimizing concept. On the other hand, construction at inner-city sites always faces shortages of space, and requirements for space vary from phase to phase during the project duration(Fikadu et al. 2023). Therefore, an integrated CSLP method is necessary and the objective function(s) should be formulated based on the lean construction's workflow optimization principle. So, the proposed framework for integrated CSLP for an inner-city building construction site is proposed to be dynamic or phased and able to use inner space (built floors) and plinth area. So, the structure of the proposed multi-objective function for the integrated CSLP is based on the workflow in the construction site and is modeled to optimize the possible shortest movement path of the work crew as shown in Figure 4.

Therefore, as shown in the parameters of the multi-objective functions for the Productivity and safety optimization problem, the objective of the model is to minimize: the crew travel distance from the working area to each connected facility and the interaction distance between different facilities, as well as to reduce the likelihood of accidents happening to

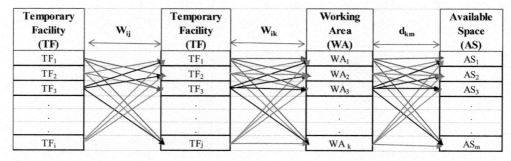

Figure 4. Lean construction workflow-based CSLP variables matrix (Facilities, working area, and available space) (after Tsegay et al. (2023)).

improve the safety level as formulated below.

$$\text{Minimize} \quad \text{Objective} \quad \text{Function} = \sum_{i=1}^{n}\sum_{j=1}^{m}\sum_{k=1}^{n} W_{ij}W_{ik}d_{ijk} \quad (1)$$

Where n, m and k are the numbers of facilities, available space, and working areas respectively, dijk is the distance of workflow path (the sum of dij and dkm), dij is the distance between temporary facility i and j, dkm is a constant distance of the available space from the working area location, and Wij is the closeness weight between facility i and j, and Wik is the closeness weight between facility i and working area k; where the values of the closeness weights (Wij and Wik) are qualitatively determined to account for influencing factors such as safety, traveling costs, trip frequency or other user-defined areas in qualitative or quantitative approaches.

4.2 Dynamo based building information modeling for integrated construction site layout planning

According to Autodesk(2022), Building Information Modeling (BIM) is defined as a holistic process of creating and managing information for a built asset based on an intelligent model that integrates structured, multi-disciplinary data to produce a digital representation of lifecycle, from planning and design to construction and operations(Autodesk 2022). BIM has revolutionized the field of construction by enabling the creation, management, and sharing of detailed digital representations of building projects. To ensure the success of a construction project by creating short and fast transport routes, clarity, and a safe work environment there is a need for efficient use of technology. So, beyond its applications in design and coordination, BIM is increasingly being utilized for integrated construction site layout planning.

This approach leverages the rich data and collaborative capabilities of BIM to optimize the arrangement of temporary facilities, equipment, and resources on the construction site. Hence, BIM with its four-dimensional (4D) has introduced an Integrated Site Planning System by integrating the project data: schedules, 3D models, resources, and site spaces using 4D CAD technology and providing 4D graphical visualization capability for construction site planning(Ma et al. 2005; Yu et al. 2016). Thus, keeping in mind the benefits of having integrated project data and graphical visualization capacity to CSLP from the 4DBIM, an optimization decision need additional consideration of incorporating optimization methods with the 4D CAD technology(Amiri et al. 2017).

The use of metaheuristic algorithms such as genetic algorithms is well-known in construction site layout planning studies. For this, Revit has a Dynamo add-in which is a Playground application that runs by combing the graphical approach in Revit with the powerful algorithmic techniques in scripting using Python script(Whitlock et al. 2021). Besides, the main advantage of using Dynamo: users don't need to have a high level of programming syntax scripting, it is used to integrate optimization methods to the 4DBIM that are written in the Python script(Mohammed Fathy et al. 2022). Hence, for this framework Dynamo is proposed to integrate the GA optimization method from the Pymoo library: a multi-objective optimization framework in Python(Blank & Deb 2020).

Therefore, the following Figure 5 shows the details of the framework for the integrated CSLP process: data organization and optimization goal(s) formulation, identifying space requirement variables and measuring available space based on project phases, computing values of influencing factors for both project productivity and safety optimizations, objective function formulation based on the workflow matrix of the objective function parameters, and integrating project data(4DBIM) and optimization method(GA) using Dynamo and finally find the optimum solution to the formulated function (eq.1) for site layout plan.

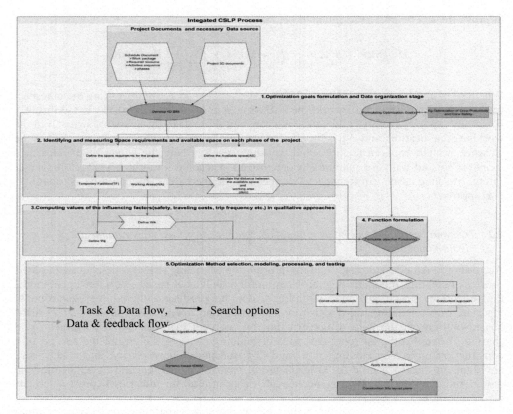

Figure 5. Proposed framework for integrated CSLP process.

5 CONCLUSION

This study has developed a framework of integrated CSLP process for inner-city sites by applying the lean construction workflow optimization concept to optimize the travel cycle of the work crew from the working area to TFs within the site by considering all possible alternative free space sources based on the project schedule phases. Thus, the formulated objective function (equation 1), the optimization process will be done with the support of mathematical optimization methods such as genetic algorithms and synchronized into the 4D BIM document by dynamo script as shown Figure 5. In this study, the efficiency of the developed methods was not evaluated and the developed framework did not formulate specific objective function(s). So, researchers can use this framework as a guideline for testing the effectiveness of the existing mathematical optimization methods and creating new method(s). similarly, practitioners can also use this framework to formulate and solve their real-world CSLP problems in construction operations.

REFERENCES

Amiri, R., Sardroud, J.M. and Soto, B.G. De (2017) 'BIM-based Applications of Metaheuristic Algorithms to Support the Decision-making Process: Uses in the Planning of Construction Site Layout', *Procedia Engineering*, 196(June), pp. 558–564.

Autodesk (2022) *Keeping Pace with the Modern Construction Industry*.

Björnfot A, Jongeling R (2007) 'Application of Line-of-balance and 4D CAD for Lean Planning', *Construction Innovation*, 7(2), pp. 200–2011.

Blank, J. and Deb, K. (2020) 'Pymoo: Multi-Objective Optimization in Python', *IEEE Access*, 8, pp. 89497–89509.

Chambers, L. (2000) *The Practical Handbook of Genetic Algorithms: Applications, Second edition, the Practical Handbook of Genetic Algorithms: Applications, Second Edition*.

Chavada, R.D. et al. (2012) 'A Framework for Construction Workspace Management: A Serious Game Engine Approach', *Computing in Civil Engineering*, 25(4), pp. 57–64.

Elbeltagi, E. and Hegazy, T. (2001) 'A Hybrid AI-Based System for Site Layout Planning in Construction', *Computer-Aided Civil and Infrastructure Engineering*, 16(2), pp. 79–93.

Ghanim, A.A.H.S.B.F. (2020) 'Construction Site Layout Planning Problem: Past, Present and Future', *Expert Systems with Applications*

Hammad, A.W.A. et al. (2021) 'The Use of Unmanned Aerial Vehicles for Dynamic Site Layout Planning in Large-scale Construction Projects', *Buildings*, 11(12).

Haupt, R.L. and Haupt, S.E. (2004) *Practical Genetic Algorithms, Practical Genetic Algorithms*.

Hosseini-nasab, H. et al. (2017) 'Classification of Facility Layout Problems: A Review Study', *The International Journal of Advanced Manufacturing Technology*, pp. 957–977.

Howell, G.A. (1999) 'What is Lean Construction – 1999', in *IGLC – 7th Conference of the International Group for Lean Construction*, pp. 1–11.

Huang, C., Wong, C.K. and Tam, C.M. (2010) 'Optimization of Material Hoisting Operations and Storage Locations in Multi-storey Building Construction by Mixed-integer Programming', *Automation in Construction*, 19(5), pp. 656–663.

Izadinia, N. and Eshghi, K. (2016) 'A Robust Mathematical Model and ACO Solution for Multi-floor Discrete Layout Problem with Uncertain Locations and Demands', *Computers and Industrial Engineering*, 96, pp. 237–248.

Jastia, N.V.K. and Kodali, R. (2015) 'Lean Production: Literature Review and Trends', *International Journal of Production Research*, 53(3), pp. 867–885.

Jongeling, R. (2006) *A Process Model for Work-flow Management in Construction*, Lulea, Sweden: Lulea University of Technology.

Keller, A.A. (2018) 'Elements of Mathematical Optimization', in *Mathematical Optimization Terminology*, pp. 1–12.

Ma, Z., Shen, Q. and Zhang, J. (2005) 'Application of 4D for Dynamic Site Layout and Management of Construction Projects', *Automation in Construction*, 14(2005), pp. 369–381.

Maghfiroh, M.F.N. et al. (2023) 'Cuckoo Search Algorithm for Construction Site Layout planning', *IAES International Journal of Artificial Intelligence*, 12(2), pp. 851–860.

Mohammed Fathy, M., Elsaid Elbeltagi, E. and Elsheikh, A. (2022) 'Dynamo Visual Programming-Based Generative Design Optimization Model for Construction Site Layout Planning. (Dept. C)', *MEJ. Mansoura Engineering Journal*, 46(4), pp. 31–42. A

Moradi, N. and Shadrokh, S. (2019) 'A Simulated Annealing Optimization Algorithm for Equal and Un-equal Area Construction Site Layout Problem', *International Journal of Research in Industrial Engineering*, 8(2), pp. 89–104.

Moraes, L.B. de, Parpinelli, R.S. and Fiorese, A. (2022) 'Application of Deterministic, Stochastic, and Hybrid Methods for Cloud Provider Selection', *Journal of Cloud Computing*, 11(1), pp. 1–23.

Ning, X., Lam, K.C. and Lam, M.C.K. (2011) 'A Decision-making System for Construction Site Layout Planning', *Automation in Construction*, 20(4), pp. 459–473.

Oral, M. et al. (2018) 'Construction Site Layout Planning: Application of Multi-objective Particle Swarm Optimization', *Teknik Dergi/Technical Journal of Turkish Chamber of Civil Engineers*, 29(6), pp. 8691–8713.

Page, M.J. et al. (2021) 'The PRISMA 2020 Statement: An Updated Guideline for Reporting Systematic Reviews', *Systematic Reviews*, 10(1), p. 89.

Prayogo, D. et al. (2020) 'A Novel Hybrid Metaheuristic Algorithm for Optimization of Construction Management Site Layout Planning', *Algorithms*, 13(117), pp. 1–19.

Rao, S.S. (2020) *Engineering Optimization: Theory and Practice*. 5th edn.

Rybkowski, Z.K., Forbes, L.H. and Tsao, C.C.Y. (2020) 'The Evolution of Lean Construction Education at US-based Universities', *Lean Construction*, 193, pp. 387–407.

Sadeghpour, F. and Andayesh, M. (2015) 'The Constructs of Site Layout Modeling: An Overview', *Canadian Journal of Civil Engineering*, 42(3), pp. 199–212.

Song, X. et al. (2018) 'Conflict Resolution-motivated Strategy Towards Integrated Construction Site Layout and Material Logistics Planning: A Bi-stakeholder Perspective', *Automation in Construction*, 87(October 2017), pp. 138–157.

Thomas, H.R. and Ellis, R.D. (2017) *Construction Site Management and Labor Productivity Improvement: How to Improve the Bottom Line and Shorten Project Schedules*, American Society of Civil Engineers.

Tsegay, Fikadu G, Mwanaumo, E. and Mwiya, B. (2023) 'Mathematical Optimization Methods for Inner-city Construction Site Layout Planning: A Systematic Review', *Asian Journal of Civil Engineering*

Tsegay, Fikadu G., Mwanaumo, E. and Mwiya, B. (2023) 'Construction Site Layout Planning Practices in Inner-city Building Projects: Space Requirement Variables, Classification and Relationship', *Urban, Planning and Transport Research*, 11(1), pp. 1–21.

Vicente A et al. (2023) *Lean Construction 4.0 Driving a Digital Revolution of Production Management in the AEC Industry*. 1st edn, Routledge Tylor & Francis Group. 1st edn. London and New york.

Whitlock, K. et al. (2021) '4D BIM for Construction Logistics Management', *CivilEng*, 2(2), pp. 325–348.

Wong, C.K., Fung, I.W.H. and Tam, C.M. (2010) 'Comparison of Using Mixed-Integer Programming and Genetic Algorithms for Construction Site Facility Layout Planning', *Journal of Construction Engineering and Management*, 136(10), pp. 1116–1128.

Yu, Q., Li, K. and Luo, H. (2016) 'A BIM-based Dynamic Model for Site Material Supply', *Procedia Engineering*, 164(June), pp. 526–533.

Zolfagarian, S. and Irizarry, J. (2014) 'Current Trends in Construction Site Layout Planning', in *Construction Research Congress 2014: Construction in a Global Network*, pp. 1723–1732.

Digital twin technology in health, safety, and wellbeing management in the built environment

A.O. Onososen*, I. Musonda, T. Moyo & H. Muzioreva
Centre for Applied Research + Innovation in the Built Environment (CARINBE), Faculty of Engineering and the Built Environment, University of Johannesburg, South Africa

ABSTRACT: Digital twin technology is an innovative approach that creates a virtual replica of a physical asset, providing real-time data and insights that can improve safety and health management in the construction industry. This can help to identify potential safety hazards, improve communication and collaboration among construction workers and stakeholders, and enhance the overall safety and health management of construction projects. This paper aims to evaluate the impact and applications of digital twin technology on enhancing safety and health management in the construction industry. A comprehensive overview of the technology, its benefits and challenges of its successful implementation in the construction industry will be provided. The findings reveal future trends and opportunities for digital twin technology in the construction industry, highlighting the potential for the technology to transform the way safety and health management is performed in the construction industry.

Keywords: Digital twin technology, Safety and health management, Construction industry, Internet of Things (IoT), Sensors.

1 INTRODUCTION

The built environment has a significant impact on the health, safety, and wellbeing of its occupants. The management of these aspects has become a critical concern for many stakeholders, including building owners, operators, and occupants (Musakwa & Moyo 2020). In recent years, digital twin technology has emerged as a promising tool for enhancing health, safety, and wellbeing management in the built environment (Opoku et al. 2021). A digital twin is a virtual replica of a physical asset, system, or process that allows for real-time monitoring, analysis, and optimization (Hou et al. 2021).

Building Information Modelling provides procedures, technologies, and data schemas that enable a standardised semantic representation of building components and systems. However, the concept of a Digital Twin conveys a more holistic socio-technical and process-oriented characterisation of the complex artefacts involved by leveraging the synchronicity of the cyberphysical bi-directional data flows (Boje et al. 2020). This is accomplished by leveraging the synchronicity of the cyberphysical bi-directional data flows. Furthermore, Building Information Modeling is lacking in semantic completeness in areas such as control systems, integrating sensor networks, social systems, and urban artefacts that go beyond the scope of buildings. Because of this, a holistic, scalable semantic approach that takes into account dynamic data at different levels is required.

Digital twin technology has been widely applied in various industries, including manufacturing, aerospace, and healthcare. In the built environment, digital twin technology has

---

*Corresponding Author

been used for building performance optimization, energy management, and predictive maintenance (Armeni et al. 2022). More recently, digital twin technology has been applied to enhance health, safety, and well-being management in the built environment. Digital twin technology has the potential to improve health management in the built environment by providing real-time monitoring and analysis of indoor air quality, temperature, humidity, and other environmental factors (Laubenbacher et al. 2022). Digital twin technology can also enhance safety management in the built environment by providing real-time monitoring and analysis of potential hazards, such as fire, smoke, and gas leaks (Dallel et al. 2023). This paper evaluates the impact and applications of digital twin technology on enhancing health, safety, and wellbeing management in the built environment. This study provides an overview of the relevant literature on the application of digital twin technology in the built environment.

2 RESEARCH METHODOLOGY

This study utilized a review approach that relied on content analysis, a widely accepted and effective method of analyzing and synthesizing past research findings. Content analysis has found broad application in engineering and construction management due to its versatility and ability to justify research findings in various contexts. The Scopus database was used to identify top-tier publications that explored the use of gamification in construction health, safety, and wellbeing. These publications were carefully selected from reputable academic journals covering topics such as engineering, construction management, science, technology, safety, and human factors. The selection aimed to ensure the relevance and credibility of the articles to the study's objectives.

Figure 1. Research approach.

3 FINDINGS AND DISCUSSIONS

3.1 *Integration of digital twin technology in health, safety, and wellbeing management*

The construction industry is one of the most hazardous sectors, with high rates of accidents, injuries, and fatalities. The management of health, safety, and wellbeing (HSW) in the construction industry is critical to prevent accidents and ensure the wellbeing of workers and occupants. Digital twin technology has emerged as a promising tool for improving HSW management in the construction industry.

3.1.1 *Integration challenges*
Despite the potential benefits of digital twin technology in HSW management in the construction industry, several challenges need to be addressed for successful integration. One of the main challenges is the lack of standardization and interoperability of digital twin platforms and applications (Malik *et al.* 2021). Different vendors may use different data models, protocols, and interfaces, which can hinder data sharing and integration. Standardization efforts, such as the Industry Foundation Classes (IFC) can help overcome this challenge. More integrated means of sharing construction data have surfaced ever since the introduction of the Industry Foundation Classes (IFC), and these approaches have now been

adopted by the construction industry as a whole. Simply the expansion of IFC has had a significant effect on the way in which current tools and methodologies are produced in the context of research and development (Boje et al. 2020). Another challenge is the privacy and security of data generated by digital twin systems. Digital twin systems can generate massive amounts of data that need to be stored, processed, and analyzed. This data can include sensitive information, such as worker health records, equipment performance, and project schedules. Ensuring the privacy and security of this data is crucial to prevent data breaches, cyberattacks, and unauthorized access. Adopting data encryption, access control, and audit trails can help mitigate these risks (Mantha et al. 2021). The integration challenges identified are summarized below;

Table 1. Integration of digital twin technology in health, safety, and wellbeing management.

S/N	Integration Challenges	Description	Sources
1.	Data Integration	One of the main challenges of implementing digital twin technology in health, safety, and wellbeing management is the integration of data from various sources, such as sensors, electronic medical records, wearables, and other devices. Ensuring the accuracy and consistency of this data is crucial for the success of the digital twin system.	You & Feng 2020
2.	Interoperability	Digital twin technology requires the integration of various systems and devices, which can be challenging due to interoperability issues. Different systems may use different data formats or communication protocols, making it difficult to exchange data between them.	Ozturk 2020
3.	Cybersecurity	The integration of digital twin technology in health, safety, and wellbeing management requires the secure handling of sensitive data. Ensuring the confidentiality, integrity, and availability of this data is essential to prevent unauthorized access or data breaches.	Mantha et al. 2021
4.	Ethics and Privacy	Digital twin technology raises ethical and privacy concerns, particularly regarding the use of personal health data. Ensuring that the use of this data is ethical and complies with privacy regulations is crucial to gain public trust in the technology.	Dallel et al. 2023
5.	Complex System Design	The design of digital twin systems for health, safety, and well-being management is complex due to integrating various systems, devices, and data sources. Ensuring that the system design is effective and efficient is crucial to ensure its success.	Armeni et al. 2022
6.	Data Management:	The volume, variety, and velocity of data generated by digital twin systems can be overwhelming, making it challenging to manage this data effectively. Ensuring that the data is stored, processed, and analyzed efficiently is essential to gain insights and make informed decisions	Malagnino et al. 2021
7.	Stakeholder Engagement	The successful implementation of digital twin technology in health, safety, and well-being management requires the engagement of various stakeholders, including healthcare providers, patients, regulators, and technology vendors. Ensuring that all stakeholders are involved in the design, implementation, and operation of the system is crucial to gain their support and acceptance.	Laing 2020
8.	Skills and Knowledge	The implementation of digital twin technology in health, safety, and wellbeing management requires specialized skills and knowledge in areas such as data analytics, system design, cybersecurity, and ethics. Ensuring that the workforce has the necessary skills and knowledge is essential to ensure the success of the technology.	Zanni 2013
9.	Cost	The implementation of digital twin technology can be expensive, requiring investments in hardware, software, and personnel. Ensuring that the costs are justified by the benefits and that technology is cost-effective is crucial to gain the support of decision-makers.	Ozturk 2020
10.	Regulatory Compliance	The implementation of digital twin technology in health, safety, and wellbeing requires compliance with various regulations, such as HIPAA and GDPR. Ensuring that the system complies with these regulations is essential to prevent legal and financial liabilities.	Chen et al. 2021

3.2 Impact of digital twin technology on health, safety, and wellbeing management

3.2.1 Impact on safety & wellbeing management

Site safety should be monitored using management tools, as the risks to safety vary in space and time. Digital twin technology can enhance safety management by providing real-time monitoring and analysis of potential hazards, such as falls, collisions, and fires. The system can also provide real-time alerts to workers and supervisors to prevent accidents. BIM is seen as the starting point for DT, acting as a semantically rich 3D reference model for the DT to use in various applications. The BIM shell is enriched with time and sensor data, to formulate parallel offline and online simulations for energy, safety, human comfort and wellbeing (Boje et al. 2020). Digital twin technology can also improve safety management by providing predictive maintenance. The system can schedule maintenance activities before equipment failure occurs, reducing downtime and preventing accidents. Digital twin technology can improve wellbeing management by providing real-time monitoring and analysis of worker behavior, preferences, and feedback.

Table 2. Impact of digital twin technology on health, safety, and wellbeing management.

S/N	Impact	Description	Sources
1.	Improved Safety	Digital twin technology can improve safety by identifying and mitigating potential hazards in real-time. This can be particularly useful in high-risk environments such as healthcare facilities, manufacturing plants, and construction sites.	Panteli et al. 2020
2.	Enhanced Health Monitoring	Digital twin technology can enable real-time health monitoring of patients, employees, and other stakeholders. This can improve early detection of health issues, facilitate remote care, and reduce the burden on healthcare facilities.	Pereira et al. 2021
3.	Predictive Maintenance	Digital twin technology can enable predictive maintenance of equipment and infrastructure, reducing downtime and improving safety. This can be particularly useful in healthcare facilities where the failure of critical equipment can have severe consequences.	Volk et al. 2014
4.	Real-time Decision Making	Digital twin technology can enable real-time decision making based on data insights. This can improve the efficiency of health, safety, and wellbeing management by enabling quick and informed decisions	Laubenbacher et al. 2022
5.	Improved Efficiency	Digital twin technology can improve the efficiency of health, safety, and wellbeing management by automating processes and reducing manual intervention. This can save time, reduce costs, and improve the accuracy of data analysis.	Malik *et al.* 2021
6.	Personalized Care	Digital twin technology can enable personalized care by providing real-time health data insights. This can enable healthcare providers to tailor their care to individual patients, improving outcomes and patient satisfaction	Hou et al. 2021
7.	Enhanced Collaboration	Digital twin technology can enable enhanced collaboration between healthcare providers, patients, and other stakeholders. This can improve the sharing of information, reduce errors, and improve communication	Akanmu et al. 2021
8.	Improved Patient Experience	Digital twin technology can improve the patient experience by enabling remote care, reducing waiting times, and providing personalized care. This can improve patient satisfaction and reduce the burden on healthcare facilities	Hou et al. 2021
9.	Enhanced Training	Digital twin technology can enhance training by enabling realistic simulations of health, safety, and wellbeing scenarios. This can improve the skills of healthcare providers, reduce errors, and improve safety	Bermudez et al. 2022
10.	Improved Regulatory Compliance	Digital twin technology can improve regulatory compliance by enabling real-time monitoring of compliance with regulations such as HIPAA and GDPR. This can reduce the risk of legal and financial liabilities and improve stakeholder trust.	Dallel et al. 2023

3.2.2 *Challenges and limitations*

Despite the potential benefits of digital twin technology on HSW management, several challenges and limitations need to be addressed. One of the main challenges is the lack of standardization and interoperability of digital twin platforms and applications. Another challenge is the cost and complexity of implementing and maintaining digital twin systems. Digital twin systems require significant investments in hardware, software, and skilled personnel. The systems also require ongoing maintenance, updates, and training, which can increase the total cost of ownership. Moreover, the complexity of digital twin systems may require additional resources, such as data scientists and cybersecurity experts.

3.3 *Applications for digital twin technology in enhancing health, safety, and wellbeing management*

3.3.1 *Applications for health & safety management*

The Architecture, Engineering, and Construction industry is transitioning into the digital age, and as a result, the processes involved in the design, construction, and operation of built assets are becoming increasingly influenced by technologies dealing with value-added monitoring of data from sensor networks, management of this data in secure and resilient storage systems underpinned by semantic models, as well as the simulation and optimization of engineering systems. Digital twin has several applications for enhancing health management, including personalized healthcare, real-time monitoring, and optimization of environmental factors, and disease surveillance (Hou et al. 2021). For example, the Covid-19 outbreak has highlighted the potential of digital twin technology in tracking and controlling disease spread. This can collect and analyze data from multiple sources, such as hospitals, airports, and social media, to identify potential outbreaks and inform public health responses. Digital twin technology has several applications for enhancing safety management, including real-time monitoring and analysis of potential hazards, predictive maintenance, and worker training.

Table 3. Applications.

S/N	Applications	Description	Sources
	Workplace Health & Safety Simulation	Digital twins can be used to simulate and analyze workplace safety scenarios, identifying potential hazards and developing strategies to minimize risks.	Getuli et al. 2020
	Emergency Responses	Digital twins can be used to simulate emergency scenarios, enabling responders to test and refine their response plans.	Kim et al. 2021
	Human-Machine/Robot Interaction on site	Digital twins can be used to simulate and optimize Human-Machine/Robot Interaction on site helping workers learn to collaborate and improve expertise in safe interaction	Boje et al. 2020
	Occupants & Workers Mental Health	Digital twins can be used to monitor and analyze mental health conditions, enabling earlier intervention and treatment.	Hou et al. 2021
	Safety Planning and Risk Mitigation	Digital twins can be used to simulate and analyze construction sites and identify potential safety hazards, enabling construction companies to develop strategies to mitigate risks	Hou et al. 2021
	Site Optimization	Digital twins can be used to optimize construction site layouts, helping to improve efficiency and reduce waste	Boje et al. 2020
	Equipment and Material Optimization:	Digital twins can be used to monitor and optimize the use of construction equipment and materials, helping to reduce waste and increase efficiency.	Taher *et al.* 2023
	Quality Assurance	Digital twins can be used to monitor and ensure the quality of construction materials and processes, identifying potential defects and ensuring compliance with building codes and regulations.	Dallel et al. 2023
	Maintenance and Repair	Digital twins can be used to monitor and predict the maintenance needs of buildings and construction equipment, helping to reduce the risk they pose on safety and wellbeing.	Hou et al. 2021

(continued)

Table 3. Continued

S/N	Applications	Description	Sources
	Remote Safety & Health Monitoring	Digital twins can be used to remotely monitor construction sites, enabling real-time safety monitoring and control of construction processes.	Dallel et al. 2023
	Worker Health and Safety	Digital twins can be used to monitor worker health and safety, identifying potential risks and hazards and developing strategies to mitigate them.	Dallel et al. 2023
	Safe Project Collaboration:	Digital twins can be used to facilitate safe collaboration and communication between project stakeholders, enabling more efficient and effective decision-making and teaming	Hou et al. 2021
	Health and Safety Training and Education:	Digital twins can be used to simulate construction processes and provide Health & Safety training and education for workers, helping to improve safety and efficiency on construction sites.	Dallel et al. 2023

4 CONCLUSION

Digital twin technology has the potential to enhance health, safety, and wellbeing management in the built environment. The application of digital twin technology can provide real-time monitoring, analysis, and optimization of indoor environmental factors, potential hazards, and occupant behavior. This study evaluates the impact and applications of digital twin technology on enhancing health, safety, and wellbeing management in the built environment through a systematic literature review. The results of this review provides insights into the effectiveness and limitations of digital twin technology in improving health, safety, and wellbeing management in the built environment.

Digital twin technology has the potential to enhance HSW management in the construction industry by providing real-time monitoring, analysis, and optimization of environmental factors, potential hazards, and worker behavior. The integration of digital twin technology in the construction industry faces several challenges, such as standardization and interoperability, privacy and security, and data management. Overcoming these challenges requires collaboration among stakeholders, including vendors, regulators, and industry associations. The successful integration of digital twin technology in HSW management in the construction industry can improve the safety, productivity, and wellbeing of workers and occupants, and reduce the overall costs of accidents and injuries. Digital twin has the potential to significantly impact HSW management by providing real-time monitoring, analysis, and optimization of environmental factors, potential hazards, and worker behavior. This can improve health management by providing personalized and proactive healthcare and real-time monitoring of environmental factors.

ACKNOWLEDGEMENT

The authors would like to appreciate the Centre for Applied Research + Innovation in the Built Environment (CARINBE), Faculty of Engineering and the Built Environment, University of Johannesburg, for financial support.

REFERENCES

Akanmu, A. A.; Anumba, C. J.; Ogunseiju, O. O. Towards Next Generation Cyber-Physical Systems and Digital Twins for Construction. *J. Inf. Technol. Constr.* 2021, 26 (June), 505–525. https://doi.org/10.36680/j.itcon.2021.027.

Armeni, P.; Polat, I.; De Rossi, L. M.; Diaferia, L.; Meregalli, S.; Gatti, A. Digital Twins in Healthcare: Is It the Beginning of a New Era of Evidence-Based Medicine? A Critical Review. *J. Pers. Med.* 2022, 12 (8). https://doi.org/10.3390/jpm12081255.

Bermúdez I Badia, S.; Silva, P. A.; Branco, D.; Pinto, A.; Carvalho, C.; Menezes, P.; Al-meida, J.; Pilacinski, A. Virtual Reality for Safe Testing and Development in Collaborative Robotics: Challenges and Perspectives. *Electron.* 2022, 11 (11), 1–14. https://doi.org/10.3390/electronics11111726.

Boje, C.; Guerriero, A.; Kubicki, S.; Rezgui, Y. Towards a Semantic Construction Digital Twin: Directions for Future Research. *Autom. Constr.* 2020, 114 (March), 103179. https://doi.org/10.1016/j.autcon.2020.103179.

Chen, X.; Chang-Richards, A. Y.; Pelosi, A.; Jia, Y.; Shen, X.; Siddiqui, M. K.; Yang, N. Implementation of Technologies in the Construction Industry: A Systematic Review. *Eng. Constr. Archit. Manag.* 2021. https://doi.org/10.1108/ECAM-02-2021-0172.

Dallel, M.; Havard, V.; Dupuis, Y.; Baudry, D. Digital Twin of an Industrial Workstation: A Novel Method of an Auto-Labeled Data Generator Using Virtual Reality for Human Action Recognition in the Context of Human–Robot Collaboration. *Eng. Appl. Artif. Intell.* 2023, 118 (December 2022), 105655. https://doi.org/10.1016/j.engappai.2022.105655.

Getuli, V.; Capone, P.; Bruttini, A.; Isaac, S. BIM-Based Immersive Virtual Reality for Construction Workspace Planning: A Safety-Oriented Approach. *Autom. Constr.* 2020, 114 (March), 103160. https://doi.org/10.1016/j.autcon.2020.103160.

Hou, L.; Wu, S.; Zhang, G. K.; Tan, Y.; Wang, X. Literature Review of Digital Twins Applications in Construction workforce Safety. *Appl. Sci.* 2021, 11 (1), 1–21. https://doi.org/10.3390/app11010339.

Kim, J. I.; Li, S.; Chen, X.; Keung, C.; Suh, M.; Kim, T. W. Evaluation Framework for BIM-Based VR Applications in Design Phase. *J. Comput. Des. Eng.* 2021, 8 (3), 910–922. https://doi.org/10.1093/jcde/qwab022.

Laing, R. Built Heritage Modelling and Visualisation: The Potential to Engage with Is-sues of Heritage Value and Wider Participation. *Dev. Built Environ.* 2020, 4 (April), 100017. https://doi.org/10.1016/j.dibe.2020.100017.

Laubenbacher, R.; Niarakis, A.; Helikar, T.; An, G.; Shapiro, B.; Malik-Sheriff, R. S.; Sego, T. J.; Knapp, A.; Macklin, P.; Glazier, J. A. Building Digital Twins of the Human Immune System: Toward a Roadmap. *npj Digit. Med.* 2022, 5 (1). https://doi.org/10.1038/s41746-022-00610-z.

Malagnino, A.; Montanaro, T.; Lazoi, M.; Sergi, I.; Corallo, A.; Patrono, L. Building In-formation Modeling and Internet of Things Integration for Smart and Sustainable Environments: A Review. *J. Clean. Prod.* 2021, 312 (May), 127716. https://doi.org/10.1016/j.jclepro.2021.127716.

Malik, A. A.; Brem, A. Digital Twins for Collaborative Robots: A Case Study in Human-Robot Interaction. *Robot. Comput. Integr. Manuf.* 2021, 68 (November 2020), 102092. https://doi.org/10.1016/j.rcim.2020.102092.

Mantha, B. R. K.; García de Soto, B. Cybersecurity in Construction: Where Do We Stand and How Do We Get Better Prepared. *Front. Built Environ.* 2021, 7 (May 2021), 1–13. https://doi.org/10.3389/fbuil.2021.612668.

Musakwa, W. and Moyo, T., 2020. Perspectives on Planning Support Systems and E-planning in Southern Africa: Opportunities, Challenges and the Road Ahead. Handbook of Planning Support Science.

Onososen, A. O.; Musonda, I.; Ramabodu, M. Construction Robotics and Human – Ro-bot Teams Research Methods. *Buildings* 2022, 12 (1192), 1–33.

Opoku, D. J.; Perera, S.; Osei-kyei, R.; Rashidi, M. Digital Twin Application in the Construction Industry: A Literature Review. *J. Build. Eng.* 2021, 40 (April), 102726. https://doi.org/10.1016/j.jobe.2021.102726.

Ozturk, G. B. Trends in Interoperability in Building Information Modeling (BIM) Research: A Scientometric Analysis of Authors and Articles. *A/Z ITU J. Fac. Archit.* 2020, 17 (3), 169–183. https://doi.org/10.5505/itujfa.2020.79026.

Panteli, C.; Kylili, A.; Fokaides, P. A. Building Information Modelling Applications in Smart Buildings: From Design to Commissioning and beyond A Critical Review. *J. Clean. Prod.* 2020, 265, 121766. https://doi.org/10.1016/j.jclepro.2020.121766.

Pereira, V.; Santos, J.; Leite, F.; Escórcio, P. Using BIM to Improve Building Energy Ef-ficiency – A Scientometric and Systematic Review. *Energy Build.* 2021, 250. https://doi.org/10.1016/j.enbuild.2021.111292.

Taher, A. H.; Elbeltagi, E. E. Integrating Building Information Modeling with Value Engineering to Facilitate the Selection of Building Design Alternatives Considering Sustainability. *Int. J. Constr. Manag.* 2021, 0 (0), 1–16. https://doi.org/10.1080/15623599.2021.2021465.

Volk, R.; Stengel, J.; Schultmann, F. Building Information Modeling (BIM) for Existing Buildings - Literature Review and Future Needs. *Autom. Constr.* 2014, 38 (March), 109–127. https://doi.org/10.1016/j.autcon.2013.10.023.

You, Z.; Feng, L. Integration of Industry 4.0 Related Technologies in Construction Industry: A Framework of Cyber-Physical System. *IEEE Access* 2020, 8, 122908–122922. https://doi.org/10.1109/ACCESS.2020.3007206.

Zanni, M.-A.; Soetanto, R.; Ruikar, K. Exploring the Potential of BIM-integrated Sustainability Assessment in AEC. In *Sustainable Building Conference*; 2013.

Infrastructure, investment and finance-trends and forecasts

Creating safe spaces: Scaling up public transport infrastructure investments in the city of Johannesburg

R.C. Kalaoane & T. Gumbo
Department of Town and Regional Planning, Faculty of Engineering and the Built Environment, University of Johannesburg, South Africa

A.Y. Kibangou
CNRS, Inria, Grenoble INP, Gipsa-Lab, University Grenoble Alpes, Grenoble, France

W. Musakwa
Department of Geography, Environmental and Energy Studies, Faculty of Science, University of Johannesburg, South Africa

I. Musonda
Department of Construction Management and Quantity Surveying, Faculty of Engineering and the Built Environment, University of Johannesburg, South Africa

ABSTRACT: Efficient public transport systems contribute to spatial ordering and help alleviate socio-economic problems in societies. Ostensibly, the state of the public transport infrastructure in the city of Johannesburg poses a higher risk of safety, compromising the tremendous efforts made to encourage a shift to public transport to achieve sustainability. However, there is limited research on safety in the minibus taxi sector. Pursuant to that, this paper aims to fill this gap by adopting a case study research design to assess commuters' satisfaction on safety and security in minibus taxis in Johannesburg. The study findings elucidate that lack of institutional capacity and limited infrastructure investments in minibus taxi has created environments that are vulnerable to crime, often influencing the feeling of being unsafe. The study recommends improving institutional capacity and resource allocation to achieve the envisioned sustainable and efficient public transport for all.

Keywords: Public transport infrastructure, safety, security, sustainability

1 INTRODUCTION

Efficient public transport is undoubtedly one of the leading indicators of a sustainable and resilient city (Mohmand *et al.* 2021). Sustainable public transport is enshrined with an efficient infrastructure that enables access to economic activities, ensures safety and supports the national economy (Marusin *et al.* 2019; Ogryzek *et al.* 2020). Nevertheless, in most African cities, public transport is faced with enormous challenges emanating from a lack of sufficient infrastructure to support existing public transport (Helon & Ejem 2021). The inefficient public transport infrastructure (PTI) has resulted in road fatalities (Bhavan 2019), and increased criminal activities (Datta & Ahmed 2020), jeopardizing the liveability of cities.

Sustainable public transport needs financial and spatial resources (Broniewicz & Ogrodnik 2020), thus the need for more investments in infrastructure for public transport. South Africa's innovative investments in PTI were realized through the establishment of Bus Rapid Transit and Gautrain (Gumbo *et al.* 2022). Despite the existence of these public transport systems, Moyo *et al.* (2022) elucidate that the majority of South Africans rely heavily on minibus taxis. Indeed Olvera *et al.* (2020) acknowledge that the existing mass transit is to some extent unable to give full user satisfaction. The minibus taxi industry is often neglected in policy and strategic plans (Klopp & Cavoli 2019). Subsequently, minibus taxis have emerged as a hotspot for sexual violence, robberies, murders, and road fatalities (Booysen et al. 2021) making South Africa a dangerous place to live in. Thus, this study sought to analyse satisfaction of minibus taxi users on safety and security in the city of Johannesburg.

2 PUBLIC TRANSPORT INFRASTRUCTURE

PTI is a framework that supports public transport (Marusin et al. 2019). These are roads, pavements, bridges, public transport services that take place on them, stops and stations (Taylor 2021). PTI is deemed to be an asset, for example, the design and development of Roman roads still influence modern landscapes in many cities around the world (Dalgaard et al. 2018). Like any other asset, PTI depreciates with time due to climate conditions, duration and these require maintenance (Taylor 2021). Maintenance of PTI is important for the safety, efficiency and operations of PTI (Magazzino & Mele 2021; Profillidis & Botzoris 2019). Public transport infrastructure is closely linked to safety. The inefficient PTI is associated with criminal activities, and increased spread of diseases like COVID-19 (Dong et al. 2021). Thus, safety is the primary concern in development plans for public transportation (Friman et al. 2020).

2.1 Investing in public transport infrastructure

Transport infrastructure investment, according to Bangaraju et al. (2022) is "an important component of infrastructure investment which covers the development of new and old construction, upgradation activity related to roads, railways, inland waterways, seaports, and airports". Indeed, progressive development for countries manifests itself through investment in PTI. Research on PTI is continually growing, given its link to countries' economic growth (Muvawala et al. 2021).

Indeed, construction projects create employment opportunities (Hlotywa & Ndaguba 2017), alleviate poverty, and contribute to the overall growth of the cities (Hanyurwumutima & Gumede 2021). Consequently, investment in PTI also improves safety in cities. When the road infrastructure is improved, users' perceived safety increases, which may also encourage NMT (Gbban et al. 2023). Othman & Ali (2020) carried out empirical research in Malaysian cities and the results yielded that increased improvements in road infrastructure have enhanced the quality of life in Malaysia. This makes investment in PT a major contributor to the overall sustainability and resilience of cities.

The world today has shifted to more innovative implementations in the transportation sector. Safety concerns in PT are addressed through developments of safe PT networks, and mass transit (Pozoukidou & Chatziyiannaki 2021). The new public transport systems are smarter, safer and more innovative driven by technological improvements (Pradhan et al. 2021). In Cape Town, Infrastructural investments in minibus taxis included improvements in sanitation facilities, office space and the use of solar energy for conservation purposes (Schalekamp & Klopp 2018). Schalekamp & Klopp (2018) further state that the maintenance of facilities is the biggest challenge.

3 MATERIALS AND METHODS

The case study research design was adopted for this study (Maree 2016). Case study design aided to understand experiences of minibus taxi user and identifying key issues affecting safety and security within the sector. Following this, a mixed method approach also known as 'third methodological orientation' used in a study to draw the strengths of both quantitative and qualitative in addressing the aim of the study (Sahin & Öztürk 2019). So as to provide the complete picture of the research problem. Using a cluster sampling technique (Birago et al. 2017), the city was divided into four clusters representing the North, South, East and west and the sample was randomly selected from each cluster. This was done to reduce bias and make the sample inclusive. A questionnaire was used to assess satisfaction of minibus taxi users from March to April 2020. 219 questionnaires were distributed to minibus taxi users and data was collected during different time series to reduce time bias. For secondary sources, policy documents and plans for CoJ's PT were collected and analyzed using thematic analysis. Data collected from questionnaires was analyzed using SPSS. Tools used to analyze data were correlation coefficient and frequency distribution. The correlation coefficient analyzed the strength of the relationship between safety and quality of infrastructure for this study.

4 FINDINGS AND DISCUSSIONS

4.1 *State of the roads in South Africa*

The city of Johannesburg (CoJ) is one of the largest beneficiaries in South Africa for PTI, which increased from 184m to 1.5b (National Treasury 2018). However, road fatalities statistics are worrying. In 2021 the country recorded rate of 26 deaths per 100 000, which makes it one of the highest in the world (National Land Transport Strategic Framework 2023–2028). According to the transport department, recorded road fatalities increased from 427 to 456 from 2020 to 2022. The respondents also support this by stating that avoiding potholes in the city has become a sport. Out of 100 000 potholes that were repaired in the city, 4, 400 were reported in Sandton alone and the number keeps rising (Road safety annual report 2019). These also highlight the institution's capacity to address the poor state of infrastructure in Johannesburg.

4.2 *Frequency of use of minibus taxis*

The study revealed that minibus taxis are used frequently by daily commuters in the city indicated in Table 1. These results align with Moyo *et al.* (2022), indicating that minibus taxis are the dominant public transport mode in Johannesburg.

Table 1. Frequency of use of minibus taxis.

		How often do you use a minibus taxi			
		Frequency	Percent	Valid Percent	Cumulative Percent
Valid	Very frequently	120	54.8	55.6	55.6
	Frequently	39	17.8	18.1	73.6
	Occasionally	43	19.6	19.9	93.5
	Rarely	13	5.9	6.0	99.5
	Never	1	.5	.5	100.0
	Total	216	98.6	100.0	
Missing	System	3	1.4		
Total		219	100.0		

4.3 *Infrastructure implications on the safety of minibus taxi users*

The respondents were asked to rate their satisfaction with minibus taxis regarding their safety level and infrastructure quality. Then, a correlation analysis was carried out by

Table 2. The relationship between safety and infrastructure in minibus taxis.

		Correlations		
			LEVEL OF SAFETY	QUALITY OF INFRASTRUCTURE
Spearman's rho	Level of safety	Correlation Coefficient	1.000	.642[**]
		Sig. (2-tailed)	.	<.001
		N	219	219
	Quality of infrastructure	Correlation Coefficient	.642[**]	1.000
		Sig. (2-tailed)	<.001	.

[**]. Correlation is significant at the 0.01 level (2-tailed).

computing Spearman's rho. The correlation, as depicted in Table 2, is significant. Results shows a strong relationship between safety and infrastructure in minibus taxis.

Safety is linked with increased criminal activities, the poor state of roads, and the lack of supporting infrastructure at stations, drop-off locations, and around the city. Indeed, most of the crimes occur around MTN taxi rank (Figure 1), also called the most notorious place in Johannesburg, according to the respondents.

Figure 1. MTN minibus taxi rank.

One of the respondents was quoted as saying:

"This place needs a serious upgrade. Not only is it congested but crime continues to rise and wreak havoc."

The place has created many debates over the past years, with social media pages created to advise on strategic ways to survive in the city, especially around the taxi ranks. Social media has become an essential tool in our lives that often inform our decisions and behavior (Padayachee 2016) thus, it was interesting to analyze information from these social media pages. The following commandments are extracted from social media pages.1. English is prohibited, 2. No asking for directions, 3. Smiling even after being robbed, and finally, walking at the same pace as the people in that place, not too fast or too slow, as this also attracts pick-pocketers and robbers. Some of the respondents who frequently use minibus taxis also confirmed the rules from social media pages. Throughout the interviews, respondents were unanimous that the safety aspects also involved littering around minibus taxi stations (Figure 2).

Figure 2. A portion of the stations turned into a dumping site.

Littering creates a sense of unsafety in places. Often cleanliness is associated with a sense of security (Özkan & Yilmaz 2019). Unclean places often attract criminal activities, which is the case in Johannesburg. Additionally, the area's lack of security patrol and lighting has perpetuated safety concerns. The infrastructure deficit has resulted in more challenges in the public transport industry, affirmed by the daily realities of respondents' experience.

5 LESSONS LEARNED AND FURTHER RESEARCH

Infrastructure investment requires extensive financial resources, which can be difficult for developing countries like South Africa but allows long-term cost savings (Zhao *et al.* 2022). Developing countries can achieve envisioned sustainable public transport through investment in infrastructure. The study significantly contributes to the existing body of knowledge on public transportation studies. It gives great insights which can be derived from the findings of this study. This study highlights the importance of investing in the minibus taxi infrastructure to increase safety and security among passengers. Johannesburg creates a wide range of opportunities when investing in minibus taxi infrastructure, including access to funding opportunities and the potential to change the negative perception in this sector. This has the potential of attracting new users to use public transport rather than motor vehicles because people opt for safe public transport. Investments can be made through various means. ICT has increased safety around other public transport systems, which can translate to real-time scheduling, payment methods and cyber security around minibus taxi ranks.

Additionally, integrating safety and security in transport planning should be factored into transportation policies and legislation. The study provides evidence-based insights to encourage stakeholders' collaboration and urges for efficient allocation of resources in transport planning while safety and security are prioritized. Current investment policies are ineffective in restoring safety and security in and outside minibus taxis. Existing transport policy should be based on investment efforts to improve the state of minibus taxis in the CoJ. This study acknowledges the critical role of the conventional mode of public transport in the city. However, the solutions lie in both policy and plans for the city where maintenance of roads and upgrading of infrastructure for minibus taxis have to be at the fore frontiers of integrated transportation plans instead of rail as indicated by the transport policies in the CoJ. Policymakers should consider investing in minibus taxi infrastructure with complementary efforts to overcome other barriers threatening the overall public transport system. The study could be further expanded using machine learning techniques to develop models that will detect which variables are essential when investing in public transport infrastructure in African cities.

6 CONCLUSION

This study contributed to the existing literature on the importance of investing in public transport infrastructure which has multiplier effects on the country's socioeconomic development. Safety is the most crucial aspect of people's daily lives. This study demonstrated the relationship between safety and infrastructure investments in minibus taxis. The infrastructure deficit in minibus taxis in Johannesburg has threatened the user's safety, affecting citizens' overall liveability. This is because this mode of public transport is used by the majority daily but is often neglected. This insight originates from discussions of this study.

REFERENCES

Abduljabbar, R., Dia, H., Liyanage, S. & Bagloee, S.A. 2019. Applications of Artificial Intelligence in Transport: An Overview. *Sustainability*, 11(1): 189.

Bangaraju, T., Aneja, P. & Mishra, B. 2022. Transport Infrastructure Investment: A Comparative Analysis Between China, Russia & India. *Earth and Environmental Science*, 1084: 1–10.

Bhavan, T. 2019. The Economic Impact of Road Accidents: The Case of Sri Lanka. *South Asia Economic Journal*, 20(1): 124–137.

Birago, D., Mensah, S.O. & Sharma, S. 2017. Level of Service Delivery of Public Transport and Mode Choice in Accra, Ghana. *Transportation Research Part F*, 46: 284–300.

Booysen, M.J., Abraham, C.J., Rix, A.J. & Ndibatya, I. 2021. Walking on Sunshine: Pairing Electric Vehicles with Solar Energy for Sustainable Informal Public Transport in Uganda. *Energy Research and Social Science*: 1–27.

Broniewicz, E. & Ogrodnik, K. 2020. Multi-criteria Analysis of Transport Infrastructure Projects. *Transportation Research Part D*, 83: 2–15.

Dalgaard, Carl-Johan, Nicolai Kaarsen, Ola Olsson, & Pablo Selaya. 2018. "Roman Roads to Prosperity: Persistence and Non-Persistence of Public Goods Provision." *University of Copenhagen, Mimeo*, 1: 1–50.

Datta, A. & Ahmed, N. 2020. Intimate Infrastructures: The Rubrics of Gendered Safety and Urban Violence in Kerala, India. *Geoforum*, 110: 67–76.

Dong, H., Ma, S., Jia, N. & Tian, J. 2021. Understanding Public Transport Satisfaction in Post COVID-19 Pandemic. *Transport Policy*, 101: 81–88.

Friman, M., Lättman, K. & Olsson, L.E. 2020. Public Transport Quality, Safety, and Perceived Accessibility. *Sustainability*, 12(9): 1–14.

Gbban, A.M., Kamruzzaman, M., Delbosc, A. & Coxon, S. 2023. The Wider Barrier Effects of Public Transport Infrastructure: The Case of Level Crossings in Melbourne. *Journal of Transport Geography*, 108: 1–11.

Hanyurwumutima, L.K. & Gumede, S. 2021. An Analysis of the Impact of Investment in Public Transport on Economic Growth of Metropolitan Cities in South Africa. *Journal of Transport and Supply Chain Management*, 15: 1–11.

Helon, C.R. & Ejem, E.A. 2021. Road Transport Management and Customer Satisfaction in Nigeria: A Study of Imo State Transport Company, Nigeria. *European Journal of Hospitality and Tourism Research*, 9(2): 16–27.

Hlotywa, A. & Ndaguba, E.A. 2017. 'Assessing the Impact of Road Transport Infrastructure Investment on Economic Development in South Africa'. *Journal of Transport and Supply Chain Management*, 11(1): 1–12.

Klopp, J.M. & Cavoli, C. 2019. Mapping Minibuses in Maputo and Nairobi: Engaging Paratransit in Transportation Planning in African Cities. *Transport Reviews*, 39(5): 657–676.

Magazzino, C. & Mele, M. 2021. On the Relationship Between Transportation Infrastructure and Economic Development in China. *Research in Transportation Economics*, 88: 1–9.

Maree, K (ed.) 2016. *The First Step in Research*. 2nd edition. Pretoria: Van Schaik Publishers

Marusin, A. Marusin, A. & Ablyazov, T. 2019. Transport Infrastructure Safety Improvement Based on Digital Technology Implementation. In *International Conference on Digital Technologies in Logistics and Infrastructure (ICDTLI 2019)*: 348–352. Atlantis Press.

Mohmand, Y.T., Mehmood, F., Mughal, K.S. & Aslam, F. 2021. Investigating the Causal Relationship Between Transport Infrastructure, Economic Growth and Transport Emissions in Pakistan. *Research in Transportation Economics*, 88: 1–9.

Moyo, T., Mbatha, S., Aderibigbe, O.O., Gumbo, T. & Musonda, I., 2022. Assessing spatial variations of traffic congestion using traffic index data in a developing city: lessons from Johannesburg, South Africa. *Sustainability*, 14(8809): 1–16.

Muvawala, J., Sebukeera, H. & Ssebulime, K. 2021. Socio-economic Impacts of Transport Infrastructure Investment in Uganda: Insight from Frontloading Expenditure on Uganda's Urban Roads and Highways. *Research in Transportation Economics*, 88: 1–18.

Ogryzek, M., Adamska-Kmieć, D. & Klimach, A. 2020. Sustainable Transport: An Efficient Transportation Network-case Study. *Sustainability*, 12(19): 1–14.

Olvera, L.D., Plat, D. & Pochet, P. 2020. Looking for the Obvious: Motorcycle Taxi Services in Sub-Saharan African Cities. *Journal of Transport Geography*, 88: 1–14.

Othman, A.G. & Ali, K.H. 2020. Transportation and Quality of Life. *Planning Malaysia*, 18: 35–50.

Padayachee, K. 2016. Internet-mediated Research: Challenges and Issues. *SACJ*, 28(2): 25–42.

Pradhan, R.P., Arvin, M.B. & Nair, M. 2021. Urbanization, Transportation Infrastructure, ICT, and Economic Growth: A Temporal Causal Analysis. *Cities*, 115: 1–19.

Profillidis, V.A. & Botzoris, G.N. 2019. Transport Demand and Factors Affecting It. In: Romer, B (ed.). *Modeling of Transport Demand: Analyzing, Calculating, and Forecasting Transport Demand*. Amsterdam: Elsevier.

Sahin, M.D. & Öztürk, G. 2019. Mixed Method Research: Theoretical Foundations, Designs and Its Use in Educational Research. *International Journal of Contemporary Educational Research*, 6(2): 301–310.

Schalekamp, H. and Klopp, J.M. 2018. Beyond BRT: Innovation in Minibus-taxi Reform in South African Cities. *37th Annual Southern African Transport Conference*, Pretoria 9–12 July 2018. Pretoria: SATC, 664–675.

South Africa. 2018. *National Treasury Report*. Pretoria: Government Printer.

Taylor, M.A.P. 2021. Transportation Infrastructure. In: Romer, B (ed.). *Climate Change Adaptation for Transportation Systems*. Amsterdam: Elsevier.

Zhao, J., Greenwood, D., Thurairajah, N., Liu, H.J. & Haigh, R. 2022. Value for Money in Transport Infrastructure Investment: An Enhanced Model for Better Procurement Decisions. *Transport Policy*, 118: 68–78.

Influence of public sector built environment professionals on infrastructure delivery in the Eastern Cape

A. Mntu, K. Kajimo-Shakantu & B. du Toit
Department of Quantity Surveying and Construction Management, University of the Free State, Bloemfontein, South Africa

ABSTRACT: South Africa has relatively a high-rate infrastructure delivery backlogs which impact the lives of many citizens at the grassroots level. Factors such as corruption, poor management style and unskilled workers have been regarded as contributing components towards backlogs. There are various social infrastructure projects planned to enhance the infrastructure; however, it is crucial to complete them within the allocated budget and timeframe to optimize the investment and provide efficient social infrastructure for the citizens. This study explores the role of built environment professionals in mitigating social infrastructure delivery delays that result in backlogs in the Eastern Cape province within South Africa. The study established that there is a correlation between skills, continuous professional development, corruption, management style and infrastructure backlogs. The study's significant findings will be used to help decision-makers at the national, provincial, and local levels improve their decision-making in order to eliminate social infrastructure deficiencies in government departments.

Keywords: Built Environment, Infrastructure delivery, Construction professional

1 INTRODUCTION

Effective infrastructure plays a significant role in South Africa's' (SA's) economic growth as it contributes to the rise in productive capacity, sustains development, eradicates; poverty, inequality and high rates of unemployment (Kumo 2012; Younis 2014). Therefore, the provision of adequate infrastructure is an important and critical activity of the South African public sector. To support this, Mbanda & Chitiga (2013) states that infrastructure investment is a space of serious importance in confidently placing South Africa on a high growth path. Looking at the past three decades; restricted advancement has been made with broad disparity and segregated societies acquired from the past government administration and prevalent spatial strategies (Adams *et al.* 2015; Booysen 2003; Tregenna & Tsela 2012).

The South African government has devoted tremendous financial assets in the recent years towards both economic and social infrastructure improvement with the expectation to catalyse economic activity (Presidency 2014). Even so, most of the social infrastructure is still poorly located and with an alarming infrastructure deficit, especially in provinces such as the Eastern Cape (Presidency 2014). Reuters (2019) further mentions that due to the lack of ability and approach of government employed, built environment professionals', in advancing their methods to support integrated development as well as fighting poor integration of development efforts between Government and the private sector, is detrimental to the delivery of these social infrastructure ventures. This is also accompanied by a devastating incoordination of exerted efforts (IDP 2011).

It cannot be that twenty (20) years into democracy, social infrastructure is still taking shape in the former homelands, even so with an increased number of backlogs (National

Treasury 2014). This is particularly so for infrastructure that provides for social needs and development that exists in South Africa such as health, roadways, communication systems, education, water and sewer lines, and transportation services. However, it must be known that social infrastructure does not encompass the provision of social services, such as that of teachers at a school or health workers at a clinic (Eastern Cape Infrastructure Plan 2016). The Eastern Cape economy's growth has been characterized by upward and downward cycles from 1996 to 2020, resulting in a combination of possible poverty-level increases and reductions (Bhorat & Van Der Westhuizen 2012). The province is showing signs of economic recovery from the past regression due to apartheid government; this is characterized by improved consumption levels last seen before the onset of the global financial crisis. The factors affecting infrastructure delivery have been delineated into three categories: bureaucratic factors, organisational factors and technical factors.

In view of the above, this paper seeks to examine if the Building Environment Professionals responsible for social infrastructure delivery in the Eastern Cape are equipped to deliver sustainable social Infrastructure project delivery. In addition, it also attempts to determine if whether the plans put in place have produced any tangible outcomes with a view to improve social infrastructure service delivery in the province. This will aid the social infrastructure stakeholders and any other interested persons to consider the proposed strategies and examine their impact towards continuous development of social infrastructure investment initiatives. To provide answers to these questions, the paper has the following objectives:

- To examine existing infrastructure plans and strategies put in place by Government to achieve better delivery of social infrastructure services in the EC.
- To examine if the BE Professionals responsible for social infrastructure delivery in the Eastern Cape are equipped to deliver sustainable social Infrastructure project delivery. Furthermore, on the roles, responsibilities and the contributions of various built environment stakeholders are also discussed.

The rest of the paper is set out as follows: in section 2 literature on the factors affecting social infrastructure investment and service delivery are reviewed. In section 3 the methodology and data used in the research paper are discussed. In section 4 we report the results and in section 5 we discuss the results and draw conclusions.

2 LITERATURE REVIEW

2.1 *The Provincial Strategic Framework (PSF)*

Statistics South Africa (2019) and ECPDP (2014) mentions that Eastern Cape Provincial Strategic Framework (PSF) is put in place is for responding to the need for transformation, aiming at auctioning the government plan and aligning the government resources to carry out the mandate of service delivery. The PSF was formulated and approved in June 2009 by the Provincial Executive Council. It appeals on the discoveries of the Provincial Growth and Developmental Plan (PGDP) review of 2009. The PGDP goal is to achieve its 10-year target; and it includes the alleviation of the high rates of unemployment, illiteracy and poverty in the EC. The PGDP further acknowledges the results of the legacy of colonialism and apartheid rule on the high backlogs in social and economic infrastructure predominantly in the underdeveloped areas. The report however displays the noticeable progress in social infrastructure development, though much progress is still required. The report also recognized other challenges that contribute to these backlogs, one of which is the lack of integrated infrastructure plans such as funding. It is evident that the existing allocations are somewhat not anywhere close to meeting the need, which is problematic.

The provincial spatial development plan directly affects the allocation of social infrastructure. It defines the province' settlement patterns, which will provide direction of where the shortfall of infrastructure facilities rests (ECPDP 2014). The settlement patterns are dictated by outward migration from rural areas to urban areas caused by lack of economic and employment opportunities, retrenchments in the formal agricultural sector (Presidency 2011). This inflicts a great responsibility on the provincial government to review its social infrastructure plans (Statistics South Africa 2019).

2.2 Roles and responsibilities of agents in Social Infrastructure development and delivery

According to Windapo & Cattell (2011) the built environment and its participants play a huge role in the outcome of the building industry. The delivery and non-delivery of the industry is hence a result of the industry's environment and its participants' joint efforts (Sahoo *et al.* 2012). The efforts or roles played by the involved stakeholders varies as it can be received or expressed in various ways, namely:

i) Providing well thought proposals of solutions to challenges that face the industry,
ii) Enabling access and use of required technology,
iii) Availing the required strategic managerial skills; as well as
iv) Provision of skilled craftsmen.

The abovementioned roles are equally paramount as assist in conceptualization and effective management of infrastructure developments and their delivery, which is of fundamental importance to the development of the country as well as the community that the built environment serves.

Zheng *et al.* (2018) suggest that the steps taken in a project environment will be of great help in reducing and or elimination the common delay causing factors. Efficient initial planning will help in defining a concrete scope and reduce possibility of scope creep and rework thus reducing conflicts. A well-defined WBS will ensure that each individual is aware of his responsibilities and accountabilities and hence better co-operation and decision making can be expected. Also, deadlines for deliverables can be expected to be honoured. Vendor's evaluation and then selection based on the type of job will help in building a good vendor database ensuring better co-operation and project performance. Regular tracking and monitoring of project progress using a project management software will ensure better collaboration, timely identification of possible issues and help in catering to the critical activities encountered throughout the project lifecycle. The importance of a clear and continuous communication is upmost in a project. An open and transparent communication framework helps share issues and experiences, ensures trust of all stakeholders and individuals from different organizations work as a team leading to better collaboration, reduced conflicts and most of all a better project delivery (Zhang *et al.* 2015).

3 RESEARCH METHODOLOGY

This research was conducted as a mono-method study using a survey as the choice of research strategy. "Survey research provides a quantitative or numeric description of trends, attitudes, or opinions of a population" (Creswell 2014:42) which will be useful in establishing the causes-and-effect and determining the most effective intervention strategy. A quantitative mono-method utilising survey questionnaires was selected because it allows for the efficient capturing of numeric and experimental data, and testing hypotheses (McCusker & Gunaydin 2014) – making it most appropriate for the nature of this study. Therefore, the study naturally adopted a deductive approach (Creswell 2014; Saunders & Thornhill 2012).

A four-part questionnaire comprising short questions designed to be answered in a rating-style system (Likert scale) was developed and administered over e-platforms for participants. The questions were related to the delivery of social infrastructure in the Eastern Cape as well as the roles and capabilities of the responsible BE professionals. The literature review identified the different factors that affect the delivery of social infrastructure in South Africa and other countries, as well as the challenges met in the delivery social of infrastructure. The identified factors were tested in the context of the Eastern Cape through the survey. The quantitative data extrapolated from the study provides a descriptive representation of the general perception of the extent to which these factors affect delivery backlogs and the rollout of infrastructure projects in the Eastern Cape. The literature explored also highlights a link between infrastructure delivery and the professionalization of disciplines within the Built Environment (Legoabe 2017). In investigating this link, the questionnaire was designed to capture the level of experience, competency and skill of BE professionals involved in government infrastructure projects in the Eastern Cape.

Purposive sampling was used in this study to identify participants based on expert knowledge and experience in the concepts articulated in this research. The sample population for this study was selected because it best approximates the characteristics of the overall population (Singh & Masuku 2014). The target population comprised of individuals that were (1) currently employed in a provincial government Department in Eastern Cape or had previously been employed in a provincial government Department and (2) was registered in some capacity with at least one of the regulatory bodies for BE professions in South Africa falling under the Council of the Built Environment (namely SACAP, SACPCMP, SACQSP and ECSA). The targeted participants were spread across four Eastern Cape provincial government Departments with infrastructure portfolios: namely, Department of Public Works and Infrastructure (DPWI), Department of Education (DoE), Department of Health (DoH), and Department of Human Settlements (DHS).

The sampling strategy adopted for the survey questionnaire was non-probability sampling given that it was more efficient and relatively easier to implement in a quantitative study. Convenience sampling from the aforementioned database of professional/candidate employees was used. This strategy was appropriate for the study when giving consideration to the time and geographical parameters.

Non-probability sampling for survey methods requires a minimum of 5-25 participants (Saunders & Thornhill 2012). However, with a population size of 108 (extracted from the database obtained by the researcher), a margin of error of 10% and a confidence level of 95%, a minimum sample size of 65 participants was established using Microsoft Excel. Degu & Yigzaw (2006) note that a larger sample size is more desirable. The desired sample size was 50-70 participants which was achieved.

The data collected was analysed by use of descriptive statistics in order to generate research findings for the study.

4 RESEARCH RESULTS PRESENTATION AND ANALYSIS

This study was undertaken with the aim of establishing the factors that cause the backlog in social infrastructure delivery in the Eastern Cape Province. This undertaking was conducted by means of a survey questionnaire of sixty-five (65) research participants whom had either been (1) currently employed in a provincial government Department in Eastern Cape or had previously been employed in a provincial government Department and (2) was registered to some extent (i.e. professional registration or candidate professional registration) under four (4) South African councils falling under the Council of the Built Environment; namely SACAP, SACPCMP, SACQSP and ECSA. Respondents were spread across four Eastern Cape provincial government Departments with infrastructure portfolios: namely, Department of Public Works and Infrastructure (DPWI), Department of Education (DoE), Department of Health (DoH), and Department of Human Settlements (DHS).

The survey questionnaire was structured into four delineated sections which correspond to the three objectives of this study, to clearly establish the patterns per objective of the data collected. The raw data was analysed through by the use of descriptive statistical analysis in order to develop the findings of this study per objective. The findings from the survey questionnaire of the above objectives were, resultantly, used to inform the research question for this study and the research aim to establish the factors affecting infrastructure delivery backlog in the Eastern Cape.

4.1 *Demographic analysis*

This section of the research questionnaire consisted of questions pertaining to the participants' professional background and was aimed at identifying the variation in the demographic of the participants. The results depicted a broad range of respondents in relation to their professional registration status. A total of thirteen (13) research respondents identified as being registered Candidates under the South African Council for the Project and Construction Management Profession, accounting for 21.67% of the overall sample. One (1) respondent identified as a Candidate registered under ECSA, accounting for a total of 1.67% of the total sample. The results are indicative of a widespread dispersion of the data collected about the mean of respondents insofar as their council registration status. Although the data presents candidates/professionals across multiple disciplines in government Departments (current or previously employed), the shortage or complete lack of skills in government entities is a prevalent discussion in literature (IMIESA 2017). This has been shown to directly contribute to government Departments' failure to provide social infrastructure in accordance with its mandate (IMIESA 2017).

The findings show that the engineering discipline makes up the lowest population of BE professionals practicing in government entities reinforcing the assertion that engineers find working for government entities unappealing (IMIESA 2017). On the other hand, the data also indicates that the majority of the respondents are registered with SACPCMP. It can, therefore, be inferred that government Departments have the capacity to effectively implement infrastructure projects.

In analysing the participants' work experience, it was found that most participants (62.29%) fall within the 0–10-year bracket. This, therefore, indicates a shortage of experienced individuals in government to inform and improve existing processes in order to mitigate the existing backlog and drive projects toward more successful outcomes (IMIESA 2017). On the other hand, the value of younger professionals tends to be more innovative and adaptable to alternative methods than professionals who have been practicing for an extensive number of years. The use of alternative construction, in addition to conventional building methods could, ultimately, improve service delivery and produce more sustainable infrastructure (Wienecke 2010).

4.2 *Infrastructure delivery plans and strategies in the day-to-day administration and implementation of infrastructure projects*

Table 1 presents numeric breakdown of the data as rated by the participants. As indicated, a total of 58.83% and 57.15% of the participants felt a general level of dissatisfaction in the implementation of the National Development Plan and the Eastern Cape Provincial Growth and Development Plan respectively. As mentioned by the Presidency (2011), this reflects a concern as existing infrastructure plans and policies are paramount as they play a pivotal role in providing direction to the infrastructure planning processes for ensuring realistic outcomes in addressing the existing challenges. Notably, a staggering total of 68.83% of the participants felt a general dissatisfaction in the implementation of plans and strategies of the National Infrastructure Plan in the day-to-day administering of infrastructure projects.

ECPDP (2014) in chapter two states that the plans are there to assist in allocation of social infrastructure, by identifying a province settlement plan. It is evident from the participants responses that the plans are in place, however, the implementation is lacking and therefore, does not produce the tangible outcomes needed. This is of great concern as the above results reveal a lack of application in the plans and strategies set by the government to alleviate backlogs which may have, in turn, been impacted to the abovementioned bureaucratic factors – a need for intervention in the implementation and administration of the plans and policies in place is required (Levy 2016). Furthermore, as seen in Table 2, 72.55% of participants perceive plans and strategies to be partially/slowly progressing, which could indicate that no improvement can be expected in the foreseeable future.

Table 1. Level of satisfaction matrix.

#	Question	Completely dissatisfied	Mostly dissatisfied	Somewhat dissatisfied	Neutral	Somewhat satisfied	Mostly satisfied	Completely satisfied
1	National Development Plan 2030 in building new infrastructure	15.69%	19.61%	23.53%	27.45%	7.84%	3.92%	1.96%
2	Satisfaction of funding for infrastructure portfolios in the Eastern Cape	17.65%	25.49%	25.49%	11.76%	13.73%	1.96%	3.92%
3	Equitable and fair distribution of economic infrastructure	20.41%	16.33%	20.41%	30.61%	6.12%	6.12%	0.00%

Table 2 indicates the extent to which the infrastructure plans and strategies are progressing in the Eastern Cape. At a mean of 2.20, 72.55% of research participants generally perceive the Eastern Cape to be partially progressing in the implementation of infrastructure plans and strategies.

Table 2. Plans and strategies level of progressiveness matric.

#	Answer	%
1	Completely/quickly progressing	3.92%
2	Partially/slowly progressing	72.55%
3	Not progressing	23.53%
	Total	100%

4.3 Competences of professionals responsible for social infrastructure delivery in the Eastern Cape

The data collected indicated the extents to which the respondents are equipped to deliver sustainable social infrastructure projects. By measuring the extent to which participants have gained, maintained and/or improved in their Project Management proficiencies, the results would assist in determining the level of expertise of the participant and, by extension, the level of skills capacity within provincial Department infrastructure portfolios. The data indicated a general level of agreement of the participants on the acquiring, maintaining and improvement of project management proficiencies. Furthermore, with mean values ranging from 5.04 to 5.98, participants conveyed a general level of agreement that the projects they have undertaken while in their respective provincial departments have allowed for acquisition, maintaining and improvement of their project management competencies. The findings indicate that the participants possess the technical know-how and—to a degree—DPWI, DoE, DoH, and DHS have employed level of skills capacity to implement and manage infrastructure projects as per the SACPCMP project management proficiencies. The findings further prove that the EC BE professionals are combatting most barriers as they are aware of the importance of social infrastructure delivery to the success of the mandate of the government they serve (Bhattachaya et al.).

The data collected on skills development of the participants indicated the following: A total of 79.17%, 65.22%, 61.70% and 59.57% of the participants had experienced or been afforded annual performance reviews, study bursaries and construction computer software training respectively. The findings demonstrate continued investment by provincial departments in professional and skills development within infrastructure portfolios. It is evident from the responses of the participants that the Eastern Cape Provincial Departments do attempt to better equip BE Professionals with skill development opportunities to achieve better delivery of social infrastructure services.

The data collected had demonstrated a general perception that the Eastern Cape Provincial Departments do have the capacity to undertake infrastructure projects. Skills development opportunities exists within the Eastern Cape Provincial Departments as seen in the data collected. Furthermore, more attentive efforts need to be put into enforcing government plans and strategies in order to meet infrastructure delivery mandates.

5 DISCUSSIONS AND IMPLICATIONS

The study explored the existing backlog in the delivery of social infrastructure investment in the Eastern Cape along with the factors that contribute to this problem. The main question that needed to be addressed was why a backlog in social infrastructure investment and development exists in the Eastern Cape. In addressing this question, the factors that affect social infrastructure delivery; the adoption of existing Government plans and strategies; as well as the proficiency and development of BE professionals in Government Departments were evaluated to determine how they contribute to the social infrastructure delivery backlog.

The plans in place were established with the aim of facilitating and introducing prompt service delivery in the province as well as evaluating the methods and processes that are in place for infrastructure delivery. The factors contributing to the backlog in social infrastructure delivery in the EC, according to their appropriateness as perceived by the research participants are (1) payment delays to service providers, (2) corruption and maladministration and (3) long processes of administering infrastructure. Further, the results (Table 1) reveal a possible lack of application in the plans and strategies set by the government to alleviate backlogs. As stated by Guney (2017), may have, in turn, been impacted to the abovementioned bureaucratic factors. Further, this inflicts a great responsibility on the provincial government to review its social infrastructure delivery plans (Presidency 2011).

The literature and survey (Table 1) have therefore indicated that the plans are there. Government departments need to impose continuous professional development and reviews on their infrastructure delivery employees to ensure prompt service delivery and proper payment budgeting for cash flow forecasting purposes.

This objective was answered, because the questionnaire findings, showed that participants conveyed a general level of agreement that the projects they have undertaken while in their respective provincial Departments have allowed for acquisition, maintaining and improvement of their project management competencies. The findings, as supported by Pereira & Andraz (2013) indicate that the participants possess the technical know-how and—to a degree—DPWI, DoE, DoH, and DHS have employed level of skills capacity to implement and manage infrastructure projects as per the SACPCMP project management proficiencies. Government must be encouraged to continue promoting professional development and to ensure that this is done for the other sections of the department(s) that form as support structure to service delivery, such as SCM and finance.

6 CONCLUSIONS AND RECOMMENDATIONS

The findings suggest that the questionnaire survey respondents are aware that there are backlogs in the provision of social infrastructure delivery in the province. The payment delays to service providers, corruption and maladministration and long processes of administering infrastructure projects were identified as contributing to the failure in delivering quality services on time and within budget to the communities. Public administration and financial management when it come to the province's development, delays the process of rolling services due to the lack of qualified personnel to emphasize effective decision-making regarding performance of services. Based on the information gathered, it is evident that the Eastern Cape Provincial Departments possess the necessary capabilities to carry out infrastructure projects. The data also indicates that there are avenues for skill development within these departments. However, it is crucial for more focused and determined actions to be taken in order to effectively implement government plans and strategies, thereby ensuring the successful completion of infrastructure delivery mandates. From the outlined findings of this study the following recommendations are made:

Institutions for innovations be included in the planning process of infrastructure government departments to promote research projects that are likely to link continuous development, management of time, cost and quality and performance reviews to infrastructure delivery as much as possible.

Government be encouraged to produce policies and strategies (especially around procurement) to promote the correct methods of construction and, Champion for research in government departments that are mandated with social infrastructure delivery so as to interact with innovative institutions in order to tap into information readily available about successful methods of social infrastructure development and delivery, expenditure management and time management for infrastructure process administration.

This study forms a basis for future research as it was limited to the backlogs of social infrastructure delivery in the EC Government Departments.

A study to establish the effectiveness of current plans, incentive and support structures in place to promote and implement social infrastructure project delivery in the province and in the country.

Government departments should align performance reviews of professionals in infrastructure delivery portfolios with the respective professional bodies' Identification of Works deliverables.

Government department must further invest in equipping all service delivery personnel (i.e., SCM personnel, BE professionals, legal services and finance department personnel) with integrated project delivery tools in order to achieve Infrastructure Delivery Management System (IDMS) objectives and mandates.

REFERENCES

Adams, C., Gallant, R., Jansen, A. & Yu, D. (2015). Public Assets and Services Delivery in South Africa: Is it Really a Success? *Development Southern Africa*, 32(6), 697–710.
Bhattacharya, A., Oppenheim, J. and Stern, N., 2015. *Driving Sustainable Development Through Better Infrastructure: Key Elements of a Transformation Program*. (Report No. 91). Brookings: Brookings Publication
Bhorat, H. and Van Der Westhuizen, C (2012) Poverty, *Inequality and the Nature of Economic Growth in South Africa*. DPRU Working Paper 12/151.
Booysen, F le R. (2003a). The Extent of, and Possible Explanations for, Provincial Disparities in Progress on Reconstruction and Development in South Africa. *Development Southern Africa*, 20(1), 21–48.
Creswell, J. W. (2014). *Research Design: Qualitative, Quantitative and Mixed Methods Approaches* (4th ed.). Thousand Oaks, CA: Sage.
DeFranzo, S. E. (2011). *What's the Difference Between Qualitative and Quantitative Research?* Available At: https://www.snapsurveys.com-blog-qualitative-andqualitative-research

Degu, G. and Yigzaw, T. (2006): *Research Methodology*. Ethopia: USAID.

Eastern Cape Provincial Development Plan. (2014). *Championing Sustainable Development: Case Studies from South Africa*. District Municipality, Eastern Cape.

Guney, T., 2017. Population Growth and Sustainable Development in Developed-developing Countries: An IV (2SLS) Approach. *Suleyman Demirel University*, 22(4): 637–662.

Kumo, W. (2012). *Infrastructure Investment and Economic Growth in South Africa: A Granger Causality Analysis. African Development Bank*, Working paper No. 160.

Legoabe, R.S. (2017). Barriers to Professionalising Municipal Civil Engineers. *IMIESA*. 61DeFranzo64.

Levy, E., 2016. *The Lack of Investment into Infrastructure in the Developing World*.

Masuku, M.M. & Jili, N.N. 2019. Public Service Delivery in South Africa: The Political Influence at Local Government Level. *Journal of Public Affairs*, 19(4):1935.

Mbanda, V. and Chitiga, M. (2013). *Growth and Employment Impacts of Public Economic Infrastructure Investment in South Africa: A Dynamic CGE Analysis. Poverty and Economic Policy Network Working Paper*.

McCusker, K. and Gunaydin, S., 2014. Research Using Qualitative, Quantitative or Mixed Methods and Choice Based on the Research. *Perfusion*, 30(7), pp.537–542.

Pereira, A. & Andraz, J. (2013). On the Economic Effects of Public Infrastructure Investment: A Survey of the International Evidence. *Journal of Economic Development*, 38(4), 1–37

Presidency (2014). Accelerated Shared Growth Initiative for South Africa: A Summary. *The Presidency*, Pretoria.

Reuters. 2019. *Global Halal Market 2018 Consumption Analysis, Health Benefits, Production Growth, Regional Overview and Forecast Outlook Till 2025*. Available at: https://www.reuters.com/brandfeatures/venture-capital/article?id=74528 [Accessed 14 June 2020].

Sahoo, P., Dash, R.K. and Nataraj, G. (2012) China's Growth Story: The Role of Physical and Social Infrastructure – *Journal of Economic Development*, 37(1), 53–75.

Saunders, M. and Thornhill, A. (2012). *Research Methods for Business Students*. Pearson Education Limited. Essex.

Singh, A. and Masuku, M. (2014). Sampling Techniques & Determination of Sample Size in Applied Statistics Research: an Overview. *International Journal of Economics, Commerce and Management*. 2(11). 1–22.

South Africa. National Treasury. (2014). *Towards Sustainable Settlements: Case Studies from South Africa*. Pretoria.

Statistic South Africa. 2019. *Public Sector Capital Expenditure Continues to Fall*. [Online]. Available from: http://www.statssa.gov.za/?p=12705 [Accessed 15 July 2020].

Tregenna, F. & Tsela, M. (2012). Inequality in South Africa: The Distribution of Income, Expenditure and Earnings. *Development Southern Africa*, 29(1), 35–61.

Windapo, A.O. and Cattell, K.S. (2011). *Research Report: Mapping the Path to Becoming a Grade 9 Contractor*. Available at: www.cidb.org.za/Documents/KC/cidb_Publications [Accessed on 18 April 2021].

Zhang, S.; Gao, Y.; Feng, Z.; Sun, W. PPP Application in Infrastructure Development in China: Institutional Analysis and Implications. *Int. J. Proj. Manag.* 2015, *33*, 497–509.

Zheng, S.; Xu, K.; He, Q.; Fang, S.; Zhang, L. Investigating the Sustainability Performance of ppp-type Infrastructure Projects: A Case of China. *Sustainability* 2018, *10*, 4162.

Factors affecting public sector infrastructure delivery in the Eastern Cape, South Africa

A. Mntu, K. Kajimo-Shakantu & T. Mogorosi
Department of Quantity Surveying and Construction Management, University of the Free State, Bloemfontein, South Africa

ABSTRACT: Most social infrastructure in South Africa still poorly located and there is a huge infrastructural deficit, especially in provinces such as the Eastern Cape which is one of the poorest in the country. This study examines reasons for social infrastructure delivery delays that result in backlogs in the Eastern Cape province, South Africa. A quantitative approach utilising a survey questionnaire was adopted. Key factors identified included corruption and maladministration, lack of governance and poor delivery management practices, long cumbersome administration, and payment delays. There was also a perception that technical-tiered factors have a less direct causal effect on infrastructure backlog compared to factors categorized under bureaucratic and organizational tiers. There is need for enhanced collaboration among various stakeholders and capacity development for better decision making to address infrastructural backlogs.

Keywords: Backlog, Social infrastructure, Infrastructure delivery, Public sector construction

1 INTRODUCTION

Effective infrastructure plays a significant role in the South African economic growth as it contributes to increased productive capacity, sustains development, eradicates poverty, inequality and high rates of unemployment (Kumo 2012). Thus, the provision of adequate infrastructure is an important and critical activity for the South African public sector. To support this, Mbanda & Chitiga (2013) state that infrastructure investment places South Africa on a high growth path. Looking at the past three decades, good progress has been made to address historical disparity and segregated societies acquired from the past government administration and prevalent spatial strategies (Adams *et al.* 2015; Booysen 2003a, 2003b; Tregenna & Tsela 2012).

The South African government has invested heavily in recent years towards both economic and social infrastructure improvement with the expectation to catalyse economic activity (Presidency 2014). Despite this, the vast majority of social infrastructure is still poorly located and there is a huge infrastructure deficit, especially in provinces such as the Eastern Cape (Presidency 2014). Reuters (2019) further states that due to lack of ability and approaches used by government-employed built environment professionals, in advancing their methods to support integrated development and poor integration of development efforts between government and the private sector, are detrimental to the delivery of social infrastructure ventures. There is also poor integration of stakeholder efforts (IDP 2011).

Twenty years into democracy, social infrastructure is still underdeveloped in former homelands, with increasing backlogs (National Treasury 2014). This is particularly so for infrastructure that provides for social needs and development that exists in South Africa such as; health, roadways, communication systems, education, water and sewer lines, and transportation services. However, social infrastructure does not encompass the provision of social services (Eastern Cape Infrastructure Plan 2016). The Eastern Cape is located in the South-eastern part of South Africa and is divided into six district municipalities, namely: Alfred Nzo, Sarah Baartman, Joe Gqabi, Chris Hani, O.R. Tambo, and Buffalo City; and a metropolitan area called Nelson Mandela Bay (NIP 2012). More than 60 percent of the province is rural and comprises the former homelands of Transkei and Ciskei (ECIDP 2016). The Eastern Cape Provincial Infrastructure Plan 2030 outlines 11 Provincial Strategic Projects (PSPs') namely;

1. Catalytic (Mega) Projects,
2. Small Town Development (4 Towns Identified, one of the priorities is Port St Johns),
3. Urban Settlements Infrastructure,
4. Water & Sanitation,
5. Energy and Electricity,
6. Agriculture and Agro-logistics,
7. Education Infrastructure,
8. Health Infrastructure,
9. Transport Infrastructure (Roads),
10. Information and Communications Technology (ICT), and
11. Enabling Interventions.

There is a lack of readily available funds to implement all the above 11 PSPs, thus a need to "prioritize the priorities." (Infrastructure South Africa 2020). The extent of the infrastructure deficit across various sectors is huge. For example, just for educational infrastructure, the Eastern Cape Department of Education carried out a provincial-wide school conditions assessment between the years 2013 and 2015 to ascertain the extent of school infrastructure deficits in the province. The assessment identified five thousand five hundred and eight (5508) schools out of which 2440 rooms were found to be inappropriate for learning. The province had 687 Grade R classrooms against the demand of 6580, indicating a deficit of 5893 units. Moreover, the schools in the province would need 5506 multi-purpose classrooms against the 334 that were available, creating a shortage of 5172 units of classrooms. There was also a deficit of 3303 rooms as the province has only 797 libraries and computer rooms against a minimum of 4100. According to Fengu (2015), an estimated budget of about R52 billion over time is needed to fix Eastern Cape schools in order to comply with acceptable national regulations. South Africa currently has a R900 billion water infrastructure backlog. The Eastern Cape is said to be the worst affected province as it will require over R120 billion to address water infrastructure and bulk water supply to its total populace (SABC 2020). The study seeks to examine the reasons for the backlog of infrastructure delivery in the Eastern Cape. This will aid social infrastructure stakeholders and other role players to identify factors affecting, and strategies to mitigate challenges constraining infrastructure investment initiatives and implementation.

2 LITERATURE REVIEW

2.1 *Factors affecting social infrastructure development backlog in the Eastern Cape*

Statistics South Africa (2019) states that high success levels in many countries are generally attained through investment in basic public infrastructure as this boosts social and economic development, namely: sanitation, water, housing, education facilities and transportation. However, the majority of social infrastructure in South Africa is still poorly located and compounded by an alarming infrastructure deficit, especially in provinces such as the Eastern Cape (Presidency 2014). The National Treasury (2012, 2013, 2014) has demonstrated plans to keep prioritising the development of infrastructure to address existing backlogs. This is achieved by allocating financial support to regions requiring economic development.

Furthermore, Perieira & Andraz (2013) state financial assets invested in infrastructure and development are in addition focused on regions of economic development.

The Eastern Cape is one of the poorest provinces in South Africa (Eastern Cape Infrastructure Plan 2016). This needs intervention as social infrastructure contributes directly to addressing fundamental needs, such as; water, sanitation and electricity and the provision of good quality education and health services (Eastern Cape Infrastructure Plan 2016). The Eastern Cape economy's growth has been characterized by upward and downward cycles from 1996 to 2020, resulting in a combination of possible poverty-level increases and reductions (Bhorat & Van Der Westhuizen 2012). It is important to adopt efficient methods to achieve successful infrastructure investment in the Eastern Cape relating to spatial aspects of future infrastructure demand, while undoing apartheid geography (ECPDP 2014). In addition, improved infrastructure can contribute to the achievement of universal access to basic social services and help to unlock the province's economic potential to contribute more to the gross domestic product (GDP) (South African Government 2015).

According to DBSA (2012) and South Africa National Treasury (2014), infrastructure delivery and quality are high on the government agenda but the following are variables where government intervention is critical to ensure that quality social infrastructure delivery is achieved by:

i) Providing enough resources towards social infrastructure;
ii) Ensuring effective leadership in social infrastructure delivery portfolios;
iii) Promoting good social delivery practices;

The above issues reveal that challenges with social infrastructure investment delivery system emanate from insufficient resources availed for infrastructure development. These include:

i) Best fitting basic delivery tools,
ii) Lack of the required training for employees employed under infrastructure delivery portfolios,
iii) Long administrative processes that may delay allocation of resources required on urgent basis, poor infrastructure delivery management by managers and ineffective leadership provided by office heads (DBSA 2012).

SA National Treasury (2014) opines that the above challenges need a programmatic approach, which requires the public sector to make an initiative towards provision of basic social infrastructure development and maintenance as a starting point. According to Watermeyer & Philips (2020), public sector should pay particular attention to the following areas:

i) Provide new and maintain existing social infrastructure such that inappropriate structures are replaced;
ii) Introduce technical teams or programme to manage the development and maintenance of social infrastructure and;
iii) Source necessary support in areas identified as needing attention within the social infrastructure development and maintenance programme (DBSA 2012).

Factors affecting infrastructure delivery have generally been delineated into three categories: bureaucratic factors, organisational factors and technical factors. According to Jacobs; Aigbovboa, Thwala & Dithebe (2021) bureaucratic factors comprise corruption, maladministration, lack of governance and delivery management practices; Organisational factors lack of management capacity, external funding, monitoring and checking processes, payment delays to service providers and poor procurement practices; technical factors identified are defects in identification, assessment, and preparation of infrastructure projects, lack of training prorammes for employees, frequent changes in project teams and long processes of administering infrastructure projects.

3 METHODOLOGY

This research was conducted as a mono-method study using a survey. It adopted a deductive approach (Creswell 2014; Saunders *et al.* 2009, 2012). A four-part questionnaire comprising short questions was developed and administered via e-platforms. The literature review identified factors that affect the delivery of social infrastructure in South Africa and other countries, as well as the challenges met in the delivery social of infrastructure. The identified factors were tested in the context of the Eastern Cape through the survey. The quantitative data extrapolated from the study provides a descriptive representation of the general perception of the extent to which these factors affect delivery backlogs and the rollout of infrastructure projects in the Eastern Cape.

The target population comprised of individuals who were (1) currently employed in a provincial government Department in Eastern Cape or had previously been employed in a provincial government Department and (2) registered with at least one of the regulatory bodies for BE professions in South Africa falling under the Council of the Built Environment (namely SACAP, SACPCMP, SACQSP and ECSA). The targeted participants were spread across four Eastern Cape provincial government Departments with infrastructure portfolios: namely, Department of Public Works and Infrastructure (DPWI), Department of Education (DoE), Department of Health (DoH), and Department of Human Settlements (DHS).

Convenience sampling from the aforementioned database of professional/candidate employees were used. Non-probability sampling for survey methods requires a minimum of 5–25 participants Saunders *et al.* (2012). However, with a population size of 108 (extracted from the database obtained by the researcher), a margin of error of 10% and a confidence level of 95%, a minimum sample size of 65 participants was established by the use of Microsoft Excel.

The data collected was analysed by the use of descriptive statistics. in order to generate research findings for the study.

4 RESULTS

The results are presented and discussed in this section.

4.1 *Section A: Demographic analysis*

Of about 107 e-questionnaires sent out, 65 useable responses were received, which gave a response rate of 60%. The respondents were spread across four Eastern Cape provincial governments with infrastructure portfolios namely; Department of Public Works and Infrastructure (DPWI), Department of Education (DoE), Department of Health (DoH) and Department of Human Settlements (DHS). Figure 1 depicts a range of respondents' professional registration statuses.

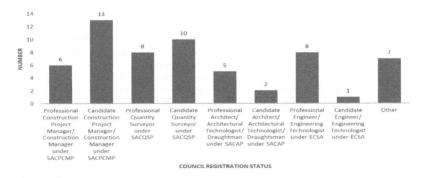

Figure 1. Respondents' professional registration status.

The results indicate a mean of 5 and standard deviation of 2.73 over the nine (9) categories ranging from professional / candidates under the South African professional councils and "Other". A total of thirteen (13) respondents identified as being registered Candidates under the South African Council for the Project and Construction Management Profession, accounting for 21.67% of the overall sample. One (1) respondent identified as a Candidate registered under ECSA, accounting for a total of 1.67% of the total sample. The results are indicative of a widespread dispersion of the data collected about the mean of respondents insofar as their council registration status.

The respondents' work experience (Figure 2) shows that the majority of them fall within the 5 – 10 year bracket.

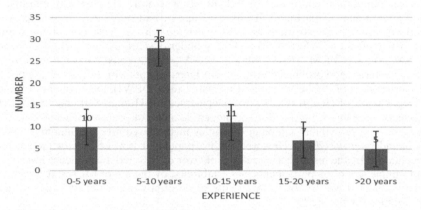

Figure 2. Respondents' years of work experience.

The results depict a skewed distribution of the data towards the lesser years of experience in government departments with fewer respondents identifying with more than twenty years of experience, suggesting that the demographic of the sample is largely new insofar as the years of experience in the Department.

4.2 *Factors affecting infrastructure delivery in the Eastern Cape*

A number of identified factors were assessed in terms of their level of appropriateness insofar as they relate to the Eastern Cape provincial government departments under study. The factors were ranked in order of importance (1 = most important and 12 = least important) as shown in Tables 1 and 2.

Table 1. Likert scale level of appropriateness data matrix.

#	Question	Absolutely inappropriate	Inappropriate	Slightly inappropriate	Neutral	Slightly appropriate	Appropriate	Absolutely appropriate
1	Corruption and maladministration	34.00%	12.00%	6.00%	4.00%	6.00%	10.00%	28.00%
2	Lack of governance and delivery management practices	24.49%	16.33%	4.08%	14.29%	8.16%	12.24%	20.41%
3	Poor procurement management practices and lack of skills and expertise in SCM units	16.33%	16.33%	12.24%	14.29%	10.20%	20.41%	10.20%

(*continued*)

Table 1. Continued

#	Question	Absolutely inappropriate	Inappropriate	Slightly inappropriate	Neutral	Slightly appropriate	Appropriate	Absolutely appropriate
4	Defects in identification, assessment and preparation of infrastructure projects	8.33%	18.75%	14.58%	12.50%	10.42%	20.83%	14.58%
5	Lack of management capacity	16.33%	20.41%	6.12%	8.16%	16.33%	16.33%	16.33%
6	Frequent changes in project teams	10.64%	14.89%	6.38%	27.66%	19.15%	17.02%	4.26%
7	Lack of external funding	10.42%	12.50%	18.75%	27.08%	10.42%	18.75%	2.08%
8	Payment delays to service providers	24.49%	10.20%	8.16%	4.08%	4.08%	16.33%	32.65%
9	Lack of monitoring and checking processes	14.29%	14.29%	10.20%	14.29%	10.20%	30.61%	6.12%
10	Lack of training programmes of employees under infrastructure portfolios	12.00%	20.00%	12.00%	8.00%	12.00%	20.00%	16.00%
11	Long processes of administering infrastructure projects	14.29%	24.49%	6.12%	10.20%	4.08%	18.37%	22.45%
12	Skills shortage; high vacancy rates of professionally registered officials	16.33%	22.45%	10.20%	4.08%	16.33%	14.29%	16.33%

Payment delays to service providers was perceived as being an "absolutely appropriate" factor that affects infrastructure delivery in the Eastern Cape at rate of 32.65%. The results indicate that payment delays may be at the forefront of infrastructure backlogs in the Eastern Cape and have a direct impact on the completion of infrastructure projects, as perceived by the research participants. These results support the findings of Khumalo et al. (2017) which highlight key hindrances to infrastructure delivery in the country.

At a rate of 22.45% of respondents, long processes of administering infrastructure projects were considered the third most appropriate factor affecting Eastern Cape infrastructure delivery. This may include planning and implementing processes of infrastructure projects by project teams and initiators from the time of inception of infrastructure projects to the commissioning of said projects to end users. These results lend credence to the argument that poor planning (Khumalo et al. 2017; Jacobs 2021) and ineffective financial and procurement strategies (Jacobs 2021; Watermeyer & Phillip 2020) as well as a lack of delivery management pose significant challenges to successful outcomes for infrastructure projects (Khumalo et al. 2017; Watermeyer & Phillip 2020) and ultimately hinder sustainable infrastructure delivery (Jacobs et al. 2021).

Observations of trends in how respondents ranked each factor indicated a generally lower importance placed on factors categorised under the technical tier, including (1) defects in assessment, identification and preparation of infrastructure projects, (2) frequent changes in project teams, (3) lack of training programmes for employees under infrastructure portfolios, and (4) long processes of administering infrastructure projects. The rate of respondents ranking Technical-tiered factors in first place ranging from 6.0% to 0.00%. These results indicate a general perception by the participants that the technical-tiered factors have less of a direct causal impact on infrastructure backlog in the Eastern Cape in comparison to the factors categorised under the bureaucratic and organisational tiers. The reasons for this perception were not investigated further at this point of the study.

An observed trend included that factors delineated under the "bureaucratic" tier were within the top five most appropriate factors in the context of the Eastern Cape. This reinforces the findings of Hollands (2019) and Watermeyer & Phillips (2020) that corruption is one of main causal factors of infrastructure backlog. Similarly, Gqaji et al. (2016) cited bureaucracy as a hindering factor to infrastructure delivery. The results further shows that lack of governance and delivery management practices was perceived as being the fourth most appropriate factor affecting infrastructure delivery in the Eastern Cape, thereby supporting the findings of Ncube & Tullock (2015) regarding the lack of accountability structures and handlers of performance accountability roles.

Table 2. Ranking of importance according to quartiles.

Question 1	1	2	3	4	5	6	7	8	9	10	11	12
	1st Quartile			2nd Quartile			3rd Quartile			4th Quartile		
Corruption and Maladministration	52.00 %	20.00 %	6.00 %	2.00 %	6.00 %	0.00 %	2.00 %	0.00 %	2.00 %	2.00 %	4.00 %	4.00 %
Poor procurement management practices and lack of skills and expertise in SCM units	10.00 %	16.00 %	20.00 %	8.00 %	12.00 %	10.00 %	8.00 %	2.00 %	8.00 %	0.00 %	4.00 %	2.00 %
Defects in identification, assessment and preparation of infrastructure projects	6.00 %	6.00 %	10.00 %	14.00 %	8.00 %	12.00 %	8.00 %	14.00 %	12.00 %	10.00 %	0.00 %	0.00 %
Lack of management capacity	2.00 %	6.00 %	14.00 %	10.00 %	22.00 %	14.00 %	8.00 %	10.00 %	6.00 %	6.00 %	0.00 %	2.00 %
Frequent changes in project teams	0.00 %	2.00 %	0.00 %	2.00 %	6.00 %	12.00 %	4.00 %	16.00 %	10.00 %	12.00 %	26.00 %	10.00 %
Lack of governance and delivery management practices	12.00 %	34.00 %	14.00 %	10.00 %	8.00 %	6.00 %	6.00 %	8.00 %	2.00 %	0.00 %	0.00 %	0.00 %
Payment delays to service providers	10.00 %	6.00 %	10.00 %	14.00 %	6.00 %	8.00 %	14.00 %	8.00 %	4.00 %	12.00 %	6.00 %	2.00 %
Lack of monitoring and checking processes	0.00 %	0.00 %	2.00 %	16.00 %	6.00 %	2.00 %	12.00 %	22.00 %	14.00 %	18.00 %	6.00 %	2.00 %
Lack of training programmes of employees under infrastructure portfolios	2.00 %	2.00 % 8.00 %	6.00 %	4.00 %	12.00 %	4.00 %	8.00 %	16.00 %	12.00 %	16.00 %	10.00 %	
Long processes of administering infrastructure projects	2.00 %	0.00 %	6.00 %	6.00 %	8.00 %	8.00 %	20.00 %	6.00 %	4.00 %	10.00 %	16.00 %	14.00 %
Skills shortage; high vacancy rates of professionally registered officials	4.00 %	6.00 %	4.00 %	10.00 %	10.00 %	10.00 %	8.00 %	2.00 %	6.00 %	10.00 %	8.00 %	22.00 %
Lack of external funding	0.00 %	2.00 %	6.00 %	2.00 %	4.00 %	6.00 %	6.00 %	4.00 %	16.00 %	8.00 %	14.00 %	32.00 %

Ranking of factors categorised in the organisational tier were observed to be generally evenly dispersed over the four quartiles. Poor procurement management practices and lack of skills and expertise in supply chain management (SCM) units was ranked the highest in the organisational tier and in third place overall, after (1) corruption and maladministration, and (2) lack of governance and delivery management practices, with 46.00% of the research participants' rankings dispersed within the first quartile. 26.00% of the participants ranked payment delays to services providers in the first place and a total participant ranking of 54.00% within the first two quartiles—28.00% which ranked payment delays to service providers within the second quartile. Lack of management capacity was ranked by 22.00% of the participants within the first quartile, with a total of 68.00% of respondents ranking said factor within the first and second quartile. The results indicate a generally medium to high importance placed on the above factors categorised under the Organisational tier as it relates to their impact on infrastructure delivery in the Eastern Cape. Holland (2019) highlights corruption as being a multi-faceted cause of development plans having a stagnant rate of implementation and the lack of proper tools to bring visions to fruition. In the case of the Eastern Cape, the province has been plagued with cases of missing fixed asset registers and inadequate valuations of property assets over years (Hollands 2019). Furthermore, findings by the Auditor General have cited corruption within the provincial government as one of contributors to the issues faced by the province's infrastructure delivery portfolios (Hollands 2019), which could arguably dispel the perception that this factor is irrelevant in affecting service delivery in the Eastern Cape.

5 CONCLUSIONS AND RECOMMENDATIONS

The study explored the existing backlog in the delivery of social infrastructure investment in the Eastern Cape and the factors contributing to this problem. The main factors contributing to the backlog were found to be; (i) payment delays to service providers, (ii) corruption and maladministration and (iii) long processes of administering infrastructure; and (iv) lack of application of plans and strategies set by the government to alleviate backlogs. Public administration and financial management when it comes to the province's development delays the process of rolling services due to a lack of qualified personnel to emphasize effective decision making regarding performance of services. The study only focused on the Eastern Cape and also the use of convenience sampling poses as limitations of the study. Notwithstanding, while the results cannot be generalized, they provide useful insights into factors that affect the delivery and implementation of infrastructure in the Eastern Cape in general.

It is recommended that institutions for innovations be included in the planning process of infrastructure government departments to promote research projects that are likely to link continuous development, management of time, cost and quality and performance reviews to infrastructure delivery as much as possible. Further, efforts should be directed to creating champions for research in government departments that are mandated with social infrastructure delivery must so as to interact with innovative institutions in order to tap into information readily available about successful methods of social infrastructure development and delivery, expenditure management and time management for infrastructure process administration. It is also recommended that capacity development be prioritised as well as partnerships with the private sector. Technological innovations are also likely to assist government departments to deal with various challenges which constrain infrastructure investment initiatives, development and implementation.

REFERENCES

Adams, C., Gallant, R., Jansen, A. & Yu, D. 2015. Public Assets and Services Delivery in South Africa: Is it Really a Success? *Development Southern Africa* 32(6): 697–710.

ADB (Asian Development Bank). 2012. *Infrastructure for Supporting Inclusive Growth and Poverty Reduction in Asia*. Mandaluyong City, Philippines: Asian Development Bank.
Available from: http://www.statssa.gov.za/?p=12705 [Accessed 15 July 2020].

Behrens, R. & Watson, V. 2012. *Making Urban Places: Principles and Guidelines for Layout Planning*. Cape Town: University of Cape Town.

Bhattacharya, A., Oppenheim, J. & Stern, N,. 2015. *Driving Sustainable Development Through Better Infrastructure: Key Elements of a Transformation Program*. (Report No. 91). Brookings: Brookings Publication

Bhorat, H. & Van Der Westhuizen, C. 2012. *Poverty, Inequality and the Nature of Economic Growth in South Africa*. DPRU Working Paper 12/151.

Booysen, F le R. 2003a. The Extent of, and Possible Explanations for, Provincial Disparities in Progress on Reconstruction and Development in South Africa. *Development Southern Africa*. 20(1): 21–48.

Booysen, F le R. 2003b. Urban and Rural Inequalities in Health Care Delivery in South Africa. *Development Southern Africa*. 20(5): 659–674.

Creswell, J. W. 2014. *Research Design: Qualitative, Quantitative and Mixed Methods Approaches* (4th ed.). Thousand Oaks, CA: Sage.

Development Bank of Southern Africa. 2012. *The State of South Africa's Economic Infrastructure: Opportunities and challenges 2012*.

Eastern Cape Infrastructure Plan. 2016. *Towards sustainable settlements: case studies from South Africa*. Eastern Cape Province.

Eastern Cape Provincial Development Plan. 2014. *Championing Sustainable Development: Case Studies from South Africa*. District Municipality, Eastern Cape.

ESCAP, U. N. 2010. *ESCAP M&E SYSTEM: Monitoring & Evaluation Overview and Evaluation Guidelines*. New York: United Nations Publications.

Gqaji, A., Proches, C. & Green, P. 2016. Perceived Impact of Public Sector Leadership on Road Infrastructure Service Delivery. *Investment Management and Financial Innovations*. 13(3): 51–59.

Hollands, G. 2019. *Corruption in Infrastructure Delivery: South Africa*. Figshare.

Infrustracture South Africa (2020) *Eastern Cape Infrastructure Engagement Report* https://infrastructuresa.org/wp-content/uploads/2021/12/Eastern-Cape-Infrastructure-Engagement-Report-_Final-Rev-0.pdf (Accessible, 01.05.2023)

Jacobs, J., *et al.* 2021. Factors Affecting Sustainable Infrastructure Delivery in South Africa – A Case of Gauteng Province. *IOP Conference Series: Materials Science and Engineering*. IOP Publishing.

Khumalo, M.J., Choga, I. and Munapo, E. 2017. Challenges Associated with Infrastructure Delivery. *Public and Municipal Finance*. 6(2): 35–45.

Kumo, W. 2012. *Infrastructure Investment and Economic Growth in South Africa: A Granger Causality Analysis*. African Development Bank.

Mbanda, V. and Chitiga, M. 2013. *Growth and Employment Impacts of Public Economic Infrastructure Investment in South Africa: A Dynamic CGE Analysis*. Poverty and Economic Policy Network.

Ncube, M. & Tullock, Z. 2015. Accountability in Infrastructure Delivery in South Africa: The Case of the Local Government Sphere. *Public and Municipal Finance*. 4(2): 46–55.

Pereira, A. & Andraz, J. (2013). On the Economic Effects of Public Infrastructure Investment: A Survey of the International Evidence. *Journal of Economic Development*, 38(4): 1–37

Presidency (2014). *Accelerated Shared Growth Initiative for South Africa: A Summary*. The Presidency, Pretoria.

Saunders, M. & Thornhill, A. 2012. *Research Methods for Business Students*. Pearson Education Limited. Essex.

Saunders, M. Lewis, P. & Thornhill, A. 2009. *Research Methods for Business Students*. 5th Edition, Pearson Education Limited, England.

South Africa. 2014. *Infrastructure Development Act, Act 23 of 2014*. Pretoria: Government. Printers.

South Africa. National Treasury. 2014. *Towards Sustainable Settlements: Case Studies from South Africa*. Pretoria.

South Africa. Office of the Premier Eastern Cape. 2004. *Provincial Growth and Development Plan 2004 – 2014*. Bhisho.

South Africa. The Presidency. 2011. *National Development Plan*. Pretoria: Government Printer.

Statistic South Africa. 2019. *Public Sector Capital Expenditure Continues to Fall*. [Online].

Straub, S. 2010. Infrastructure and Development: A Critical Appraisal of the Macro-level Literature. *Journal of Development Studies*, 47(5): 683–708.

Tregenna, F. & Tsela, M. (2012). Inequality in South Africa: The Distribution of Income, Expenditure and Earnings. *Development Southern Africa*, 29(1), 35–61.

Watermeyer, R. & Phillips, S. 2020. Public Infrastructure Delivery and Construction Sector Dynamism in the South African Economy. *NPC Economic Series*.

Importance of proactive road maintenance over reactive road maintenance in developing countries: A case of Malawi

B.F. Mwakatobe, W. Kuotcha & I. Ngoma
Malawi University of Business and Applied Sciences, Chichiri, Blantyre, Malawi

S. Zulu
Leeds Beckett University, Leeds, UK

ABSTRACT: Road infrastructure has become a popular transportation infrastructure worldwide because it connects most parts of the world. Regardless of its importance, road maintenance has been neglected. Therefore, this study aims to unravel the importance of proactive maintenance over reactive maintenance of roads. This study adopted a quantitative approach whereby data were collected through an observation checklist which was done on 31 recently maintained and upgraded Blantyre City's roads. The study reveals that the reactive maintenance adoption in most parts of Malawi has worsened urban road stock condition which requires a quick response and this has adversely affected the country's economy. Hence, requires the adoption of proactive maintenance to boost the declining economy. Therefore, this study provides useful practical and policy insights into the factors that lead to the deferring of maintenance other than budget constraints and competing demands because it is a common problem in both developed and developing countries.

Keywords: Road, Reactive maintenance, Proactive maintenance, Economy, Budget limitation

1 INTRODUCTION

Road networks have become a popular mode of transport worldwide since they connect places starting from buildings, villages, cities, and countries. However, currently there has been an influx of cars which has triggered the large traffic load which in turn has led to the inevitability of pavement damage. Unfortunately, damaged roads have adversely been damaging the vehicles which induce a significant loss of people's lives and properties (Wang et al. 2018). Besides, damaged roads have cascading effects on Green House Gas (GHG) emission due to traffic disruption and fuel consumption. Also, it leads to delays to the road users which in turn affects countries' economies (France-mensah et al. 2019; Mohammadi et al. 2020). Consequently, with any deficiency in the road, many households and businesses incur unexpected cost (Mcneil et al. 2014). The lack of attention given to road users, as evident from the postponement of road maintenance, contributes to the failure of numerous road projects. This failure stems from road users feeling undervalued as their needs are not promptly met (Hartmann & Hietbrink 2013; Mohammadi et al. 2020).

According to Gupta et al. (2011) and Mohammadi et al. (2020), the vitality of good preservation of road network is widely known and hence proper measures should be taken in preserving existing roads such as having frequent performance evaluation of existing roads in order to earlier identify the faults. Moreover, numerous road authorities across the globe face a significant challenge in inspecting roads using state-of-the-art technology. Therefore, the primary approach continues to be manual inspection, which requires a considerable amount of time to identify damages. Consequently, the delay in addressing these issues leads to increased maintenance expenses, as highlighted by Zhang (2006). The purpose of this study is to unravel the importance of proactive over reactive maintenances in urban roads in developing countries. Correspondingly, according to American Association of State Highway and Transportation Officials (AASHTO), pavement performance is defined as "the ability of the pavement to serve the demand of the traffic in the existing conditions." In other words, it is the prediction of the road serviceability from the time after construction to the desired lifetime. Nevertheless, the deterioration of pavement can be caused by several factors such as traffic, age, environment, material properties, poor or delayed maintenance, pavement thickness, strength of pavement and subgrade properties which affect pavement mechanical characteristics (Gupta et al. 2011). These factors have the impact on performance of pavement but in this paper proactive and reactive road maintenance will be discussed.

Therefore, this paper contributes to the existing body of knowledge on road maintenance by discussing on the importance of proactive over the reactive maintenance which is backed up by prior plans to the operation and maintenance of the paved roads in urban areas. Proactive and reactive maintenances are very important to be discussed since they are the ones that can predict and enable the longevity of the road infrastructure. This is because due to restricted budget, road authorities have resorted to reactive road maintenance, resulting in inadequate upkeep of the roads. Consequently, the roads fail to meet the necessary maintenance standards, leading to more frequent and costly repairs for the authorities. More so, this study will enable developing countries to improve their economy and avoid unnecessary spending on road infrastructure that are caused by deferred maintenance. Following this introductory section, section 2 focuses on the existing literature that offers an understanding of the subject matter. Section 3 outlines the methodology employed to gather the data. In section 4, the results are presented and discussed in corroboration to the extant literature. Finally, section 5 offers the conclusion, along with the implications and recommendations arising from the research.

2 LITERATURE REVIEW

2.1 *Road maintenance*

The quest for national development has resulted in significant demand for improved road networks that serve the intended purpose and reduce inconveniences that might be caused by the poorly preserved roads. Having well preserved roads require the proper budg*et a*llocation and avoidance of unnecessary wastes (Mohammadi et al. 2020). According to Elbagalati *et al.* (2018) and Khadka *et al.* (2017), there are some studies which have been done to come up with recommendations and solutions which will be enhancing the road networks condition and safety considering budget limitations. There has been an increased demand for road maintenance worldwide. Unfortunately, budget limitations have hindered the response towards such demand (Guevara et al. 2017). Therefore, the procrastinated maintenance has in turn accelerated to the rise of the future maintenance costs and triple the rehabilitation cost (Gultom et al. 2017; Yepes et al. 2016). This has worsened most road networks conditions in such a way that they do not need to be maintained but reconstructed.

2.2 *Reactive maintenance and proactive maintenance*

Reactive maintenance is maintenance performed upon the occurrence of a fault in a system, like in this study, it is a response to the failure of proper road functioning. While proactive maintenance is sometimes referred to as preventive maintenance, it allows actors in the infrastructure sector to be creative in dealing with potential risks that could lead to the failure of a road network system (Cheng et al. 2020; Eriksson et al. 2019). As a result, reactive maintenance is maintenance by surprise because it is generally done under high pressure and unplanned, whereas proactive maintenance is incorporated in earlier planning before things get out of hand and it is performed at regular intervals, allowing professionals to analyze the consequences of each maintenance that has ever been performed, allowing them to plan for a better solution for the other maintenance to come (Salih et al. 2016).

Reactive maintenance is rarely duplicated to other projects since there are no traces of what exactly was done because the maintenance staff is under intense pressure to ensure that the road system returns to function. This is because maintenance is performed after a disruption has occurred, causing the team to struggle to remedy the issue (Eriksson et al. 2019). As a result, documenting unscheduled maintenance is extremely difficult. Proactive maintenance, on the other hand, is a reliable maintenance approach that can be reused by other actors in the infrastructure sector because it is a well-documented approach that has been tested and evaluated based on the outcomes that have been proven through application (Cheng et al. 2020; Eriksson et al. 2019). Similarly, proactive maintenance extends the service lifetime of road networks by predicting failure and resolving it before it occurs (Chen et al. 2016; Salih et al. 2016). As a result, it is a long-term plan that allows the road sector to anticipate the future (Mohammadi et al. 2020).

Furthermore, according to a World Bank analysis of 85 developing countries, it is projected that around $45 billion in road stock has been lost owing to a lack of proactive maintenance, which could have saved if $12 billion had been spent on proactive maintenance (Salih et al. 2016). This also emphasizes the need of proactive road maintenance because it is the only technique capable of addressing the problem at its source. While studies have indicated that proactive maintenance is crucial (Salih et al. 2016; Yepes et al. 2016), many developing countries have not implemented it due to budget constraints, causing maintenance to be postponed (Guevara et al. 2017). Other studies have found that many developing countries practice reactive maintenance (Cheng et al. 2020; Eriksson et al. 2019). Therefore, the importance of proactive over reactive maintenance in developing countries with limited budgets remains uncertain.

3 RESEARCH METHODOLOGY

This study used a quantitative method, with data collected using an observation checklist (Creswell 2015). Data was taken from 31 newly maintained and upgraded roadways in Malawi's Blantyre City. Blantyre was chosen as a case study because it is Malawi's second largest city, business, and industrial centre. The check-list, on the other hand, included the use of a Likert scale that varied from very poor, poor, fair, good, and very good, reflecting the observed condition of the roads (Kothari 2004). In addition, varied road conditions were photographed to back up the observation checklist results because images can reveal the reality of whatever was commented to avoid the researcher's prejudices during the observation.

3.1 *Data analysis*

The analysis was carried out using STATA 15, and the inferential analysis was carried out using the Chi-square test of independence. Inferential statistics used data collected from observed roads to describe and infer about them. The Chi-square test with a 5% level of significance was employed in inferential analysis to identify the relationship between two variables having two or more categories. The Chi-square test was employed in this study to examine the relationship between road conditions, maintenance type, and maintenance quality. Because the quality of maintenance work is a dependent variable that is classified as Very poor, Poor, Good, Fair, and Very good, as well as an independent variable (road conditions and type of maintenance).

It is computed using the formula;

$$\chi^2 = \sum_i^r \sum_j^c \frac{(f_y - e_y)^2}{e_y} \sim \chi^2(c-1)(r-1) \qquad (i)$$

Where χ^2 = chi square, f_y = means the observed frequency (the observed counts in the cells), (c-1) (r-1) shows the degree of freedom

$\sum_i^r \sum_j^c$ = sum of ith row and j^{th} column respectively.

e_y = means the expected frequency But

$$\text{Expected frequencies} = \frac{Row\ sum \times Colum\ sum}{Grand\ total} \qquad (ii)$$

As depicted in the formula, the Chi-Square statistic is based on the difference between what is actually observed in the data and what would be expected if there was truly no association between the variables.

With r rows and c columns in the contingency table, the test statistic has a Chi-square distribution with

(r −1)(c −1) degrees of freedom provided that the expected frequencies was two or more for all categories (Calvin & Long 1998).

Hypothesis

H0: $r = 0$ (There is no association between dependent and independents variables).

Ha: $r \neq 0$ (There is association between dependent and independents variables).

Cramer's V

Cramer's V is a measure of association between two categorical variables, giving a value between 0 and 1 (inclusive). Cramer's V is the most popular of the Chi-square-based measures of nominal association because it gives good norming from 0 to 1. The study employed Cramer's V to examine the strength of the association between quality of maintenance works and (type of maintenance and road's condition) as predictor/independent variables.

The interpretation of the strength of the association between variables based on Cramer's V is as follows:

- If V = 0, the variables are not associated.
- If 0 < V < 0.25, the association between the variables is weak.
- If 0.25 ≤ V < 0.5, the association between the variables is moderate.
- If 0.5 ≤ V < 0.75, the association between the variables is strong.
- If V ≥ 0.75, the variables are perfectly associated.

4 RESULTS AND DISCUSSION

4.1 *Road condition*

Table 1 presents the results of an analysis of the association between the condition of roads and the quality of maintenance work. The quality of maintenance works is the dependent variable, while road conditions are the independent variables. The table shows the frequency counts of the quality of maintenance works for each level of road conditions. The expected values are also provided in brackets. The Chi-square test of independence was conducted to test the null hypothesis that the quality of maintenance works is independent of the condition of roads. The test yielded a Chi-square value of 52.5327 with a p-value of 0.013, which is statistically significant at the 5% level since p-value 0.013 is less than 0.05 at 5% level of significance. The Cramer's V value is 0.4213, indicating a moderate association between the two variables.

Table 1. The association between the condition of roads and the quality of maintenance work.

road's conditions (I.V)	Very Poor	Poor	Fair	Good	Very good	Total	χ^2	P-value	Cramer's V
Drainage	1(1.4)	4(2.0)	2(1.1)	0(1.3)	0(1.1)	7(7.0)	52.5327	0.013**	0.4213
Potholes	0(1.8)	7(2.6)	0(1.5)	2(1.7)	0(1.5)	9(9.0)			
Road edges	3(1.8)	0(2.6)	1(1.5)	3(1.7)	2(1.5)	9(9.0)			
Silting	2(1.8)	0(2.6)	0(1.5)	2(1.7)	5(1.5)	9(9.0)			
Traverse Cracks	1(1.6)	2(2.3)	3(1.3)	0(1.5)	2(1.3)	8(8.0)			
Thermal cracking	2(1.8)	3(2.6)	0(1.5)	4(1.7)	0(1.5)	9(9.0)			
Distress Patches	2(1.6)	1(2.3)	2(1.3)	1(1.5)	2(1.3)	8(8.0)			
Rutting	2(1.4)	1(2.0)	2(1.1)	2(1.3)	0(1.1)	7(7.0)			
Fatigue	2(1.6)	3(2.0)	2(1.3)	0(1.5)	1(1.3)	8(8.0)			
Total	15(15.0)	21(21.0)	12(12.0)	14(14.0)	12(12.0)	74(74.0)			

(..) in brackets are the expected values
**Significant at 5% level

Overall, the results indicate that the quality of maintenance work is related to the road condition of drainage. Bad quality maintenance works are more probable on roads with very bad and poor drainage than on roads with fair, good, and very good drainage. This supports the views of Nora & Reddy (2018) and Zumrawi (2016), who contend that the longevity of paved roads is dependent on the existence of effective drainage systems. However, there is no substantial relationship between maintenance work quality and other road characteristics such as potholes, road margins, silting, transverse cracks, thermal cracking, distress patches, rutting, and fatigue. These findings are useful for policymakers and road authorities in prioritizing maintenance work for specific road conditions in order to increase overall quality of roads.

Blantyre's drainage systems have been badly maintained. As a result, throughout the rainy season, both the city council and residents have been performing reactive maintenance, similar to a fire-fighting technique (Eriksson et al. 2019). Furthermore, the blockage of drainage systems has resulted in prolonged road wetness, which gradually weakens the pavement, and with time, due to the passage of vehicles in weak road networks, the roads develop cracks, then small potholes, and finally large potholes, which have a negative impact on road users and the environment at large. As a result, any flaw in a road system is expected, and many issues could have been avoided if the city council had implemented proactive measures (Cheng et al. 2020) in the road sector which tackle problems at early stages.

Figure 1. Showing the flooded road due to inadequate drainage system.

Drainage infrastructure acts as a road protector when it comes to water; consequently, if more attention was given to it, most of Blantyre city's roads would have been kept safe from regular damages, particularly during the rainy season. However, drainage systems have been taken for granted, resulting in road damage during the rainy season. This is because, as a metropolis, many residential and institutional plots are paved, leaving only relatively tiny areas for water penetration into the ground. As a result, there is a lot of surface runoff on roadways, and if the drainage structures aren't well maintained, floods in the city are unavoidable, as illustrated in Figure 1, which leads to road surface damage. Subsequently, this necessitates purposeful measures such as involving other sectors in its plans, particularly proactive plans such as physical planners who can regulate the land uses for the plots that will be included in building permit requirements, such as having a large percentage of land that allows water penetration into the ground compared to one that will be paved, which resists water penetration. Additionally, involving diverse disciplines in maintenance plans will lead to the consideration of sustainability in social, ecological, and economic terms (Eriksson et al. 2019).

However, there is no substantial relationship between maintenance work quality and other road characteristics such as potholes, road margins, silting, transverse cracks, thermal cracking, distress patches, rutting, and fatigue. Still, the city council must include these in their plans, particularly a proactive maintenance plan, because, over time they cause road deterioration (Figure 2a), which increases road user costs, and in order to include sustainability as explained by Eriksson et al. (2019), as well as user satisfaction (Mohammadi et al. 2020; Wheat, 2017) and user safety (Havránek et al. 2016) in maintenance activities.

Table 2 shows the association between the type of maintenance and the quality of maintenance works in the roads. The dependent variable is the quality of maintenance works, while the independent variable is the type of maintenance. The results of the study were analyzed using the Chi-square test and Cramer's V coefficient. The findings reveal that there is a significant association between the quality of maintenance works and the type of maintenance performed on the roads. The p-value for the Chi-square test is less than 0.05, indicating that the results are statistically significant. The Cramer's V coefficient is 0.3493, which suggests a moderate effect size.

Table 2. The association between types of maintenance and quality of maintenance works.

Type of maintenance (I.V)	Very Poor	poor	Fair	Good	Very good	Total	χ^2	P-value	Cramer's V
Drainage and shoulder improvement	0(2.6)	6(3.7)	6(2.1)	0(2.5)	1(2.1)	13(13.0)	27.0824	0.008**	0.3493
Upgrading of asphalt surface	5(4.1)	2(5.7)	1(3.2)	6(3.8)	6(3.2)	20(20.0)			
Dualization	8(4.9)	8(6.8)	2(3.9)	4(4.5)	2(3.9)	24(24.0)			
Shoulder improvement	2(3.4)	5(4.8)	3(2.8)	4(3.2)	3(2.8)	17(17.0)			
Total	15(15.0)	21(21.0)	12(12.0)	14(14.0)	12(12.0)	74(74.0)			

Association between type of maintenance and quality of maintenance works.(..) in brackets are the expected values**Significant at 5% level

(a)   (b)

Figure 2. The effect of untreated sources of road deterioration.

Looking at the individual categories, drainage and shoulder improvement have the strongest correlation with the quality of maintenance work. This category's expected values vary from 2.1 to 3.7, whereas the observed values range from 0 to 6. This difference shows that the quality of drainage and shoulder enhancement is inadequate and needs to be improved. Figure 3 depicts a stretch of road from Bangwe to St.Patricks with deteriorating shoulders and no drainage system.

Figure 3. Bangwe to St.Patricks Road with poor shoulders and without a drainage system.

Upgraded asphalt surface, Dualization, and shoulder improvement, on the other hand, have a moderate association with the maintenance work quality. The expected and observed values for these categories are relatively close, indicating that the quality of maintenance work for these types of maintenance is fair to good.

In summary, the study's findings indicate that there is a considerable relationship between the type of maintenance and the quality of road maintenance work. Drainage and shoulder improvement have the strongest relationships with poor quality of maintenance works due to reactive maintenance, while asphalt surface upgrading, dualization, and shoulder improvement have moderate associations with the quality of maintenance works. These findings suggest that there is a need to improve the quality of road maintenance works by implementing a proactive maintenance strategy, notably for drainage and shoulder improvement.

As shown in Figure 3, the road has poor shoulders and no drainage system, which has led to road deterioration over time. Whenever water gets a chance to flow along that road, whether from rain or human activities, it penetrates to the sub-grade and with time the road carrying capacity is being weakened, which has a cascading effect on the country's economy due to delays, increased road user costs, and health problems due to emissions. The challenge is that, as a result of the lack of proactive maintenance and the use of reactive maintenance when there is practically total failure (Guevara et al. 2017), the road infrastructure has suffered tremendous deterioration infrastructure in the eyes of operators, resulting in indirect costs to most governments, such as an increase in demand for foreign currencies which are used for ordering fuel, which is triggered by an increase in demand for fuel due to rough roads (Wang et al. 2020). Also, the lack of drainage systems along roads is a huge mistake, especially in these days when the effects of climate change are so severe. Other areas have adapted to climate change on how they operate, maintain, upgrade, and expand infrastructure networks (Neumann et al. 2021). This must also be done in underdeveloped countries because the entire world is experiencing climate change, including intense rains and acidic rains, both of which have negative consequences on roadways. With heavy rains now, proactive adaptation measures such as the provision of drainage systems are required to preserve the roads (Wang et al. 2020).

Table 3. The summary of importance of proactive road maintenance.

Importance of proactive road maintenance	Reference
It deals with the root-cause of the road network failure	Cheng et al. (2020); Eriksson et al. (2019)
It involves prior plans	Cheng et al. (2020); Eriksson et al. (2019); Salih et al. (2016)
It extends the lifetime of the road network	Chen et al. (2016); Salih et al. (2016)
Reduce maintenance cost	Gultom et al. (2017)

5 CONCLUSION

Because it is difficult to mobilize funds to finance reactive maintenance whenever total failure occurs in the road system due to deferring maintenance works for other competing demands, this paper has explored much on the importance of proactive maintenance over reactive maintenance under the limited budget. However, it is preferable to respond to a projected failure because the cost increases as the problem expands in size. Numerous governments have recognized the need of repairing minor failures, but shockingly few have acted on the idea, resulting in the deterioration of numerous metropolitan roads worldwide, which costs the world $30 billion yearly.

According to this study findings, adopting proactive maintenance can help the country improve its economy and save wasteful spending on road infrastructure caused by deferred maintenance. Some emergent contributions are evident through exploration of differing maintenance approaches in road works. To the best of our knowledge, the findings provide the first useful insights into the importance of proactive over reactive road maintenances in developing countries, using Malawi as a case. Equally, the implications of this study are three folds: First, they open researchers' eyes to more research on the variables that contribute to maintenance deferral other than budget restrictions and competing demands, because it is a problem not only in developing countries but also in wealthy countries. Second, by comprehending the importance of proactive maintenance, over reactive, construction practitioners will be able to apply effective maintenance practices for better performance of road works projects. Third, the finding would provide governments with policy direction with regard to proactive road maintenance principles in the construction industry.

It is recommended that in order for the road network to function correctly, developing countries adopt a proactive maintenance method, which is a reliable technique that ensures the longevity of roads because it addresses problems at an early stage. Likewise, with this strategy, governments may be more rigorous in allocating the funding set aside for road maintenance to its intended purpose, reducing budgetary excuses. As a result of investing in preventative road maintenance, future maintenance expenses and total failure of road network systems which create unnecessary interruption, will be reduced, resulting in economic benefit in developing countries.

REFERENCES

Calvin, J. A., & Long, J. S. (1998). Regression Models for Categorical and Limited Dependent Variables. *In Technometrics* (Vol. 40, Issue 1, p. 80). https://doi.org/10.2307/1271407

Chen, Y., Cowling, P., Polack, F., Remde, S., & Mourdjis, P. (2016). PT US CR. *European Journal of Operational Research*. https://doi.org/10.1016/j.ejor.2016.07.027

Cheng, J. C. P., Chen, W., Chen, K., & Wang, Q. (2020). Automation in Construction Data-driven Predictive Maintenance Planning Framework for MEP Components Based on BIM and IoT Using Machine Learning Algorithms. *Automation in Construction*, 112(January), 103087. https://doi.org/10.1016/j.autcon.2020.103087

Creswell, J. W. (2015). *Educational Research*.

Eriksson, P. E., Larsson, J., & Szentes, H. (2019). *Reactive Problem Solving and Proactive Development in Infrastructure Projects*. 1–3. https://doi.org/10.33552/CTCSE.2019.03.000558

France-mensah, J., Asce, S. M., Brien, W. J. O., Ph, D., & Asce, M. (2019). Developing a Sustainable Pavement Management Plan: Tradeoffs in Road Condition, User Costs, and Greenhouse Gas Emissions. *35*(3), 1–13. https://doi.org/10.1061/(ASCE)ME.1943-5479.0000686.

Guevara, J., Garvin, M. J., Asce, M., & Ghaffarzadegan, N. (2017). Capability Trap of the U. S. Highway System: Policy and Management Implications. *33*(4), 1–14. https://doi.org/10.1061/(ASCE)ME.1943-5479.0000512.

Gultom, T. H. M., Tamin, O. Z., Sjafruddin, A., & Pradono. (2017). The Road Maintenance Funding Models in Indonesia use Earmarked Tax The Road Maintenance Funding Models in Indonesia Use Earmarked Tax. *AIP Concfrence Proceedings*, 060009(November 2017). https://doi.org/https://doi.org/10.1063/1.5011563

Gupta, A., Kumar, P., & Rastogi, R. (2011). Pavement Deterioration and Maintenance Model for Low Volume Roads. *4*(4), 195–202.

Hartmann, A., & Hietbrink, M. (2013). Construction Management and Economics An Exploratory Study on the Relationship Between Stakeholder Expectations, Experiences and Satisfaction in Road Maintenance An Exploratory Study on the Relationship Between Stakeholder Expectations, Experiences and. December 2014, 37–41. https://doi.org/10.1080/01446193.2013.768772

Havránek, P., Valentová, V., Ĝlyiqnryi, X. D. Q. D., & Striegler, R. (2016). Identification of Hazardous Locations in Regional Road Network – Comparison of Reactive and Proactive Approaches. 14, 4209–4217. https://doi.org/10.1016/j.trpro.2016.05.392

Kothari, C.. (2004). *Research Methodology*.

Mcneil, S., Ph, D., & Asce, M. (2014). Impact of Road Conditions and Disruption Uncertainties on Network Vulnerability. 1–8. https://doi.org/10.1061/(ASCE)IS.1943-555X.0000205.

Mohammadi, A., Igwe, C., Amador-jimenez, L., Nasiri, F., Mohammadi, A., Igwe, C., Amador-jimenez, L., & Nasiri, F. (2020). Applying Lean Construction Principles in Road Maintenance Planning and Scheduling and Scheduling. *International Journal of Construction Management*, 0(0), 1–11. https://doi.org/10.1080/15623599.2020.1788758

Neumann, J. E., Chinowsky, P., Helman, J., Black, M., Fant, C., Strzepek, K., & Martinich, J. (2021). *Climate Effects on US Infrastructure: The Economics of Adaptation for Rail, Roads, and Coastal Development*.

Salih, J., Edum-Fotwe, F., & Price, A. (2016). Investigating the Road Maintenance Performance in Developing Countries. *International Journal of Civil and Environmental Engineering*, 10(4), 395–399.

Wang, Huo, N. A., Li, J., Wang, K. A. N., & Wang, Z. (2018). A Road Quality Detection Method Based on the Mahalanobis-Taguchi System. *IEEE Access*, 6, 29078–29087. https://doi.org/10.1109/ACCESS.2018.2839765

Wang, T., Qu, Z., Yang, Z., Nichol, T., & Clarke, G. (2020). Climate Change Research on Transportation Systems: Climate Risks, Adaptation and Planning. *Transportation Research Part D*, 88, 102553. https://doi.org/10.1016/j.trd.2020.102553

Wheat, P. (2017). Scale, Quality and Efficiency in Road Maintenance: Evidence for English Local Authorities. 59(June 2016), 46–53. https://doi.org/10.1016/j.tranpol.2017.06.002

Yepes, V., Torres-machi, C., Chamorro, A., & Pellicer, E. (2016). Optimal Pavement Maintenance Programs Based on a Hybrid Greedy Randomized Adaptive Search Procedure Algorithm. 3730(May). https://doi.org/10.3846/13923730.2015.1120770

Public procurement: Driver for achieving a circular economy among SME housing developers

C.P. Mukumba & K. Kajimo-Shakantu
Department of Quantity Surveying and Construction Management, University of the Free State, Bloemfontein, South Africa

ABSTRACT: Housing development construction practices significantly contribute to global challenges, including climate change, biodiversity loss and waste, and pollution. Because of its purchasing power, public procurement can stimulate sustainable practices in housing development among SMEs to transition to a circular economy. This study explores public procurement as a strategic policy tool to transition to a circular economy among SME housing developers. The study employed qualitative research using semi-structured interviews with 22 purposively selected public procurement officials and SME developers involved in housing development in Lusaka, Zambia. The collected data was analysed using thematic analysis. The findings indicate that public procurement can promote a circular economy among SME housing developers. The study shows that achieving a circular economy requires implementation of legislation and monitoring in collaboration with industry stakeholders. Further, the study findings inform policy in strategically designing public procurement to advance sustainable built environment practices in housing development.

Keywords: Housing development, small-medium enterprises, circular economy

1 INTRODUCTION

Zambia like many governments in the global south, governments in the global south Zambia struggles with the provision of affordable housing. Its housing stock is estimated at 2.5 million units, of which 64% is traditional housing while 36% [n = 800,000 units] is urban housing (Phiri 2016). However, the country continues to experience a critical housing shortage, estimated at 1.5 million, expected to rise to 3.3 million by 2030 (Ministry of Infrastructure, Housing and Urban Development [GRZ-MIHUD] 2020). Some estimates suggest that 110,000 units are required annually to close the housing backlog over the next decade (MIHUD 2020). According to the 2022 Census, 39% of Zambia's population resided in urban areas (Zambia Statistics Agency [ZamStats] 2023). Lusaka, Zambia's capital city, accommodates over two-thirds of the city's population on 20 percent residential land, with nearly 70% of Lusaka's housing stock being substandard and informal (Chisumbe *et al.* 2022). The high rate of urbanization in Lusaka reflects a countrywide increase in urbanization caused by urban economic growth resulting in migrating from rural to urban areas (Mukumba 2019).

The Zambian construction Industry [ZCI], mainly the housing sector, has been seen to be unsustainable in its current production and consumption practices in a linear economy. As construction activities require vast natural resources (Oke *et al.* 2019). This has resulted in environmental challenges ranging from excess consumption energy consumption to pollution of the surrounding environment (Zulu *et al.* 2022). According to Oke *et al.* (2019), the housing sector generates about one-third of all construction waste. In recent years, there has been scholarship debate on reducing material use and waste resulting from unsustainable housing development construction activities. To ensure that housing construction activities align with the needs of the circular economy, the government, through the recently enacted National Housing Policy [NHP], intends to provide quality housing with basic services and healthy sanitation conditions (MIHUD 2020).

The circular economy promotes different ways to current housing development practices to enhance resource efficiency while satisfying the growing housing demand (John *et al.* 2023). With a provision in the Zambian national budget for housing development, public procurement is strategically positioned as a government policy tool to transition housing development practices to a circular economy in the housing sector (MIHUD 2023; National Assembly of Zambia 2023). Though public procurement has been recognised for its role in delivering the deficit housing stock in Zambia, its potential as a driver to transition to a circular economy in housing development has not fully been exploited.

This study investigates the potential link of sustainable public procurement in achieving a circular economy among SME housing developers in Lusaka, Zambia. Lusaka was selected mainly because it has Zambia's highest urbanization rate and housing demand. Further, Lusaka consumes the most significant natural resources for housing construction. Hence its suitability as a choice for this study's preliminary investigation of how public procurement can drive a circular economy among SME housing developers.

2 SUSTAINABLE HOUSING DEVELOPMENT

2.1 *Housing development*

Housing development plays a significant role in human life quality and impacts the natural environment (Godfrey *et al.* 2019; Saliu & Akiomon 2022). From a functional point of view, a house is considered a place which provides shelter, where domestic activities such as sleeping, leisure and rest are performed (Fadlalla 2011). Hence the need for the promotion of sustainable practices in housing development to improve environmental and social performance, which can enhance the well-being of families (Obrenovic *et al.* 2020). Oktay (2002) acknowledged that housing not only satisfies the basic needs for shelter but also other sustainability requirements. As the construction phases produce a lot of pollution and harmful environmental effects, sustainability entails addressing all conditions and attitudes of sustainability within the housing projects' design, construction and maintenance phases (John *et al.* 2023). Thus, sustainable housing development enables efficient use of energy, which requires technologies and designs to deliver lower zero-carbon homes (O'Malley *et al.* 2020).

There are many ways to achieve sustainable housing development. These include: encouraging housing developers to embrace green building practices such as the use of renewable energy sources and water conservation measures and promotion of mixed-use development that combines residential and commercial facilities (Adabre & Chan 2019; Bibri *et al.* 2020). Further, an increase in density in housing developments helps to reduce the amount of land needed for development (Adabre & Chan 2019; Rauterkus & Miller 2011). According to Alemaw *et al.* (2020), investing in green infrastructure, such as green roofs and promoting green building materials such as recycled materials and low-VOC paints, helps reduce construction's environmental impacts. Furthermore, it can enhance sustainable housing development (Sharma & Jha 2021).

2.2 *Features and importance of sustainable housing development*

Sustainable housing development is Sustainable housing development also often includes features that promote walkability and access to public transportation, as well as access to green spaces and other amenities (Rauterkus & Miller 2011). It helps to reduce the environmental impact of housing, conserve resources, and create healthier living environments.

Moreover, it helps to reduce energy costs, improve air quality, and reduce water usage (Wan et al. 2019). Additionally, sustainable housing can help to create more resilient communities and reduce the risk of displacement due to climate change (Ragheb et al. 2016). Finally, sustainable housing can help to create more equitable and affordable housing options for all (Adabre & Chan 2019).

There are many ways to achieve sustainable housing development. These include: encouraging housing developers to embrace green building practices such as the use of renewable energy sources and water conservation measures and promotion of mixed-use development that combines residential and commercial (Adabre & Chan 2019; Bibri et al. 2020). Further, an increase in density in housing developments helps to reduce the amount of land needed for development (Adabre & Chan 2019; Rauterkus & Miller 2011). According to Alemaw et al. (2020), investing in green infrastructure, such as green roofs and promoting green building materials such as recycled materials and low-VOC paints, helps reduce construction's environmental impacts. Furthermore, it can enhance sustainable housing development (Sharma & Jha 2021).

3 CIRCULAR ECONOMY

3.1 *Circular economy defined*

A circular economy is as an economic system that is restorative and regenerative by design, and it is based on the principles of reuse, repair, and recycling of materials and products (MacArthur 2013). It helps reduce the resources used, creating economic opportunities and increasing sustainability (Elegbede et al. 2023). It addresses decoupling, resource efficiency, production efficiency, slower material flows rather than linear economic models, and lower resource extraction without reducing economic activity (Ari & Yikmaz 2019). Governments, businesses, and individuals are all responsible for creating and maintaining a circular economy (Ibn-Mohammed et al. 2021). Governments can create policies incentivising businesses to move towards a circular economy, while businesses can invest in more sustainable production and consumption practices (Narayan & Tidstrom 2020).

3.2 *Strategies to achieve a circular economy*

There are several strategies for achieving a circular economy. These include: *Reduce Waste-reducing* the number of materials used in production, reusing and recycling materials, and minimizing the amount of waste generated: *Design for durability and longevity:* that can be easily repaired, reused, and recycled; *Increasing product efficiency*: by using renewable energy sources, reducing energy consumption, and using more efficient production processes. (Ibn-Mohammed et al. 2021). Transitioning to circular has its own cost associated with it. Generally, the cost of transitioning to a circular economy includes investments in infrastructure, research and development, and new business models (Tirado et al. 2022). Additionally, the cost of transitioning to a circular economy may include investments in training, education, and awareness-raising initiatives (Marino & Pariso 2020).

3.3 *Benefits of a circular economy*

The growing environmental awareness has led governments and companies to look for ways of doing their business sustainably. This calls for reusing and recycling products to ensure a slowdown in the use of natural resources (Elegbede et al. 2023; Ibn-Mohammed et al. 2021). The world's population is growing, and the demand for raw materials is growing. However, the supply of crucial raw materials is limited (Jones et al. 2020). Recycling raw materials mitigates the risks associated with supply, such as price volatility, availability and import dependency (Tirado et al. 2022). Moving towards a more circular economy could increase competitiveness, stimulate innovation, boost economic growth and create jobs

(Ogunmakinde *et al.* 2022). Redesigning materials and products for circular use would also boost innovation across different sectors of the economy (Friant *et al.* 2021).

4 POTENTIAL OF PUBLIC PROCUREMENT TO DRIVE A CIRCULAR ECONOMY

4.1 *Conceptualising public procurement*

Public procurement is the purchase of goods and services by government departments and state enterprises using public funds to fulfil their functional objectives (Organisation for Economic Co-operation and Development [OECD] 2020). Globally, governments spend an estimated US$ 9.5 trillion in public contracts every year, and many developing countries, including Ghana, Botswana and Zambia, represent approximately 15–22% of the Gross Domestic Product [GDP] (World Bank 2021). As such, public procurement is crucial for the government's services delivery because of its spending volume (Lazaroiu *et al.* 2020). Besides its primary objectives of value for money, public procurement can also meet horizontal objectives such as environmental or social ones (Kristensen *et al.* 2021). Hence governments can employ circular public procurement to procure goods, services and works with a reduced environmental impact throughout their life cycle by embedding circular public procurement [CPP] in procurement guidelines (Kristensen *et al.* 2021; Morales 2021). The government can motivate and stimulate innovation to develop green products and make them the market standard (Harland *et al.* 2019).

4.2 *Employing public procurement to achieve a circular economy*

Public procurement as the driver for achieving a circular economy can be achieved by integrating Circular Economy [CE] into Public Procurement [PP], such as including different strategies of CE in procuring entities' procurement plans and guidelines (Alhola *et al.* 2019). The most recent literature presents public procurement as a tool to promote CE (Bernstein & Vos 2021; Harland *et al.* 2019). Circular public procurement has been recognised as a critical element in the transition towards a circular economy (Kazancoglu *et al.* 2021). The importance of the circular economy in both policy and practices is that it enables public authorities to purchase works, goods or services that seek to contribute to closed energy and material loops within supply chains (OECD 2020). At the same time, minimizing and, in the best case avoiding negative environmental impacts and waste creation across their whole life cycle (European Commission 2021). The circular economy in public procurement can be operationalised by promoting circular supply chains and ensuring that different contract forms demonstrate ways of defining contracts supporting the circular economy (Kazancoglu *et al.* 2021; Kristensen *et al.* 2021). Though focusing on purchasing products and services, Alhola *et al.* (2019) argue that much consideration must be shifted towards broader perspectives, including systems and stakeholders. Bernstein & Vos (2021) posit that public procurement should consider the broader and more complex network of stakeholders. Moreover, supply chains in the procurement process entail new roles for different stakeholders (Alhola *et al.* 2019).

For public procurement to be effective as a tool for the transition to a circular economy, Kristensen *et al.* (2021) posit that it requires collaboration internally across departments and externally with partners in the supply chain. However, the lack of circular public procurement tools presents a barrier to implementing circular public procurement, as circular public procurement depends on practices and guidelines stipulated by policy (Reike & Hekkert 2017). Policy implementation can be done by setting specific requirements for procuring goods and services, such as requiring that products are made from recycled materials or designed for reuse or repair. Governments can also use their purchasing power to support the development of new circular products and services by investing in research and development or providing grants for start-ups. Finally, governments can use their procurement processes to promote transparency and traceability in the supply chain, ensuring that materials are sourced responsibly and that products are not made using unethical labour practices.

5 METHODOLOGY

This study adopted a qualitative approach to collect and analyse the data. According to Christensen *et al.* (2014), the qualitative design enables the researcher to explore, uncover salient issues and gain insight into activities that might be missed in structured surveys. Therefore, the qualitative design was deemed appropriate for the study as it enabled the researcher to elicit information on how public procurement can be a driver to achieve a circular economy among SME housing developers (Miles *et al.* 2014).

The target population for this study are public procurement personnel, SME housing developer owners and construction site managers. Due to limitations to access participants, data was collected from 22 purposively selected participants using semi-structured interviews (Christensen *et al.* 2014). Semi-structured interviews enabled the researcher to interact one-to-one with the interviewees, ask for more explanations and seek clarifications to vague answers (Teddlie & Tashakkori 2009).

Analysis of data followed the multiple steps of analysis. Data were organised and prepared for analysis (Creswell & Creswell 2018). The researcher read through all the data and coded it. Coding involves organising data by bracketing the texts and segmenting the sentences into categories (Holton 2007). Descriptions were then generated into themes for further analysis using an approach that involves identifying repetition and similarities as suggested (Bryman 2012). The analysed themes were displayed as the significant findings discussed in the results section (Creswell & Creswell 2018).

6 RESULTS

6.1 Demographic information of participants

Table 1 presents the participants' demographic information regarding gender, work experience, and occupation related to circular public procurement activities. As shown in Table 1, most respondents are males, representing 73%, and the remaining 27% are female. The highest number of respondents have above ten years of experience, represented by 36%. The

Table 1. Demographic information of participants.

Category		frequency	Percentage
Gender	Female	06	27
	Male	16	73
	Total	22	100
Work Experience			
	Less than a year	3	14
	Between 1 and 5 years	5	23
	Between 5 and 10 years	6	27
	Above 10 years	8	36
	Total	22	100
Occupation			
	Public procurement officer	4	18
	Site construction manager	8	36
	Housing developer owner	10	46
	Total	22	100

second highest number of respondents have between 5 and 10 years of experience. This is followed by respondents with experience between 1 and 5 years, representing 23%, and less than a year at 14%. Most respondents are owners of housing developing firms, represented by 46%. Site construction managers and public procurement officers are represented at 36% and 18%, respectively.

6.2 *Awareness levels of the concept of circular economy*

Awareness of the concept of circular economy was identified as the first theme. Awareness is essential in stakeholders' promotion and effective participation in the transition to a circular economy. The quality of environmental, social and economic depends on the active participation of all societal actors. Most respondents indicated levels of circular economy awareness and the significance of transitioning to a circular economy. Some of the results, as indicated by the interviewees, are presented below:

"Circular economy is important for our society and economy at large. It will help us to better look after our environment in a sustainable way" *[ppo.2]*

" ... We are very much aware of the circular economy; however, the awareness should be spread to the entire supply chain if we are collectively to achieve a circular economy in the supply chain ... " *[ppo.4]*

Other participants stated:

"Circular economy means that we are careful with our environment and how we use the materials, if possible we can try to reuse some of the material ... I think that is the basic understanding of what circular economy means ... " *[scm.7]*

"I think it's more to do with being conscious of our environments that as we build these structures, we do not destroy the environment which can negatively affect the people and how they live ... " *[hdo.5]*

" ... it is all about reducing waste, reusing and recycling materials ... " *[ppo.1: scm6: hdo.9]*.

6.3 *Role of public procurement to promote a circular economy*

The overall indication from the participant on the second emergent theme is that public procurement plays a significant role in promoting a circular economy. Research participants indicated that buying public goods and services should promote businesses that point to a circular economy. Further, the respondents indicated that public buying of goods should compel suppliers to adhere to activities that promote a circular economy.

" ... Public procurement tenders are well positioned to have clauses that promote circular economy ... " *[ppo.1]*

" ... Procurement begins with planning ... so they can plan to ensure that the right goods and services are procured that support circular economy ... " *[hdo.9: ppo.41]*

Another participant stated that though public procurement is vital for promoting a circular economy, tender specifications do not provide material reuse provisions.

" ... Public procurement can ensure that were use the materials we use in construction ... though specifications in tenders do not stipulate the reuse of construction materials ... " *[ppo.2: scm.2]*.

6.4 What role would the SME housing developers play in transitioning to a circular economy

The third theme that emerged was SME housing developers' role in transitioning to a circular economy. They stated that construction activities impact the environment, consume resources and that they should sustainably manage their waste.

"It is straightforward to see where construction activities occur in our community. Most times, waste is never managed properly … . It ends up being mixed with the environment, thereby polluting the environment. Housing developers must ensure they clean the construction sites and immediate polluted environment … " *[ppo.3: csm.5]*.

"Most construction activities affect the environment; SME housing developers should have measures of minimal environmental damage caused by construction activities … " *[ppo.3]*

"Construction of houses requires many resources such as building sand, quarry, stones … , these are obtained from the pit which is never restored. You find that there is damage to the environment. So, SME housing developers should find ways of minimizing environmental damage during construction … " *[hdo.8]*

7 DISCUSSION OF RESULTS

The results show that from the explanations provided by the participant that there is awareness of the concept of circular economy among the stakeholders. These findings corroborate those advanced by Adams *et al.* (2017), that there are awareness levels among the construction industry actors regarding environmental, social and economic aspects. The results further suggest that to most respondents, a circular economy means reducing, reusing and recycling. The findings are consistent with those reported in the literature that the awareness levels of circular economy among stakeholders are inclined toward reducing, reusing and recycling, where an economy can regenerate itself (MacArthur 2013). The results findings show a good sign for public institutions and stakeholders to push the development of the circular economy agenda. However, the new insight from the study is that awareness of the circular economy should be extended to the entire public procurement supply chain. This position is advanced by PPO.4. According to Xue *et al.* (2010), the main channels for disseminating circular economy information are radio and television, including secondary channels of government documents, magazines and newspapers.

The overall indication from the participant on the second emergent theme is that public procurement plays a significant role in promoting a circular economy. Research participants indicate that buying public goods and services should promote businesses that point to a circular economy. Further, the respondents indicate that public procurement should compel suppliers to adhere to activities that promote a circular economy. The findings support those reported in the literature that public procurement positively impacts circular consumption (Kristense *et al.* 2021). It aims to promote closed material loops and value retention which can be achieved by reusing products and materials in a circular manner. Concerning having tenders stipulate the reuse of materials as stated by PPO.2 and SCM.2, the results support those Alhola *et al.* (2019) reported that public procurement contracts can stipulate recycling criteria and conditions in contracts due to their relatively high purchasing volumes.

The findings show that the practical application of a circular economy involves enterprises and that the implementation requires legislation and supervision is observed. The results indicate that most SME firms' activities affect the environment due to construction activities. As such, construction activities should be regulated to ensure less environmental damage and reduce resource consumption. These research findings supported those reported in the literature that construction and demolition waste accounts for the most significant portion of waste that affect the environment (Ghisellini *et al.* 2018).

8 CONCLUSION

This paper examined public procurement as a driver for promoting a circular economy among SME housing developers. The research findings indicate that industry stakeholders know about the circular economy concept. However, it has to be extended to the supply chain for effective implementation. The results indicate that public procurement significantly promotes a circular economy through legislation embedded in procurement and contracts. The study identified SME housing developers as critical implementers of circular economy through improved ways of construction activities which can reduce resource consumption and damage to the environment.

The research findings indicated that establishing and transitioning to a circular economy requires the collaboration of public institutions and industry actors. Awareness of the concept of circular economy should be extended to the actors involved in practical implementation. The findings provide insight for policy-makers to design procurement policies promoting a circular economy. Whereas for SME housing developers, the study provides practical application of the circular economy through innovative construction activities that reduce resource consumption and are restorative to the environment.

REFERENCES

Adabre, M.A. & Chan, A.P. 2019. Critical Success Factors (CSFs) for Sustainable Affordable Housing. *Building and Environment (156)*:203–214.

Adams, K.T., Osmani, M., Thorpe, T. & Thornback, J. 2017 Circular Economy in Construction: Current Awareness, Challenges and Enablers. In *Proceedings of the Institution of Civil Engineers-Waste and Resource Management* 170 (1):15–24).

Alemaw, B.F., Chaoka, T.R. & Tafesse, N.T. 2020. *Modelling of Nature-based Solutions (NBS) for Urban Water Management-investment and Outscaling Implications at Basin and Regional Levels.*

Alhola, K., Ryding, S.O., Salmenperä, H. & Busch, N.J. 2019. Exploiting the Potential of Public Procurement: Opportunities for Circular Economy. *Journal of Industrial Ecology* 23(1):96–109.

Bernstein, J.M. & Vos, R.O. 2021. *SDG12-Sustainable Consumption and Production: A Revolutionary Challenge for the 21st Century*. Emerald Group Publishing.

Bibri, S.E., Krogstie, J. & Kärrholm, M. 2020. Compact City Planning and Development: Emerging Practices and Strategies for Achieving the Goals of Sustainability. *Developments in the Built Environment* (4):100021.

Bryman, A. (4th ed.) 2012. *Social Research Methods*. Oxford, Oxford University Press.

Chisumbe, S., Aigbavboa, C., Mwanaumo, E. & Thwala, W. 2022. A Measurement Model for Stakeholders' Participation in Urban Housing Development for Lusaka: A Neo-Liberal Perspective. *Urban Science*, 6(2): 34.

Christensen, L.B., Johnson, R.B. & Turner, L.A. (12th ed.) 2014. *Research Methods, Design, and Analysis*. Boston: Pearson Education.

Creswell, J.W. & Creswell, J.D. (5th ed.) 2018. *Research Design-qualitative, Quantitative, and Mixed Methods Approaches*. California: Sage.

Elegbede, I.O., Muritala, I., Afeez, Y., Majolagbe, Y., Marques, L., Duarte, C., Ndidi, N.E. & Iskilu, S.O. 2023. BS8001 Circular Economy Standard. In *Encyclopedia of Sustainable Management*: 1–8. Cham: Springer International Publishing.

Fadlalla, N. 2011. Conceptualizing the Meaning of Home for Refugees. *Spaces & Flows: An International Journal of Urban & Extra Urban Studies 1*(3).

Friant, M.C., Vermeulen, W.J. & Salomone, R. 2021. Analysing European Union Circular Economy Policies: Words Versus Actions. *Sustainable Production and Consumption* (27):337–353.

Ghisellini, P., Ripa, M. & Ulgiati, S. 2018. Exploring Environmental and Economic Costs and Benefits of a Circular Economy Approach to the Construction and Demolition Sector. A Literature Review. *Journal of Cleaner Production (178)*:618–643.

Godfrey, S., Dean, J. & Regier, K. 2019. Sustainable Housing. In *A Research Agenda for Housing*: 151–164. Edward Elgar Publishing.

Harland, C., Telgen, J., Callender, G., Grimm, R. & Patrucco, A. 2019. Implementing Government Policy in Supply Chains: an International Coproduction Study of Public Procurement. *Journal of Supply Chain Management* 55(2):6–25.

Holton, J.A. 2007. *The Coding Process and Its Challenges. The Sage Handbook of Grounded Theory* (3):265–289.

Ibn-Mohammed, T., Mustapha, K.B., Godsell, J., Adamu, Z., Babatunde, K.A., Akintade, D.D., Acquaye, A., Fujii, H., Ndiaye, M.M., Yamoah, F.A & Koh, S.C.L. 2021. A Critical Analysis of the Impacts of COVID-19 on the Global Economy and Ecosystems and Opportunities for Circular Economy Strategies. *Resources, Conservation and Recycling (164)*:105169.

John, I.B., Adekunle, S.A. & Aigbavboa, C.O. 2023. Adoption of Circular Economy by Construction Industry SMEs: Organisational Growth Transition Study. *Sustainability 15(7)*: 5929.

Jones, B., Elliott, R.J. & Nguyen-Tien, V. 2020. The EV Revolution: The Road Ahead for Critical Raw Materials Demand. *Applied Energy (280)*: 115072.

Kazancoglu, I., Sagnak, M., Kumar Mangla, S. & Kazancoglu, Y. 2021. Circular Economy and the Policy: A Framework for Improving the Corporate Environmental Management in Supply Chains. *Business Strategy and the Environment, 30*(1):590–608.

Kristensen, H.S., Mosgaard, M.A. & Remmen, A. 2021. Circular Public Procurement Practices in Danish Municipalities. *Journal of Cleaner Production (281)*:24962.

MacArthur, E. 2013. Towards the Circular Economy. *Journal of Industrial Ecology 2*(1):23–44.

Marino, A. & Pariso, P. 2020. Comparing European Countries' Performances in the Transition Towards the Circular Economy. *Science of the Total Environment (729)*:138142.

Miles, Huberman & Saldana (3rd ed.) 2014. *Qualitative Data Analysis., Methods Sourcebook*. California: Sage.

Ministry of Infrastructure, Housing and Urban Development [MIHUD]. 2020. *National Housing Policy 2020–2024. Implementation Plan*. Ministry of Housing and Infrastructure Development.

Morales, A.H. 2021. *Circular Economy and Public Procurement: Dialogues, Paradoxes and Innovation*. Aalborg Universitetsforlag.

Mukumba, C.P. 2019. Enablement Approaches to the Upgrading of Informal Settlements in Zambia. Case study of Misisi Compound, Lusaka. Unpublished Thesis. Bloemfontein: University of the Free State.

National Assembly of Zambia. 1994. *National Housing Authority Act 13 of 1994*. Available at: www.parliament.gov.zm/sites/default/files/documents/acts/National%20Housing%20Authority%20Act.pdf.

National Assembly of Zambia. 2023. *Estimates of Revenue and Expenditure [Output Based Budget]*. Available at: www.parliament.gov.zm/sites/default/files/images/publication_docs/07%20Main%20Report%20Budget%202023%20%282%29.pdf.

O'Malley, M.J., Anwar, M.B., Heinen, S., Kober, T., McCalley, J., McPherson, M., Muratori, M., Orths, A., Ruth, M., Schmidt, T.J. & Tuohy, A. 2020. Multicarrier Energy Systems: Shaping Our Energy Future. *Proceedings of the IEEE 108*(9): 1437–1456.

Obrenovic, B., Jianguo, D., Khudaykulov, A. & Khan, M.A.S. 2020. Work-family Conflict Impact on Psychological Safety and Psychological Well-being: A Job Performance Model. *Frontiers in Psychology (11)*: 475.

Ogunmakinde, O.E., Egbelakin, T. & Sher, W. 2022. Contributions of the Circular Economy to the UN Sustainable Development Goals Through Sustainable Construction. *Resources, Conservation and Recycling (178)*:106023.

Oke, A., Aghimien, D., Aigbavboa, C. & Musenga, C. 2019. Drivers of Sustainable Construction Practices in the Zambian Construction Industry. *Energy Procedia (158)*: 3246–3252.

Phiri, D. 2016. *Challenges of Affordable Housing Delivery in Zambia*. Centre for Affordable Housing Finance Africa.

Ragheb, A., El-Shimy, H. & Ragheb, G. 2016. Green Architecture: A Concept of Sustainability. *Procedia-Social and Behavioral Sciences (216)*: 778–787.

Rauterkus, S.Y. & Miller, N. 2011. Residential Land Values and Walkability. *Journal of Sustainable Real Estate 3* (1):23–43.

Saliu, I. & Akiomon, E. 2022. Sustainable Housing in Developing Countries: A Reality or a Mirage. Sustainable Housing. *IntechOpen*. DOI: 10.5772/intechopen.99060.

Sharma, N. & Jha, A.K. 2021. Overview of Eco-Friendly Construction Materials. In *Intelligent Energy Management Technologies: ICAEM 2019*: 111–118. Springer Singapore.

Teddlie, T. & Tashakkori, A. 2009. *Mixed Methods Research. Integrating Quantitative and Qualitative Approaches in the Social and Behavioral Sciences*. Los Angeles: Sage.

Tirado, R., Aublet, A., Laurenceau, S. & Habert, G. 2022. Challenges and Opportunities for Circular Economy Promotion in the Building Sector. *Sustainability 14*(3):1569.

Wan, C., Shen, G.Q. & Choi, S. 2019. Waste Management Strategies for Sustainable Development. In *Encyclopedia of Sustainability in Higher Education*: 2020–2028. Cham: Springer International Publishing

Xue, B., Chen, X.P., Geng, Y., Guo, X.J., Lu, C.P., Zhang, Z.L. & Lu, C.Y. 2010. Survey of Officials' Awareness on Circular Economy Development in China: Based on Municipal and County Level. *Resources, Conservation and Recycling, 54*(12), pp.1296–1302.

Zambia Statistics Agency [ZamStats]. 2023. *2022 Census of Population and Housing Report*. Available at: www.zamstats.gov.zm/wp-content/uploads/2023/05/2022-Census-of-Population-and-Housing-Preliminary.pdf.

Zulu, S.L., Zulu, E., Chabala, M. & Chunda, N. 2022. Drivers and Barriers to Sustainability Practices in the Zambian Construction Industry. *International Journal of Construction Management*: 1–10.

An assessment of cost and socio-economic implications of applying innovative active design principles in construction

L. Le Roux & K. Kajimo-Shakantu
Department of Quantity Surveying and Construction Management, University of the Free State, Bloemfontein, South Africa

ABSTRACT: There is limited knowledge among Quantity Surveyors in South Africa, on cost and socio-economic benefits of active design principles in construction. This study sought to establish a base level understanding of cost and socio-economic implications of active design principles to improve cost management and provide insights into the associated benefits for the construction sector. The study was exploratory and utilized interviews with purposive sampling within a case study approach. Key findings included five identified phenomena covering health and wellness through construction; interest in active design implementation amongst Quantity Surveyors; financial factors; socio-economic factors, barriers and limitations. The study concludes that despite the potential, there are inherent constraints that limits implementation in the South African construction industry. It recommends that implementation be embraced as an industry norm to add value for end-users and also the emerging role of the Quantity Surveyor in the process of driving innovative concepts in construction industry.

Keywords: Active Design Construction, Cost Management, Health, Socio-economic, Quantity Surveying

1 INTRODUCTION

Obesity is a globally increasing problem which poses numerous challenges to quality of life (Dietz & Santos-Burgoa 2020). In 2015, approximately 4 million deaths and 120 million disability-adjusted life years, worldwide, were caused by physical inactivity (Ng *et al.* 2020). Socio-ecological models of physical activity theorize that individual physical activity levels are influenced by policy, physical and socio-cultural environments, and individual personalities (Sawyer *et al.* 2017). According to Weavers (2011), the occurrence of overweight and obesity can be connected to transportation, work environment, and media resources such as television and social media. Obesity further poses a severe impact on employee productivity in terms of absenteeism, presenteeism, return on investment, increased health insurance costs and company turnover (Crous 2014).

Lopez (2012) states that the built environment provides a framework for day-to-day living. The built environment is a major contributor to individual health and well-being and represents an important pathway through which individuals are influenced. Physical activity can be promoted through good design of active-friendly buildings and neighbourhoods (Marsh *et al.* 2016). From a health and wellness viewpoint, there is growing research on the impact of interior environments on individual health (Marsh *et al.* 2016).

A possible way offered in the literature to mitigate problems associated with overweight, obesity and inadequate physical activity is the concept of active design (Sawyer *et al.* 2017). Active design is an approach to the development of buildings and communities in such a way that daily activity and healthy food choices become more attractive to the public (Engelen et al. 2016). Socio-ecological models of physical activity theorize that individual physical activity levels are influenced by policy, physical and socio-cultural environments, and individual personalities (Sawyer et al. 2017).

There are ten (10) Active Design Principles, which include active spaces, walkable communities, connected walking and cycling routes, co-location of facilities, open spaces, good quality streets, infrastructure, active buildings, management, maintenance, monitoring & evaluation as well as the promotion of activity to create healthier cities and communities. However, in the literature, there is limited information on active buildings and/or how the application of active design principles can be innovatively maximized in the construction process for cost and socio-economic benefits. This study investigates the cost and socio-economic implications of this concept in practice from a Quantity Surveying perspective. The study will provide useful insights as baselines to guide how to the concept can be better exploited to advise clients on its adoption and implementation for project cost management.

2 LITERATURE REVIEW

2.1 Active design and the built environment

Active design developed from the theory that the design of the built environment has a direct impact on individual health and wellbeing (Marsh et al. 2016). Construction consultants focus on four ways to introduce healthier lifestyles to the public namely, active recreation, active buildings, alternative transportation methods, and improved access to healthier nutrition (Finch et al. 2017). Alternative transportation can involve the design of safer, easier to use cycling routes and active buildings making use of stairs rather than elevators or lifts (Engelen et al. 2016). Interior designers and architects can increase public activity through increased stair usage via design and placement (Taemthong & Chaisaard 2019).

Finch et al. (2017) suggests that building functions be located in such a way that communal areas such as mail and lunchrooms require some walking through appealing and supportive walking routes, staircases should be placed in eyesight and be usable as well as attractive. Research argue that providing centrally visible active spaces like bicycle storing rooms, fitness centres, showers, locker rooms, and drinking fountains will encourage people to participate in physical activity (Finch et al. 2017). Standing desks or sit-stand workstations can be incorporated to reduce the long-term effects of sitting (Taemthong & Chaisaard 2019). Active design is thus a form of design making activity easier and more approachable to the public (Engelen et al. 2016). It forms part of the built environment and aims at creating opportunities for physical activity. The construction industry is still in the early stages of filling the gap between building design, construction, and health (Bernstein & Russo 2014). There are limited studies and information available regarding active design buildings, hence the opportunity presents itself to study the effect of light, air and acoustics on the health and wellbeing of individuals (Bernstein & Russo 2014).

Sustainable design challenges developers and the professional team to think outside the box. Sustainable design includes processes to reduce negative environmental impacts, promote healthy and productive environments and minimize waste (Kibert 2016). The overall concept of sustainability is divided into economic-, socio-political and environmental sustainability (Marsh et al. 2016). Green design and construction focus on the reduction of social and environmental impacts on the built environment whilst improving the quality of life for the building occupants (Cole 2019). Sustainable design and green infrastructure incorporates the interconnection between community and health, providing a way to control benefits, to build buildings that will last longer, be more efficient, reduce operating costs, increase employee productiviyty and contribute to healthy living (Kubba 2010). The elimination of waste has been used as a key driver to contribute to sustainable construction (Kibert 2016).

Implementing lean construction principles offers the construction industry the opportunity to improve sustainability by optimizing resource utilisation and human safety whilst contributing to construcion cost savings though waste minimization (Taemthong & Chaisaard 2019). Sport England and Public Health England established ten (10) principles of active design to create environments that will promote physical activity and healthy lifestyles (Finch et al. 2017). These principles can be applied to various forms of development and includes principles such as mixed land use; facilities for cyclists; easy, available, and fresh food; movement at work; parking; exterior working spaces; stairs; urban gardening; and exercise facilities (Finch et al. 2017; Xue et al. 2016).

The implementation of the above said principles in construction is associated with direct, indirect, and hidden costs (Burton 2008). These costs need to be evaluated (Taemthong & Chaisaard 2019) and managed (Russ et al. 2020) to ensure viability of implementation. The involvement of the Quantity Surveyor from the inception of such projects is required to ensure cost effectiveness. Finch et al. (2017) highlights the traditional activities of the Quantity Surveyor as single-rate approxmate estimation, cost planning, procurement advice, measurement and quantification, document preparation, cost management and cost control. The incorporation of active design principles requires effective and innovative cost- planning and control techniques. The Quantity Surveyor plays a vital role in the viability of active design implementation in the construction industry (Russ et al. 2020).

A healthy workplace is defined by two factors namely organizational culture and the physical work environment (Burton 2008; Kubba 2010; Xue et al. 2016). The loss of productivity accompanying sick days lead to the realisation that it is more effective to prevent illness than to treat it (Bernstein & Russo 2014). More firms are embracing workplace wellness through architecture (Xue et al. 2016). Smallwood (2015) notes that architects, interior designers, and associated professionals combine information and methodologies to create spaces providing new experiences and concepts. Designing for health, safety, and wellness is important in an industry that is recognised as one of the most dangerous industries world-wide (Chandramohan et al. 2022). Clients therefor should give preference to contractors with good and improved health and safety records (Cole 2019). Smallwood (2015) notes that designers must familiarize themselves with ergonomic design principles to prevent and minimize ergonomic related risks during the various phases of a project.

2.2 *Cost implications of active design*

Construction projects globally are known for exceeding the initial project budget (Odediran & Windapo 2014). Cost, time, and quality therefor remains the three parameters of project success (Russ et al. 2020). Odediran & Windapo (2014) compiled a list of factors influencing cost overruns on active design projects. These include but are not limited to; inaccurate feasibility studies and cost estimates, limited knowledge about active design amongst the professional team (Knox et al. 2013), the availability of skilled labour, frequent work scope changes, excessive workload, poor project planning and scheduling, poor project management and control, poor communication, poor contract administration, delays, inadequate contractor knowledge, and poor cost control and managing. Fonarrow et al. (2015) adds the influence of the internal and external working environment to the list. Cost is however considered to be the leading concern in the implementation of active design strategies in construction (Abidin & Azizi 2016). It is therefore important to understand the factors that will influence the cost of active design buildings, these include awareness, knowledge, financial, technical and government support (Cole 2019).

Building for health creates healthy, comfortable and economically prosperous places for employees and families (Abidin & Azizi 2016). As cost is seen as the biggest barrier to the implementation of active design strategies it is pertinent to understand the cost elements in order to recommend ways to reduce cost and make active designed buildings feasibly presentable (Abidin & Azizi 2016). Besides the contributions of such buildings to the community and end-users, there is a growing concern about the effect on the financial performance of these buildings (Taemthong & Chaisaard 2019). Active designed buildings have however been found to generate higher rental income, higher occupancy rates, longer economic lives and high profits over a longer period of time for investors (Taemthong & Chaisaard 2019). Numerous studies focus on the profit of active design building owners however, there seems to be an absence in examining the capital market, i.e. the cost of equity capital of these buildings (Abidin & Azizi 2016). According to Russ *et al.* (2020) healthy employees hold individual, organizational and societal benefits. Chronic exposure to stress impacts mental health, leading to increased absenteeism and presenteeism and decreased company turnover (Fonarrow *et al.* 2015). Recent literature highlights the following strategies to minimize and manage the costs associated with active design implementation: clear goal setting, appointment of a knowledgeable project team, education, clear expectations, budget assumptions and time management (Fonarrow *et al.* 2015)

2.3 *Active design barriers and limitations*

Even though the relationship between the built environment and physical activity is gaining momentum, financial barriers of high initial construction cost and uncertainties about future value present barriers and limitations to implementation Odediran & Windapo (2014). Fonarrow *et al.* (2015) adds that this uncertainty of future project value and increased initial construction cost could have a negative influence on capital financing. Russ *et al.* (2020) notes that active design presents a new business model resulting from increased worldwide awareness of health and well-being, hence constituting that the barriers and limitations to the implementation of active design principles need to be considered, not only from a technical point of view but also from a financial point of view.

Active Design presents the application of a new technique to the construction industry which increase cost and associated risks leading to uncertainty from a feasibility perspective (Taemthong & Chaisaard 2019). This problem constitutes a financial barrier to the implementation of active design principles in construction, followed by a possible financing barrier (Odediran & Windapo 2014).

3 METHODOLOGY

The exploratory nature of this study adopted a qualitative research approach to understand and interpret how participants perceive and react to this specific topic of active design, a relatively new topic in the construction sector, and where little is known. A two-way case study approach was used comprising two modern designed buildings in Gauteng. The case study buildings namely Building A, a national bank, and Building B, a conventional office building in the Woodmead Office Park were studied in detail to investigate and analyse the cost and socio-economic impact of the application of innovative active design strategies and the effect on employee productivity and company performance. The selection criteria for the case studies included that one (1) building needed to have incorporated active design principles and the other not (conventional building) and that they both should be located in the same area and used for commercial purposes.

The study further interviewed a total of 18 case study participants to elicit their contribution to the comprehensive understanding of the cost and socio-economic benefits accompanying the implementation of innovative active design principles and strategies. Random sampling was used to sample the participants who are users of the respective buildings, until point of saturation was reached. Further to this, five (5) industry experts were purposively selected and interviewed for their contribution to the comprehensive understanding of active design implementation in the construction industry. The data was analyzed thematically. The main analytical strategies used to analyse the data in this study encountered revisiting of collected data and examining and categorising the data to achieve the study aims. The study also made use of techniques of applied pattern matching and explanation building to analyse the obtained data, codes to identify the anchors that allowed the key points of data to be gathered, concepts that included the collection of codes of similar nature enabling grouping of the data, categories to generate theory and theory to explain the research subject.

4 RESULTS

4.1 Demographic details

The case studies and participate details are provided in Table 1

Table 1. Empirical study data.

	Building A (Case Study)	Building B (Case Study)	Other (Expert)
Case Studies	Active Design – Bank	No Active Design – Conventional Office Building	N/A
Profession	Financial / Other	Quantity Surveying	Quantity Surveying/ Construction related
Designations	4 x Employers, 3 x Employees	4 x Employers, 9 x Employees	Employees
Interviewees	7	12	5
Interviewee ID's	A1, A2, A3, AS4, AS5, AS6, AS7	B1, B2, B3 BS4, BS5, BS6, BS7, BS8, BS9, BS10, BS11, BS12	C1, C2, CS3, CS4, CS5
Interviewee age group	3 x Participants per the following age group: 30–40 4 x Participants per the following age group: 20–30	3 x Participants per the following age groups: 30–40; 40–50 4 x Participants per the following age group: 20–30 1 x Participant per the following age group: >60	2 x Participants per the following age group: 20–30 2 x Participants per the following age group: 30–50 1 x Participant per the following age group: 40–50
Interviewee gender	Female x 2 Male x 5	Female x 3 Male x 9	Female x 1 Male x 4

Source: Field Study (2020)

Table 1 details the two case studies (Buildings A and B) used by the study to investigate the main research question as well as the expert interviews conducted to broaden the understanding on the implementation of active design strategies in the construction industry. Building A represented an active designed building, designed to promote individual health and wellbeing. The building incorporates active design strategies such as a variety of eateries, shopping, and lifestyle options. The architectural design is one of hard structures to optimize productivity through the incorporation of urban green oasis. The outside area is landscaped with indigenous trees, fountains, and an outdoor lounge area. The courtyard is built entirely from locally sourced materials, with many of these materials being recycled or repurposed and is situated on top of a four-level basement parking area. Various structured and semi-structured interviews with building occupants and industry experts were conducted to study the research phenomenon. The line of inquiry of this study justified the identified case studies and selected group of participants were justified as appropriate and fit for purpose.

4.2 Presentation and discussions of results

From the coded data, five themes emerged namely; 1. Health and Wellness; 2. Interest; 3. Financial; 4. Socio-economical; 5. Implementation. These were developed further and explained as phenomenon

Phenomenon 1: Contribution to health and wellness through the construction industry. The first theme included factors influenced by and which contribute to the overall health and wellness of the building occupants (employees and employers). These findings support the literature where it was found that the linkage between the built environment and health outcomes is increasing. The empirical study addressed the effect of active design on employee health and the findings supported the sentiment by Marsh *et al.* (2016) that there is growing research on the impact of interior environments on individual health and wellness in that the respondents all value personal health and wellness and all respondents indicated interest in learning more about the impact of active design on their personal health and wellbeing. The study further found that the respondents who value a healthy and active lifestyle, are more likely to be interested in learning more about active design and the impact that the implementation of active design strategies can have on employee health and wellness.

Phenomenon 2: Interest in active design implementation amongst Quantity Surveyors and construction related professions. The second theme revolved around the growing interest in active design implementation amongst the interviewed group, with a focus on industry experts. The study found that there is a high level of interest in active design implementation across the professions selected to partake in the study. The study further established a link between the interest in – and knowledge on the concept of active design. This finding supports the problem statement and highlights the limited interest and knowledge about active design amongst Quantity Surveyors. This study findings support the literature where Bernstein & Russo (2014) found growing interest in healthcare programmes and imbursements related to physical activity.

Phenomenon 3: Financial factors related to implementation. One of the key objectives of this study was to determine the financial factors related to the implementation of active design in construction. Theme 3 was woven around knowledge of the cost related to the implementation of active design strategies in the construction industry. This also dealt with perceptions of whether the benefits of active design principles justify the associated implementation cost and drew a link between the implementation of active design and cost control. The findings support the literature in terms of direct- and indirect cost implications of active design implementation and finding that a healthy workplace can decrease company turnover and increase the company brand and employee production (Burton 2008; Crous 2014). The study revealed four predominant findings which were directly linked to the financial impact of applying innovative active design principles in construction namely; knowledge, viability, implementation, and success factors. The empirical study revealed limited knowledge regarding active design amongst Quantity Surveyors, which supports the literature by Knox *et al.* (2013). The study further strengthens the findings of Abidin & Azizi (2016) that cost is one of the leading barriers in the implementation of active design and found that the implementation of active design strategies in the construction industry could be viable if effective cost management strategies are applied. The findings indicate that the core success factors contributing to the successful implementation of active design strategies in construction includes knowledge, feasibility, and cost management

Phenomenon 4: Socio-economic response to the implementation of active design principles in the construction industry. The fourth theme developed around the socio-economic factors of the study. The study found that the socio-economic impact of applying innovative active design strategies in construction is evident and valued. It was found that employee relationships, employee/employer relationships, employee performance and productivity, positivity amongst the workforce, self-confidence, education, and the effect of buildings on the environment are positively enhanced by implementation. This phenomenon supports the findings by Kubba (2010); and Xue et al. (2015) that the implementation of active design has a positive effect on employees, their families, and surrounding environments. These findings support the literature and findings that a workplace which gives access to movement, ventilation, daylight and outdoor areas may increase employee satisfaction and productivity (Kubba 2010; Marsh et al. 2016; Weavers 2011; Xue et al. 2015).

Phenomenon 5: Barriers and limitations to Active Design adoption. The fifth theme developed around the factors which influence the implementation of active design strategies in the construction industry. It was found that the main limitation of active design implementation is limited knowledge about the concept of active design and appropriate cost management strategies amongst industry experts. This finding supports the problem statement and literature findings by Knox et al. (2013) regarding limited knowledge amongst industry professionals. The study found that the implementation of innovative active design strategies improves the overall morale of the inhabitants and contributes to overall happier employees. The findings further support the literature by Fonarrow et al. (2015), that adequate cost management strategies are needed to control and regulate implementation and avoid over expenditure on items justifying alternative methods in construction projects.

5 CONCLUSIONS AND RECOMMENDATIONS

The study identified obesity as a growing problem posing numerous challenges to the quality of life. The literature indicates the implementation of active design of buildings as a possible way to mitigate problems associated with obesity and inadequate physical activity. The literature also indicated that individual physical activity levels are influenced by policy, physical and socio-cultural environments, and individual personalities. The built environment contributes to individual health and well-being through good design of active-friendly buildings and neighbourhoods. The concept of active design is not widely explored, and its implementation fully exploited in the South African construction industry. The findings suggest that implementation of active design innovatively can be feasible if optimal knowledge and cost management systems are applied. The study findings found that the implementation of active design strategies can have a positive effect on socio-economic factors including employee relationships, employee/employer relationships, employee performance and productivity, positivity amongst the workforce, self-confidence, education, and the effect of buildings on the environment. In addition, the study found the biggest constraints to the implementation of active design strategies in the construction industry to be limited knowledge and limited available research. It emerged in the study that the implementation of active design has benefits which could outweigh constrained implementation costs.

The study concludes that despite the potential of active design, there are inherent barriers that limits implementation in the South African construction industry. It is therefore recommended that the implementation of active design strategies in the South African construction industry should be generalized as an industry norm to add value to the end-user whilst the emerging role of the Quantity Surveyor should be reconsidered as a valued contributor to establishing innovative concepts in the South African construction industry. It would also be beneficial for the construction industry to invest in in-depth feasibility studies and cost analysis for optimisation of the benefits of the application of active design principles in construction as well as raising more awareness among various industry role players such as Quantity Surveyors and clients. There is a need for increased awareness and implementation of active design strategies in the built environment. The findings of this study will create increased investment and participation from the Quantity Surveyor to promote active design as a sustainable practice in the construction industry with economic and socio-economic benefits. The possible positive socio-economic effects of active design including employee/employer relationships, performance and productivity, positive morale amongst the workforce and self-confidence, makes a case for Quantity Surveyors to think outside the box of their traditional roles in order to make meaningful contribution and add value to the process through better informed cost management principles and practices. The study highlights an important gap in the role which Quantity Surveyors could play in advancing sustainable construction principles through active design. There is need for increased training and upskilling of knowledge and skills in this area of active design and the role of construction for a better healthier built environment.

The limitation of this study is that the case studies were both based in Gauteng hence the situation might be different in other areas or provinces. It is therefore further recommended that future studies should consider case studies from different areas and provinces to see if similar findings will obtain. Notwithstanding the results do provide a baseline from which future studies could be extended on to benefit the construction industry.

REFERENCES

Abidin, N. and Azizi, N., 2016. Identification of Factors Influencing Costs in Green Projects. *World Academy of Science, Engineering and Technology International Journal of Civil and Environmental Engineering*, [online] 10(9), pp.1160–1163. Available at: <https://zenodo.org/record/1126421#.Xu3cJ2gzaM8> [Accessed 20 June 2020].

Bernstein, H.M., Russo, M.A. 2014. United States of America. Department of Health. 2015. *Beyond ROI: Building Employee Health & Wellness Value of Investment*. Eden Prairie: Op tum.

Burton, J. 2008. *The Business Case for a Healthy Workplace*. [pdf] IAPA. Available from: <http://www.iapa.ca/pdf/fd_business_case_healthy_workplace.pdf> [Accessed: 10 July 2016].

Chandramohan, A., Perera, B.A.K.S. and Dewagoda, K.G. (2020). Diversification of Profess sional Quantity Surveyors' Roles in the Construction Industry: The Skills and Competencies Re- quired. *International Journal of Construction Management*, pp.1–8. https://doi.org/10.1080/15623599.2020.1720058.

Cole, L., 2019. Green Building Literacy: A Framework for Advancing Green Building Education. *International Journal of STEM Education*, [online] 6(1), p.1. Available at: <https://stemeducationjournal.springeropen.com/articles/10.1186/s40594-019-0171-6> [Accessed 16 June 2020].

Crous, S. 2014. *Absenteeism Costs SA Employers R12bn Annually*. [Online]. South Africa: Cape Town. Available from: <http://www.bizcommunity.com/Article/196/22/113953.html> [Ac cessed: 11 July 2016].

Dietz, W. and Santos-Burgoa, C., 2020. Obesity and It's Implications for COVID-19 Mortality. *Obesity*, [Online] 28(6), pp.1005–1005. Available from: <http://onlinelibrary.wiley.com/doi/full/10.1002/oby.22818>.

Engelen, L., Dhillon, H.M., Chau, J.Y., Hespe, D., Baumann, A.E. 2016. Do Active Design Buildings Change Health Behaviour and Workplace Perceptions? 66(5), pp 1.

Finch, L., Tomiyama, A. and Ward, A., 2017. Taking a Stand: The Effects of Standing Desks on Task Performance and Engagement. *International Journal of Environmental Research and Public Health*, [online] 14(8), p.939. Available at: <http://file:///C:/Users/lknobel/Downloads/ijerph-14-00939.pdf> [Accessed 14 June 2020].

Fonarrow, G., Calitz, C., Arena, R., Baase, C., Isaac, F., Lloyd-Jones, D., Peterson, E., Pronk, N., Sanchez, E., Terry, P., Volpp, K. and Antman, E., 2015. Workplace Wellness Recognition for Optimizing Workplace Health. *Circulation*, [online] 131(20), pp.1–18. Available from: <https://www.ahajournals.org/doi/full/10.1161/cir.0000000000000206> [Accessed 8 June 2020].

Kibert, C.J. (2016) *Sustainable Construction: Green Building Design and Delivery*, 4th edn., Florida: John Wiley & Sons.

Knox, E.C., Esliger, D.W., Biddle, S.J.H., Sherar, L.B. 2013. Lack of Knowledge of Physical ac tivity Guidelines: Can Physical Activity Promotion Campaigns Do Better? 3(12), pp. 22

Kubba, S., 2010. *Green Construction Project Management and Cost Oversight*. 1st ed. [Place of publication not identified]: Butterworth-Heinenann Inc., pp.1–120.

Lopez, R., 2012. *The Built Environment and Public Health*. 1st ed. USA: John Wiley & Sons, pp.20–59.

Marsh, M., Erickson, I. and Bleckner, J., 2016. Transforming Building Industry and Health Out comes Through Social Data-supported Design. In: *Bloomberg Data for Good Exchange Confer ence*. [online] New York City, NY, USA: Bloomberg Data for Good Exchange Conference., p.1. Available at: <https://arxiv.org/abs/1609.08778> [Accessed 13 June 2020].

Ng, M., Rosenberg, M., Thornton, A., Lester, L., Trost, S., Bai, P. and Christian, H., 2020. The Effect of Upgrades to Childcare Outdoor Spaces on Preschoolers' Physical Activity: Findings from a Natural Experiment. *International Journal of Environmental Research and Public Health*, [online] 17(2), p.468. Available at: <https://www.mdpi.com/1660-4601/17/2/468> [Accessed 16 June 2020].

Odediran, S. and Windapo, A., 2014. Systematic Review of Factors Influencing the Cost Performance of Building Projects. In: *Proceedings 8th Construction Industry Development Board (cidb) Postgraduate Conference*. [online] Johannesburg, pp.32–50. Available at: <https://www.researchgate.net/publication/261296539_Systematic_review_of_factors_influencing_the_cost_performance_of_building_projects> [Accessed 12 October 2020].

Rajasekar, S., Philominathan, P. and Chinnathambi, V., 2013. *Research Methodology*. Doctor's Degree. University, Tiruchirapalli.

Russ, N., Hamid, M. and Ye, K., 2020. The Development Of Conceptual Model For Improving Sustainable Building Construction Implementation. Masters Degree. University Malaysia.

Sawyer, A., Smith, L., Ucci, M., Jones, R., Marmot, A. and Fisher, A., 2017. Perceived Office Environments and Occupational Physical Activity in Office-based Workers. *Occupational Medicine*, 67(4), pp.260–267.

Smallwood, J., 2015. Designing for Construction Ergonomics. In: *6th International Conference on Applied Human Factors and Ergonomics 2015 and the Affiliated Conferences*. [online] Port Elizabeth: Elserivier, pp.6400–6405. Available from: <https://www.sciencedirect.com/science/article/pii/S2351978915009713> [Accessed 21 June 2020].

Taemthong, W. and Chaisaard, N., 2019. An Analysis of Green Building Costs using a Minimum Cost Concept. *Journal of Green Building*, [online] 14(1), pp.53–78. Available at: <https://meridian.allenpress.com/jgb/article/14/1/53/10693/AN-ANALYSIS-OF-GREEN-BUILDING-COSTS-USING-A> [Accessed 2 August 2020].

Weavers, W. 2011. *Active Living Architecture*. Thesis (Master). Aucland: Unitec Institute of Technology.

Xue, F., Gou, Z. and Lau, S. (2016) "Human Factors in Green Office Building Design: The Impact of Workplace Green Features on Health Perceptions in High-rise High-density Asian Cities," *Sustainability*, 8(11), p. 1095. Available at: https://doi.org/10.3390/su8111095.

Construction ergonomics, health, safety and wellbeing

Psycho-social well-being programs required for construction workers in Zimbabwe

W. Mateza
Faculty of the Built Environment, National University of Science and Technology, Bulawayo, Zimbabwe

T. Moyo
Department of Quantity Surveying, Nelson Mandela University, Port Elizabeth, South Africa

ABSTRACT: Globally, psycho-social well-being is a significant challenge affecting construction workers. While the developed world has made great strides in solving this challenge, developing countries, including Zimbabwe, are lagging behind. This research, sought to determine the psycho-social well-being initiatives required for skilled construction workers in Zimbabwe. Quantitative data was collected using self-administered structured questionnaires for management personnel and interviewer-administered questionnaires for construction workers. Participants were sought from the Construction Industry Federation of Zimbabwe affiliated companies in higher categories. Relative importance index and T-test (two samples) were used for analysis. Both groups ranked counselling and team-building activities as the first and second-best initiatives. Health and safety regulations must address the absence of such initiatives. Further studies can investigate on the structuring and implementation of the initiatives. The study's main limitation was that participants needed to be drawn from all the construction company categories to ascertain the full extent of the challenge.

Keywords: Psycho-social, Well-being, Occupational Safety and Health, Developing countries, Zimbabwe

1 INTRODUCTION

The construction industry in developing countries is labour-intensive (Chigara & Moyo 2014a). Due to the high number of skilled construction workers and the nature of the work, incidences and accidents are inevitable in the construction industry (Chigara & Moyo 2014b). Liang & Shi (2021) posit that the demanding nature of the construction industry poses grave health risks to construction workers. This notion is also supported by Khan *et al.* (2020) when they reported that owners often neglect health and safety (H & S) in construction firms, leading to poor health and safety performance and unacceptably high fatality and injury rates. However, according to The International Labour Organisation (ILO), workplace well-being relates to all aspects of working life, from the quality and safety of the physical environment, to how workers feel about their work, their working environment, the climate at work and work organisation. In addition, the National Wellness Institute (NWI) (2023) defines wellness as an active process through which people become aware of and make choices toward a more successful existence.

As may be the case in other developing countries, workers in Zimbabwe are exposed to numerous occupational safety and health challenges (Chirazeni & Chigonda 2018). Research by Chipato *et al.* (2019) on the H&S practices in the Zimbabwean construction industry established that a negative H&S culture is one of the significant factors affecting the industry.

In their research, Chirazeni & Chigonda (2018) also established that workers are experiencing work-related stress and depression, and such psycho-social challenges need a well-structured response. Traditionally, worker health or occupational safety and health discipline has been fixated on worker exposure to various workplace hazards (Adams 2019). However, the field's scope has since expanded with time to embrace the concept of worker well-being, or the capability of people to address normal stresses, work productively, and achieve their highest potential. Despite the evidence of these deficits, limited research on wellness interventions has been undertaken, especially in the study area. Therefore, the main objective of this study was to determine the most significant and suitable psycho-social wellness interventions that are needed to respond to the challenges affecting construction workers in Zimbabwe. In addition, the study determined a statistically significant difference between the insights of managers and skilled construction workers. This approach enables achieving consensus on the initiatives to be implemented. Harvard Business Review (2013) stated that research shows that skilled construction workers in good physical, mental, and emotional health are more likely to deliver the best results in the workplace than skilled construction workers who are not. Also, such initiatives provide interventions to address health risks and manifest diseases and promote healthy lifestyles (Mattke *et al.* 2014). Corporations embrace employee wellness, especially in developed countries, because of this need for improved performance.

2 REVIEW OF RELATED LITERATURE

This section concentrated on the study's theoretical framework and psycho-social well-being initiatives.

2.1 *Theoretical framework*

Maslow's motivational theory focuses on a hierarchy of needs. Maslow believes that people strive to reach their maximum level of potential. He further believes that all human behaviour is directional and that there is a reason for everything a person does (Leonard & Trusty 2015). At the same time, Maslow divides needs into lower and higher levels. Lower-level needs are physiological and security needs; higher-level needs are social (belonging), esteem, and self-actualisation needs; for the broader study, three needs have been extracted and focused on. The three types of needs all point to skilled construction workers' physical and psycho-social well-being. These needs must be provided for through the use of employee wellness initiatives. Hence, the focus of the study on psycho-social well-being initiatives is a positive stride towards ensuring such needs are achieved. Moyo *et al.* (2022) believe that regardless of the size or type of workplace you manage, instigating a health and well-being initiative leads to significant improvements for the employer and skilled construction workers. In 1929, Herbert William Heinrich developed the Domino theory of accident causation, which used dominos in the sequences. The Domino theory is widely used by authorities to control the aspects causing undesirable events and to prevent errors before being made (Sabet *et al.* 2014). Applying the Domino theory to psycho-social well-being, the initiatives eliminate just one of the first four dominos, implying that the dominos will not complete the sequenced fall and resultantly avoid the challenges affecting the employees. Leonard & Trusty (2015) also highlight that it is vital for organisations to include medical plans as benefits for skilled construction workers to fulfil the welfare of the skilled construction workers. Due to the absence of such plans (Moyo *et al.* 2019), wellness initiatives are necessary. In support, Chigara & Smallwood (2019) pointed out poor H & S performance in the Zimbabwean construction industry, which needs to be addressed.

2.2 *Psycho-social well-being initiatives*

Construction workers have not been spared from the adverse psychological effects of their work. The World Health Organization (WHO) (2022) penned the guidelines on mental health at work, which state that Governments and employers have the responsibility of upholding the right to access, participate and thrive in work by providing work that seeks to simultaneously prevent skilled construction workers from experiencing unwarranted stress and mental health risks; protects and promotes workers' mental health and well-being; and supports people to fully and effectively take part in the workforce, free from stigma, discrimination or abuse. Rouhanizadeh & Kermanshachi (2021) identified the top-three factors leading to mental health issues: work pressure, emotional and physical demands, and bullying and harassment. However, mental health is more than the absence of mental health conditions. Instead, mental well-being allows people to cope with the stresses of life, realise their abilities, learn well and work well, and contribute to their communities (WHO 2022). Nwaogu & Chan (2021) support introducing wellness programmes for blue-collar workers to respond to economic hardships in developing countries. In an endeavour to address issues that affect skilled construction workers' mental well-being, several initiatives have been introduced. However, this study only considered a limited number of cost-effective initiatives that suit the economic stature of a developing country like Zimbabwe. The initiatives include peer support, social support, team-building activities, counselling, and helplines.

Peer support is an increasingly important trend in population-level mental health initiatives, supplementing mental health professional therapies (Mezzina *et al.* 2019). According to Palaniappan *et al.* (2022), "Peer" refers to a person at the same level in the occupational hierarchy and shares the social and demographic context. "Support" means providing emotional motivation and encouragement through everyday experiences and a sense of empathy. "Peer support" is a form of emotional and social support provided by people who have gone through similar experiences of mental distress and have recovered from the same (Rosenberg & Argentzell 2018). It has been argued that peers can better recognise issues associated with mental illnesses, such as oppression and social isolation (Le Boutillier *et al.* 2011). Palaniappan *et al.* (2022) also pointed out that the method of promoting peer support among migrant workers was practicable and effective in improving the mental health of migrant workers, despite these affirmations. The lack of such initiatives may indicate unstructured, informal initiatives that may cause more harm if not regulated.

Social support, as an initiative comes in various forms and from many different people in one's life. Co-workers may grant support in the workplace, while at the same time, friends and family can provide emotional or practical support in other facets of life. Gros *et al.* (2016) reported that having robust social support in times of crisis can aid in the reduction of the consequences of trauma-induced disorders. Pidd *et al.* (2012) report that high levels of general social support, workplace social support, work engagement, and communication skills were moderately associated with low levels of psychological distress. On the other hand, low levels of workplace social support were significantly associated with risky alcohol use and the frequency of cannabis use.

Team building has also proven to be of value as a psycho-social initiative. According to Wilemon & Samhain (1983), team building is the process of taking a collection of individuals with different needs, backgrounds and expertise and transforming them by various methods into an integrated, effective work unit. If adequately implemented, team building activities have got the positive effect of creating relationships between skilled construction workers where they open up to each other and also lead to a reduction in the occurrences of such adverse incidences of workplace bullying and discrimination, which in most cases lead to skilled construction workers suffering going through stress. Team building is a vital cog in the success of projects, as alluded to by Adeel *et al.* (2018), who conveyed that effective project managers ensure that they select and utilise competent teams to ensure successful projects. It is also essential to state that team-building should be implemented across all designations. A sense of one-ness is usually cultivated during team building activities, breaking down the inferiority complex that exists mainly in minority groups.

Workplace counselling services for workers experiencing psycho-social challenges can also be of assistance. Counselling permits people to speak out about what is in their minds without feeling judged. A study by Chan et al. (2020) and Bajorek & Bevan (2020) established that some companies afford their employees' confidential counselling, access to life coaches and 24/7 on-call advisors. The counsellors are equipped with skills to ease the burden the affected individuals will carry. With technological advancement, the Construction Industry Helpline created a free mobile application available on the google play store, allowing individuals to self-assess and providing links to further support if required (Chan et al. 2020). One can use such a facility without divulging his or her identity, thereby eliminating the fear of stigmatisation occurring during face-to-face counselling sessions.

Research has also established that some initiatives might require little financial commitment from the employer yet have a long-reaching impact on the employees, and these can work well in developing countries where employees have financial challenges. One of the most vital aspects of keeping the stress levels of the employees on the lower side is by making sure that their wages are paid well on time, as supported by Rani et al. (2022), who believed that late payments harm employee morale, bring workplace conflict, and also tend to force employees to pursue alternative jobs. This view is also supported by Mollo & Emuze (2020), who cited addressing psycho-social risk factors associated with construction work, such as long hours and work-life imbalance, as critical in supporting the well-being of workers. Lingard et al. (2010) believed that, in most cases, construction professionals are required to work unusual work schedules, principally on weekends, resulting in work–family conflicts and job-related psychological damages.

3 RESEARCH METHODS

This study adopted a quantitative research design method in evaluating the Psycho-social well-being initiatives required for construction workers in Zimbabwe. The participants were drawn from the project/ site managers from construction companies and skilled construction workers on construction sites. Only construction companies affiliated with the Construction Industry Federation of Zimbabwe under the A, B, and C categories were considered. These project/ site managers were selected because of the strategic nature of the information required by this survey. The choice of these categories was mainly based on the fact that they have the most structured organisations. The construction sites were identified through the National Social Security Authority records.

The Taro Yamane formula was used to calculate the sample sizes for the management and the skilled workers. From the population of 113 construction companies under Construction Industry Federation of Zimbabwe (CIFOZ) A, B and C categories, a sample size of 88 was obtained. In contrast, a sample size of 112 was obtained for the skilled workers from the 5 contractors who had a population of 156 workers. Simple random sampling was then used for selecting respondents for the managers, which involved randomly selecting managers from the population without any bias. For the skilled workers, census sampling was used whereby all skilled workers were identified and approached to participate.

The study made use of questionnaires as a tool for data collection. Two types of questionnaires were used, which are online and self-administered structured questionnaires and interviewer-administered questionnaires. Online self-administered structured questionnaires were used to collect information from the management personnel of the construction companies. Google Forms were used for creating the online questionnaire, which was then distributed electronically using several platforms, the main ones being Email, WhatsApp and LinkedIn. Saunders et al. (2016) stated that the main advantage of online questionnaires is that with the advancement in technology, surveys can now be sent online, and it can be effortless to access the information gathered is available to the researcher in real-time. Since the interviewer was not involved during the completion of the questionnaires, there was the assurance of not having interviewer error or bias occurring. The interviewer-administered structured questionnaires were used for gathering information from the skilled construction workers. According to Leedy & Ormrod (2016), structured questionnaires enable the respondents to stick to the predetermined response range avoiding wild guesses, thereby enhancing data validity and reliability, especially when one is dealing with members of the public.

Data analysis was carried out using the statistical package for social sciences (SPSS) through descriptive and inferential statistics. The relative importance index (RII) assessed importance with: 'not important' < 0.2; $0.2<$ 'of little importance' ≤ 0.4; $0.4<$ 'somewhat important' ≤ 0.6; $0.6<$ 'important' ≤ 0.8; $0.8<$ 'very important' ≤ 1 (Moyo & Chigara 2022). Cronbach's alpha coefficient was used to test the reliability of the results, which is the most common indicator used for evaluating the data's internal consistency to reflect the scale's homogeneity (Field 2014). Cronbach's coefficient alpha is acceptable when its value equals or exceeds 0.7. The value of Cronbach's alpha for the five items inferred was 0.787, meaning the values are higher than the minimum standard. Generally, this means the data is reliable. This study used the Kolmogorov-Smirnov (KS) test to test the hypothesis that the variables are normally distributed. This test is used to assess whether a variable follows a given type of distribution, making it worthwhile to test normality and other types of distributions (Razali & Wah 2011). For this study, however, the KS test was used to test whether variables follow a normal distribution with a p-value of more than 0.05. For the confidence level of 0.05, any significant value greater than 0.05 indicates that the sample distribution is normal. The t-test was used to test the null hypothesis that the difference between the means of the insights of managers and workers is zero (Field 2014). In addition, where $p \leq 0.05$, the means are significantly different.

4 RESULTS AND DISCUSSIONS

4.1 *Response rate and profile of respondents*

Eighty-nine questionnaires were distributed to randomly selected managers, mainly electronically and completed using Google Forms. This was after a successful pilot test with 5 questionnaires. Out of the 89 questionnaires for the management, 51 were successfully completed representing a response rate of 57.3%. From the sample of 112 skilled workers from 5 construction sites, 82 completed the interviewer-administered questionnaire

Table 1. Demographic profile of respondents.

Characteristic	Description	Frequency	Percentage (%)
Gender	Male	102	76.7
	Female	31	23.3
	Total	133	100
Designation	Managers	51	38.3%
	Skilled workers	82	61.7%
Educational background	National Certificate	39	29.7
	Diploma	39	29.7
	Bachelor's degree	36	27.1
	Post Grad	18	13.5
	Total	133	100
Number of years working in the construction industry	0–5 years	30	22.5
	6–10 years	12	9
	Above 10 years	91	68.5
	Total	133	100

translating to a response rate of 73.2%. Using 51 respondents for management and 82 for skilled workers, a total of 133 respondents was obtained. The mean response rate of 56% is above the expected minimum, as highlighted by Chigara & Moyo (2021), who said that the response rate for construction survey questionnaires ranges between 20 to 30%.

The demographic analysis shown in the table above shows that the majority of the respondents, 76.7% were males. In comparison, 23.3% were females cementing the view that the construction industry is male-dominated, as highlighted by Zitzman (2018). Regarding educational qualifications, the analysis shows that 13.5% of the respondents had post-graduate qualifications, followed by 27.1% with an honours degree, while national certificate and diploma holders were 29.7%. Regarding the number of years in the construction industry, showed that 68.5 had more than 10 years of experience, followed by those between 0 and 5 years at 22.5%, with the minor being between 6 and 10 years at 9%. Generally, all demographic variables were appropriately represented, which validates the study.

4.2 Psycho-social well-being initiatives required for construction workers in Zimbabwe

Table 2 below indicates the various interventions or initiatives that can be implemented in the construction sector for skilled workers regarding percentage responses on a scale of 1 (not important) to 5 (very important).

Table 2. Psycho-social well-being initiatives required for construction workers in Zimbabwe.

Initiative	Skilled workers Mean Score	Evaluation	Rank	Project/site managers Mean Score	Evaluation	Rank	T-Test of means
Counselling	0.946	Very Important	1	0.896	Very important	1	0.058
Team Building Activities	0.888	Very important	2	0.876	Very important	2	0.099
Social Support	0.872	Very important	3	0.862	Very important	3	0.052
Peer Support	0.852	Very important	4	0.830	Very important	4	0.070
Help Line (Anonymous)	0.744	Important	5	0.754	Important	5	0.050

The psycho-social employee wellness initiatives have all been found important for employee wellness at work, the most being counselling, with an RII of 0.946 for skilled construction workers and 0.896 for managers. The least important were helplines by managers and skilled construction workers, but they were still important. In broad terms, all the initiatives were important to very important, with $0.6 < $ 'important' ≤ 0.8; $0.8 < $ 'very important' ≤ 1 for both skilled construction workers and managers. From the analysis in Table 2 above, we also find out that the t-test of means has values greater than 0.05, which is the critical value. This means that the difference between the skilled construction workers' and managers' responses is not significant enough for us to reject the null hypothesis of consensus. The critical initiatives are discussed hereafter.

The results suggest that counselling was the primary initiative perceived as an intervention to the psycho-social challenges. The popularity of counselling may mainly be attributed to the fact that counsellors are well-equipped with proper training to identify the challenges individuals face. Workplace counselling can be an internal or external service to help skilled construction workers with any mental health or related issues that have arisen from, or are worsened by, work (Bajorek & Bevan 2020). The CITB (2021) outline the varying facets of counselling, pointing out that it can cater for those with challenges in financial, drug and alcohol, family and relationship matters. Also, they have countermeasures which help calm down the affected skilled construction workers, thereby reducing the effects.

The second rated initiative that can be implemented to reduce psycho-social problems among skilled construction workers is team building activities. As alluded to by Adeel *et al.* (2018), team-building activities positively affect workers' performance. Furthermore, if adequately implemented, team building activities have the positive effect of creating relationships between workers where they open up to each other and also reduce the occurrences of such adverse incidences of workplace bullying and discrimination, which in most cases lead to stress. Hence, team-building activities must be monitored and appraised if they are going to attain their objectives.

Another program that was found to be paramount to skilled construction workers was social support. Social support comes in various forms and also from many different people in one's life. For example, co-workers may grant support in the workplace, while at the same time, friends and family can provide emotional or practical support in other facets of life. This finding is in line with Gros *et al.* (2016), whose research also showed that having robust social support in times of crisis can aid in the reduction of the consequences of trauma-induced disorders. In support of this initiative, Pidd *et al.* (2012) reported that high levels of general social support, workplace social support, work engagement, and communication skills were moderately associated with low levels of psychological distress. Research also constantly identifies social support in the environment from sources such as one's supervisor or line manager as an important determining factor of employees' well-being (Bradley *et al.* 2009). Such support from other scholars strengthens the significance of social support in addressing psycho-social challenges.

Peer support emerged as the fourth-rated initiative that can also be useful in resolving psycho-social problems skilled construction workers face. Rosenberg & Argentzell (2018) define Peer support as a form of emotional and social support provided by people who have gone through similar experiences of mental distress and have recovered from the same. This has proven to be effective mainly because the peer assisting the affected one will have gone through similar experiences, and this experience enables peers to recognise better issues associated with mental illnesses, such as oppression and social isolation (Le Boutillier *et al.* 2011). Kurtz (1990) pointed out that the recipients of peer support identify well with the givers, thereby contributing to self-efficacy in handling and problem-solving skills. This is a practical and enriching initiative which works in instances where relationships can be developed over time.

This study also showed that the anonymous helpline is considered by both the management and skilled construction workers as a valuable tool in addressing the psycho-social challenges being encountered by skilled construction workers. The main reason why this program is acceptable and effective among skilled construction workers is because of the element of anonymity that it affords the skilled construction workers. One can use the helpline without divulging his or her identity, thereby eliminating the fear of stigmatisation, usually during face-to-face counselling sessions. This aligns with a study by Pirkis *et al.* (2016).

5 CONCLUSION AND RECOMMENDATIONS

The prevalence of economic hardships and the inadequacy of health and safety provisions necessitates the implementation of psycho-social wellness initiatives for skilled construction workers in Zimbabwe. Notably, the impact of COVID-19 on the construction sector was felt in the well-being side of the skilled construction workers, mainly on the psycho-social side whereby some people ended up in depression and stress after losing incomes during the lock-down period. Also, the various illness challenges had a psychological impact. This study sought to determine the psycho-social initiatives that can be implemented in Zimbabwe to solve such challenges. Both managers and skilled construction workers supported the importance, in ranking order, of five main initiatives; Counselling, Team Building Activities, Social Support, Peer Support and Anonymous Help Line. These initiatives must be implemented in close consultation with skilled construction workers and ensure they are of high quality through employing or contracting counsellors and team-building professionals that renowned bodies or institutions accredit. While it is noble to come up with these initiatives to alleviate the challenges being faced by skilled construction workers, this study recommendsbuilding a caring culture that embraces well-being by incorporating mental health awareness, health, wellness, and employee/labour relations, is essential. Screening of prospective skilled construction workers at the recruitment stage must be taken seriously to ensure those with such conditions are assisted. Further studies can investigate how these initiatives can be structured and implemented. The study's main limitation was that participants were not drawn from all the construction company categories to ascertain the full extent of the challenge.

REFERENCES

Adeel, A., Batool, S. & Ali, R. 2018. Empowering Leadership and Team Creativity: Understanding the Direct-indirect Path. *Verslas: Teorija ir Praktika/Business: Theory and Practice*, 19, 242–254.
Bajorek, Z. & Bevan, S. 2020. *Demonstrating the Effectiveness of Workplace Counselling Reviewing the Evidence for Well-being and Cost-effectiveness Outcomes*. Ahead of print. Emerald Publishing.
Chan, A. P., Nwaogu, J. M. & Naslund, J. A. 2020. Mental Ill-health Risk Factors in the Construction Industry: Systematic Review. *Journal of Construction Engineering and Management*, 146(3), 04020004.
Chigara, B. & Moyo, T. 2014a. Factors Affecting Labour Productivity on Building Projects in Zimbabwe. *International Journal of Architecture, Engineering and Construction*, 3(1), 57–65.
Chigara, B. & Moyo, T. 2021. The Impact of Covid-19 on the Construction Sector in Zimbabwe. *Journal of Engineering, Design and Technology*, 20(1), 24–46.
Chigara, B. and Moyo, T. 2014b. Overview of the Operational and Regulatory Framework for Occupational Safety and Health Management in Zimbabwe's Construction Industry. In: Emuze, F.A. & Aigbavbao, C. (Eds.). *Proceedings of TG59 'People in Construction' Conference*, 6–8 April 2014, Port Elizabeth, South Africa, pp. 150–159.
Chipato, E., Chigara, and Smallwood, J. 2019. *Health And Safety Practices in the Zimbabwean Construction Industry*. (Doctoral dissertation, Nelson Mandela University).
Chirazeni, E. & Chigonda, T. 2018. Occupational Safety and Health Challenges in Construction Industries: Case of Masvingo, Zimbabwe. *World Journal of Healthcare Research*, 2(6), 16–20.
CONSTRUCTION INDUSTRY TRAINING BOARD. 2021. *Mental Health and Well-being Research – Final Report*
Field, A. 2014. *Discovering Statistics Using IBM SPSS Statistics*. 4th Ed. Los Angeles: Sage Publications.
Gros D.F., Flanagan J.C., Korte K.J., Mills A.C., Brady K.T. & Back S.E. 2016 Relations Among Social Support, PTSD Symptoms, and Substance use in Veterans. *Psychological Addict Behaviour*. 2016;30 (7):764–770.
Harvard Business Review. 2013. *The Radical Innovation Playbook* Available at: https://store.hbr.org/product/harvard-business-review-october-2013/BR1310. Accessed on 23 March 2023
Khan, S. B., Proverbs, D. G. & Xiao, H. 2022. The Motivation of Operatives in Small Construction Firms Towards Health and Safety–A Conceptual Framework. *Engineering, Construction and Architectural Management*, 29(1), 245–261.
Kurtz, L.F. 1990. "The Self-help Movement: Review of the Past Decade of Research", *Social Work with Groups*, Vol. 13 No. 3, pp. 101–115, doi: 10.1300/J009v13n03_11.
Le Boutillier, C., Chevalier, A., Lawrence, L., Leamy, M., Bird, V. J., Macpherson, R., Williams, J & Slade, M. (2015). Staff Understanding of Recovery-orientated Mental Health Practice: A Systematic Review and Narrative Synthesis. *Implementation Science*, 10, 1–14.
Leedy, P. D. & Ormrod, J. E. 2016. *Practical Research: Planning and Design*. 11th ed. Boston: Pearson
Leonard, E.C. & Trusty, K.A. 2015. *Supervision: Concepts and Practices of Management*. Cengage Learning.

Liang, H. & Shi, X. 2022. Exploring the Structure and Emerging Trends of Construction Health Management: A Bibliometric Review and Content Analysis. *Engineering, Construction and Architectural Management*, 29(4), 1861–1889.

Lingard, H., Francis, V. & Turner, M. 2010. The Rhythms of Project Life: A Longitudinal Analysis of Work Hours and Work-life Experiences in Construction. *Constr. Manag. Econ.* 2010, 28, 1085–1098.

Mattke, S., Kapinos, K., Caloyeras, J. P., Taylor, E. A., Batorsky, B., Liu, H. & Newberry, S. 2015. Workplace Wellness Programs: Services Offered, Participation, and Incentives. *Rand Health Quarterly*, 5(2).

Mezzina, R., Rosen, A., Amering, M. & Javed, A. 2019. The Practice of Freedom: Human Rights and the Global Mental Health Agenda. *Advances in Psychiatry*, 483–515.

Mollo, L.G. & Emuze, F. 2020. The Well-Being of People in Construction. W.Leal Filho et al. (eds.), *Good Health and Well-Being*, https://doi.org/10.1007/978-3-319-69627-0_123-1

Moyo, T. and Chigara, B. 2022. Factors Affecting the Competence of Quantity Surveying Professionals in Zimbabwe: *Journal of Construction for Developing Countries*. Early view

Moyo, T., Crafford, G. & Emuze, F. 2019. Assessing Strategies for Improving the Social Security of Construction Workers in Zimbabwe: *Journal of Construction*, 12(1): 4–13

Moyo, T., Crafford, G. & Emuze, F. 2022. Significant Decent Work Objectives for Monitoring Construction Workers' Productivity Performance in Zimbabwe: *Journal of Construction for Developing Countries*, 27(1): 95–110.

National Wellness Institute. 2023. *Defining Wellness*. Available at: https://nationalwellness.org/resources/six-dimensions-of-wellness/ Accessed 01 April 2023.

Nwaogu, J. M., & Chan, A. P. 2020. 'Evaluation of Multi-level Intervention Strategies for a Psycho-logically Healthy Construction Workplace in Nigeria', *Journal of Engineering, Design and Technology*, Vol. 19(2), pp. 509–536.

Palaniappan, K., Rajaraman, N., & Ghosh, S. 2022. Effectiveness of Peer Support to Reduce Depression, Anxiety and Stress Among Migrant Construction Workers in Singapore. *Engineering, Construction and Architectural Management* (ahead-of-print).

Pidd, K., Duraisingam, V., Roche, A. & Trifonoff, A. 2017. Young Construction Workers: Sub-stance Use, Mental Health, and Workplace Psycho-social Factors. *Advances in Dual Diagnosis*.

Pirkis, J., Middleton, A., Bassilios, B., Harris, M., Spittal, M. J., Fedszyn, I. & Gunn, J. 2016. Frequent Callers to Telephone Helplines: New Evidence and a New Service Model. *International Journal of Mental Health Systems*, 10(1), 1–9.

Rani, H.A., Radzi, A.R., Alias, A.R.; Almutairi, S. & Rahman, R.A. 2022. *Factors Affecting Workplace Well-Being: Building Construction Projects. Buildings* 2022, 12, 910. https://doi.org/10.3390/ build-ings12070910.

Razali, N. M. & Wah, Y.B. 2011. 'Power Comparisons of Shapiro-Wilk, Kolmogorov-Smirnov, Lilliefors and Anderson-Daring Tests', *Journal of Statistical Modelling and Analytics*, Vol. 2(1), pp. 23–33.

Rosenberg, D., & Argentzell, E. 2018. Service Users Experience of Peer Support in Swedish Mental Health Care: A J. Psychosoc. Rehabil. Ment. Health 123 "Tipping Point" in the Care-giving Culture? *Journal of Psycho-social Rehabilitation and Mental Health*, 5(1),53–61.

Rouhanizadeh, B., & Kermanshachi, S. 2021. Causes of the Mental Health Challenges in Construction Workers and Their Impact on Labour Productivity. In *Tran-SET 2021* (pp. 16–26). Reston, VA: American Society of Civil Engineers.

Saunders, M., Lewis, P. & Thornhill, A. 2016. *Research Methods for Business Students*. 7th ed. Har-low: Pearson Education Limited.

Wilemon, D. L., and Thamhain, H. J. 1983. *Team Building in Project Management*. Project Management Institute.

Implementing construction risk management methods on private sector projects in South Africa

K. Kajimo-Shakantu & H. Nengovhela
Department of Quantity Surveying and Construction Management, University of the Free State, South Africa

F. Muleya
Department of Quantity Surveying and Construction Management, University of the Free State, South Africa
Copperbelt University, School of the Built Environment, Kitwe, Zambia

ABSTRACT: This study aimed to examine ways in which the implementation of risk management methods is carried out in the private sector. A study was undertaken involving private sector projects using a questionnaire survey among 155 randomly selected respondents in construction in the Gauteng province of South Africa representing a response rate of 34%. SPSS 21 and Microsoft excel were used to analyse data and to present descriptive statistics. Key results of the study indicate that training of the project team was most critical in implementing risk management systems effectively. Collaborative and proactive approaches, decentralised risk management system, regular review of risks, regular updates of risk register, were also found to be fundamental in successful implementation of risk management. The study recommends auditing of risk management policies and strategies, change in the organisational culture and attitude towards risk management.

Keywords: Barriers, private sector, construction risk management, implementation, improvement.

1 INTRODUCTION

This study focuses on the implementation of risk management methods in South Africa. Risk management is a process utilised to identify, analyse and respond to risks over the lifecycle of projects in order to mitigate, eliminate or control risk (Tembo 2017). The construction industry is one of the major global sectors with developing countries accounting for most risk management failures, as a result of disintegrated methods of risk management (Roslan 2015; Sharma and Goyal 2015). In the construction industry, risks have a notable impact on projects outcomes when it comes to the time (duration), cost and quality parameters of the project (Gbahabo & Ajuwon 2017; Osipova 2008; Tembo 2017) highlights that failed risk management in Southern Africa has been due to lack of knowledge and lack of skills. According to Tembo (2017) poor implementation of risk management methods remains a key factor in the failure of risk management procedures. The governments' method of managing projects has led to many project failures (Selepe 2019). Poor implementation of project risk management methods on South African government projects is said to have been caused by ignoring the need to develop core support teams that will assist in ensuring that methods are effectively implemented (Selepe 2019). A study on small projects conducted by Fischer (2015) revealed that ineffective implementation of project risk management methods was caused by lack of skills and knowledge to manage risks. This study focussed on how project risk management is implemented in construction could be improved in the Gauteng Province private sector construction projects in South Africa.

2 LITERATURE REVIEW

2.1 Construction risk management

Risk in a project context is defined as a set of circumstances or an uncertain event, that if it occurs, one or more objectives of the project will be affected (Baker et al. 2020). The exposure to a risk in a project must be managed (Baker et al. 2020). Risk management in the construction industry is outlined to plan, monitor and control the actions which are needed to prevent a project from getting exposed to risks (Morphy 2020). Risk management assists project stakeholders to meet the project's requirements and to achieve quality, time and cost objectives (Banaitiene & Banaitis 2012). Risk management is one of the most difficult aspects to handle on construction projects and therefore a project manager should study the causes of risks, effect of the risk, mitigation or elimination of the risk and the consequences of the risk on the project. (Roseke 2018). Roseke (2018) states that PMBOK (Project Management Body of Knowledge) has seven processes to risk management. The processes are: Planning risk management, identifying the risks, performing a qualitative risk analysis, performing a quantitative risk analysis, planning risk responses, implementing risk responses and monitoring risks. The risk management process has a crucial process when assessing risk, which is rating the risks (Gul et al. 2018). According to Babut et al. (2011), there are various risk rating methods. The Fine-Kinney method developed by Kinney and Wiruth in 1976 is based on three parameters, namely, the probability of a risk occurring, the frequency of being exposed to a risk and the consequence of an occurred risk (Babut et al. 2011). The Fine-Kinney method is the most used method in the world to rate risk (Babut et al. 2011). The European Commission (2011) asserts that some of the risks which can cause project failure due to poor project management are: unclear or impossible project objectives, Poor definition of the project scope, Poor estimation from the cost management team, Contractual problems, Insurance problems, Delays, Quality concerns, The usage if incomplete data to budget; and Lack of time to test.

2.2 Risk allocation

Risk allocation is defined as assigning management responsibility and assigning the liability of risks (Tembo 2017). Risk allocation should be conducted in the presence of all relevant stakeholders in order to take all their needs into consideration (Peckiene et al. 2013). Assigning risks to the wrong project stakeholders and inappropriately assigning risks to stakeholder results in risk misallocation (Tembo 2017). Risk misallocation is the act of assigning risk to a certain part without considering which party may be in the best position to evaluate, control and endure the cost or benefit from the risk should it occur (Hanna et al. 2013). When allocating risk, Baker et al. (2020) proposed that the following guidelines be considered by the parties involved in a project: Which party is in the best position to control the risk with its associated consequences? Which party is in the best position to foresee the risk? Which party is in the best position to bear the risk and Which party is in the best position to benefit or suffer the most when the risk occurs? Baker et al. (2020) further stated that in order to achieve a fair and equitable risk allocation process, a risk should be assigned to a party if the party is in control of the risk, in order to ensure efficiency.

A study by Tembo (2017) is consistent with the rationale used above to allocate risks. In order to allocate risks, one must know if the person appointed is best suited to be liable to manage the risk (Tembo 2017). The use of a suitable construction contract with altered clauses forms a useful tool to manage project risks through effective allocation of risks (Baker et al. 2020; Tembo 2017). Some risks can be allocated using certain construction contracts and the type of procurement plays an important role in how risks will be allocated. The types of procurement are explained below by Baker et al. (2020): In Traditional procurement, the contract allows for the contractor to "construct only" as the designs and drawings will be done prior to the appointment of the contractor. Between the contractor and the employer, the employer bares the risk for any design defects. In Design and build, the contractor is responsible for the design and the construction. It is the contractor's risk to ensure that the built product matches the requirements of the employer. The employer holds power to control how things are executed in this type of procurement. The employer may require a project management team to overlook the project (ACMS Group 2019). Lastly in Engineering, procurement and construction (EPC) or Turnkey, the EPC contracts allows the contractor to have full discretion on the design if the designs meet the functional requirements of the employer. The employer does not have to worry about the project management of the project as the contractor appointed is responsible for that (ACMS Group 2019). Tembo (2017) states that the risk is more for the client than the contractor on the construction management contract compared to the design and build. In the design and build contract, the greater risk is on contractor than the client. This shows that there is an agreement between Baker *et al.* (2020) and Tembo (2017). However, the construction industry prefers the traditional procurement which Choudhry & Iqbal (2013) found to be containing the following as barriers to risk management:

1. There is a lack of formal risk management systems.
2. There is a lack of collaboration in the risk management system by parties.
3. Insufficient knowledge and techniques.
4. The complexity of risk management.
5. A tendency to be reactive than be proactive.
6. Risk management which is centralised rather than decentralised.
7. Risk analysis is done instead of risk identification.
8. Risk management performed periodically and not continuously.
9. Inadequate historical data in order to perform risk trend analysis.
10. A lack in risk awareness.

Tembo (2017) further agrees with Choudhry & Iqbal (2013) that a sophisticated risk management process is a barrier to the success of risk management. The lack of knowledge of techniques is also agreed to by Tembo (2017) that it acts as barrier to successful risk management. Shunmugan & Rwelamila (2014) further indicate that the barriers to effective and efficient risk management are caused by the lack of subjectivity of the current methods. Makombo (2011) established that there is a skills gap among the professionals dealing with project risk related issues, poor scope management and minor focus given to risk management in the initiation phases of a project in South Africa. A South African study conducted by Chihuri & Pretorius (2010) found that the risk management methods were not widely used in construction and engineering projects, there is a lack of adaptation of risk management policies and a lack of implementation of the risk management policies.

3 RESEARCH METHODOLOGY

This study adopted an objective ontological perspective. This ontological viewpoint assists the researcher to easily compare data and have control over the research process (Al-Saadi 2014; Ejnavarzala 2019). This study had a non-experimental fixed design strategy (Babbie 2010). The study used a quantitative survey which utilised a structured questionnaire. This approach is useful in analysing data collected in a numerical manner. The target population of this study carried out in the second half of 2021 included Gauteng Province construction stakeholders responsible for managing projects in private sector construction projects. The target population included construction firms, consulting firms and developers. This is because much work been carried out in the public sector compared to the private sector. A total of 460 questionnaires were sent out and 155 were received. The represented a response rate of 34%. Random sampling method was used. Random sampling ensures that all the participants in the sample will have an equal chance of being selected (Shadish *et al.* 2002; Walonick 2010). Despite the response rate being 34%, the number of respondents was still high enough to obtain credible data required to provide valid results. This response rate is acceptable is supported by Cycyota & Harrison (2006)

A five-point Likert scale was used on the questionnaire. A Likert scale is a scale used to get respondents to choose the option that aligns with their views. It is often used to measure the attitudes of the respondents to find out the extent to which they disagree or agree with a particular statement (Losby & Wetmore 2012). The data was analysed statistically so that it could be quantified. Data was manually coded and presented in charts and tables produced on Microsoft Excel. Descriptive statistics was deployed to further analyse the data received. SPSS 21 was used generate mean scores. Ethical consideration comprised the researcher's conduct during the research process. One of the reasons for an ethical approach is to protect participants from the negative consequences, harm or potential harm of the research (Simushi 2017). The responses were handled with extreme confidentiality with identification details of the respondents never to be revealed or disclosed. The identity of the respondents was treated with respect and dignity with the views of the respondents being portrayed as the respondents intended them to be.

4 RESULTS PRESENTATION AND ANALYSIS

4.1 *Demographic profile*

The questionnaire was sent out in the form of electronic links. The responses received were one hundred and fifty-five (155) from four hundred and sixty (460) questionnaires sent out, which gives a response rate of thirty-four percent (34%). As stipulated in the sample size determination, a minimum of one hundred and thirty-eight (138) respondents were required to achieve a ninety percent (90%) confidence level and a margin of error of seven percent (7%). Given that the respondents went beyond the minimum requirements, a calculation based on the Krejcie & Morgan (1970) formula showed that the confidence level was raised to 91.8%. Of the respondents, the majority (78.3%) were male and the remaining 21.7% were female. The majority of respondents (one hundred and forty-seven, 94.8%) worked for the private sector and eight (5.2%) respondents worked in the public sector where they have been seconded by private sector firms. This allowed them to work on private sector projects. Among the respondents, thirty-two (20.6%) worked for consulting firms, one hundred and seven (69%) worked for construction or engineering firms, nine (5.8%) worked for manufacturing firms and seven (4.5%) worked for developers or real estate firm.

The professionals that responded in this study are indicated in Table 1.

Table 1. Grouping of respondents by profession.

Profession	Quantity	Percentage
Quantity Surveyor	31	20.00%
Civil Engineer	49	31.60%
Mechanical Engineer	3	1.90%
Electrical Engineer	12	7.70%
Structural engineer	1	0.60%
Contracts Manager	4	2.60%
Project Manager	13	8.40%
Construction Manager	12	7.70%
Architect	4	2.60%
Real Estate Developer	2	1.30%
Foreman	4	2.60%

(*continued*)

Table 1. Continued

Profession	Quantity	Percentage
Site Agent	11	7.10%
Civil Engineering Technologist	1	0.60%
Safety Manager	1	0.60%
Construction Health and Safety Officer	1	0.60%
Metallurgical Engineer	2	1.30%
Senior building inspector	1	0.60%
Design and Draughtsman	1	0.60%
Project Planner	1	0.60%
Industrial Engineer	1	0.60%
Total	**155**	**100%**

4.2 Key findings

This section presents the key results of the study. It is important to note that where a five-point Likert scale was used, the following codes were used:
1 – Strongly disagree; 2 – Disagree; 3 – Neutral; 4 – Agree; 5 – Strongly agree.

Where the mean is from 1 to 1.8, this symbolises strongly disagree, from 1.81 to 2.60 it symbolises disagree, from 2.61 to 3.40 it symbolises neutrality, from 3.41 to 4.20 it symbolises agree and 4.21 to 5 symbolises strongly agree (Benhima 2020).

4.3 Training received

Majority of the respondents (one hundred and thirteen (72.9%)) had not done any additional risk management training apart from the qualifications they hold which cover risk management. Of the 72.9% of respondents who had no relevant risk management training, they indicated that experience in the construction industry works well as a tool to manage risks. On the other hand, forty-two (27.1%) had completed their qualifications and further undertook courses or training which deal directly with project risk management. The courses or training done among the 27.1% was found to be as follows:

Figure 1 is a representation of the adequacy of the training or education obtained by the respondents to manage project risks. Further, of those that had received training, only 11.9% had received specific training in risk management.

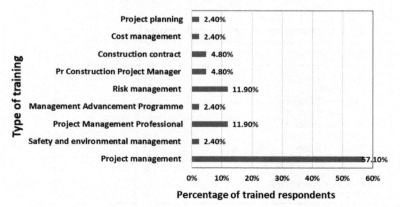

Figure 1. Training received by respondents.

4.4 Improvement to the implementation if risk management methods

Table 2 illustrates the ranked strategies which can be adopted to ensure improvement of the implementation of risk management methods. Table 2 shows that training, risk reviews, creation of risk register and reporting project risks were identified and ranked as the most important with regards to implementation of risk management in the private sector of the construction industry. This is supported by Al-Keim (2017; Olaru et al. (2014); Jongo et al. (2019) and Roseke (2018) who states that training is essential for project risk management personnel

Table 2. Improvements to the implementation of risk management methods.

Improvement to the implementation of risk management methods	Quantity	Percentage	Rank
Train the project team on project risk management	143	92.30%	1
Regularly review risks	125	80.60%	2
Create a project risk register	112	72.30%	3
Report on project risks	106	68.40%	4
Determine the likelihood and impact of risks	97	62.60%	5
Determine the response to the risk if it occurs	97	62.60%	6
Determine the cost to address the risks	79	51.00%	7

Overall, the results are in agreement with Rwelamila (2017) that effective implementation of risk management systems is pivotal to deliver successful projects. It is traditionally known that the success of a construction project depends on quality, time and cost (Banaitiene & Banaitis 2012). With the improved implementation of risk management methods, project success is guaranteed including in the private sector. Table 3 presents results on the clarity of JBCC, FIDIC, GCC and NEC construction contracts used in Gauteng province of South Africa's private sector. Each contract was rated according to which one has the most clarity with regards to risk allocation.

Table 3. Clarity of construction contracts for risk allocation.

Clearest contract for allocation of risk and responsibilities	Quantity	Percentage	Rank
Joint Building Contracts Committee (JBCC)	86	57.0%	1
New Engineering Contract (NEC)	59	39.1%	2
General Conditions of Contract (GCC)	54	35.8%	3
International Federation of Consulting Engineers (FIDIC)	22	14.6%	4

Table 3 shows that the Joint Building Contracts Committee (JBCC) was ranked first in best clarity for risk allocation and responsibilities followed by the New Engineering Contract (NEC) and General Conditions of Contract (GCC). The International Federation of Consulting Engineers (FIDIC) was rated lowest at 14.6% because it is not widely used nationally but internationally. Within the boundaries of Gauteng Province, it is hardly or never used. The selection of the four contracts above was based on the fact that these are construction standard contracts and using them assists in having a procurement process which is more efficient and competitive (Bowman's Law 2019). These contracts are mostly used in Africa (Bowman's Law 2019).

4.5 Barriers to the implementation of risk management methods

Table 4 summarises the ranked barriers to the implementation of risk management methods. The ranking is based on the mean value. Table 4 indicates that all the 10 barriers to implementation of risk management methods had a mean score exceeding 3.7 representing significance of the barriers overall. A tendency to be reactive, lack of collaboration, insufficient knowledge and techniques, lack of risk awareness and lack of formal risk management system scored a mean score of above 4.000 signifying the critical stated barriers based on respondents. The results are supported by (Azis *et al.* 2013) that effective strategic planning and proactive risk management are some of the keys to effective risk management. Further, the results are supported by (Choudhry & Iqbal 2013) whose study emphasises that effective risk management is achieved through collaboration rather than working in silos which presents a risk to successful implementation of risk management methods

Table 4. Summarised ranking of barriers to the implementation of risk management methods.

Barriers to the implementation of risk management methods	Mean	Ranking
A tendency to be reactive than be proactive.	4.2194	1
A lack of collaboration in the risk management system by parties.	4.0516	2
Insufficient knowledge and techniques to handle risks	4.0516	2
A lack in risk awareness.	4.0452	3
A lack of formal risk management systems.	4.0194	4
Risk management performed periodically and not continuously.	3.9484	5
Risk management which is centralised rather than decentralised.	3.8516	6
The complexity of risk management.	3.7548	7
Inadequate historical data in order to perform risk trend analysis.		
Sufficiency of risk training or education obtained to manage project risks	3.7161	8

5 DISCUSSION

The study place emphasis on the subject that risk management challenges also exist in the private sector and therefore not be ignored. The traditional focus of construction risk management has mainly been on the public sector. This study has revealed the private sector in the construction industry equally requires attention and effort in attaining high standards of risk management as seen in Tables 2 and 4.

6 CONCLUSIONS AND RECOMMENDATION

Based on the findings, this study concludes that poor implementation of risk management methods leads to project failure in the private sector. The study further concludes that training with regards to risk management is needed to equip the project team on how to manage risks. A comprehensive risk management framework needs to be developed and explained to all the employees. It is crucial to have every stakeholder's participation because the success or failure of a project affects the whole company or organisation. In a risky environment such as the construction industry, it is important for employees to feel that management cares for them to be fully dedicated to their duties. This is part of a proactive strategy approach to risk management. Intensive planning needs to be done to ensure that all risk aspects are prepared for. The use of a contract that is not suitable for the type of construction project at hand can have drastic consequences when allocating risks. Implementation of risk management methods is also an important factor when drafting a risk management strategy. Lastly, the study recommends that risk management factors equally require full attention in the private sector just like the public sector which has for many years received more attention and interest.

REFERENCES

ACMS Group. 2019. *What is the Difference Between Turnkey and Design-Build Construction?* [Online]. Available at: <https://acmsgroup.com/2019/10/what-is-the-difference-between-turnkey-and-design-build-construction/> [Accessed 15 March 2021].

Al-keim, A. 2017. *Strategies to Reduce Cost Overruns and Schedule Delays in Construction Projects.* Dissertation (PhD). Minnesota: Walden University.

Al-Saadi, H. 2014. *Demystifying Ontology and Epistemology in Research Methods.* Thesis (PhD). Shefield: University of Sheffield

Azis, A.A.A., Memon, A.H., Rahman, I.A. & Karim, A.A. 2013. Controlling Cost Overrun Factors in Construction Projects in Malaysia. *Research Journal of Applied Sciences, Engineering and Technology* [online]. 5(8). Available at: <https://www.researchgate.net/publication/265966383_Controlling_Cost_Overrun_Factors_in_Construction_Projects_in_Malaysia> [Accessed 13 May 2021].

Babbie, E.R. 2010. *The Practice of Social Research.* 12th edition. Belmont, CA: Wadsworth Cengage

Babut, G., Moraru, R. and Cioca, L. 2011. "Kinney-Type Methods": Useful or Harmful Tools in the Risk Assessment and Management Process? [Online]. *Proceedings of the 5th International Conference on Manufacturing Science and Educations, Romania: International Conference on Manufacturing Science and Education,* pp. 315–318. Available at: <https://www.researchgate.net/publication/221691651_Kinney-type_methods_useful_or_harmful_tools_in_the_risk_assessment_and_management_process> [Accessed 11 September 2019].

Baker, E., Hill, R. and Hakim, I. 2020. *Allocation of Risk in Construction Contracts.* [Online]. Available at: <https://www.whitecase.com/publications/alert/allocation-risk-construction-contracts> [Accessed 15 March 2021].

Banaitiene, N. and Banaitis, A. 2012. Risk Management in Construction Projects, *Risk Management – Current Issues and Challenges.* Available from: <https://www.intechopen.com/books/risk-management-current-issues-and-challenges/risk-management-in-construction-projects> September, DOI: 10.5772/51460. [Accessed 4 March 2021].

Benhima, M. (2021). Moroccan English Department Student Attitudes Towards the Use of Distance Education During COVID-19: Moulay Ismail University as a Case Study. *International Journal of Information and Communication Technology Education (IJICTE)*, *17*(3), 105–122.

Bowman's Law. 2019. *Guide – Construction Contracts in South Africa.* [Pdf]. Available at: <https://www.bowmanslaw.com/wp-content/uploads/2016/12/Guide-Construction-Contracts-Digital.pdf> [Accessed 12 November 2021].

Chihuri, S. and Pretorius L. 2010. Managing Risk for Success in a South African Engineering and Construction Project Environment. *South African Journal of Industrial Engineering* [online], 21(2). Available at: <http://sajie.journals.ac.za/pub/article/download/50/44> [Accessed 22 May 2020].

Choudhry, R. M., & Iqbal, K. 2013. Identification of Risk Management System in Construction Industry in Pakistan. *Journal of Management in Engineering* [online], 29(1). Available at: <https://ascelibrary.org/doi/full/10.1061/%28ASCE%29ME.1943-5479.0000122> [Accessed 15 March 2021].

Cycyota, C.S. and Harrison, D.A., 2006. What (not) to expect when surveying executives: A meta-analysis of top manager response rates and techniques over time. *Organizational Research Methods*, 9(2), pp.133–160.

Ejnavarzala, H. 2019. Epistemology–Ontology Relations in Social Research: A Review. *Sociological Bulletin*, 68(1) pp. 94–104. DOI: https://doi.org/10.1177/0038022918819369.

European Commission. 2011. *FWC Sector Competitiveness Studies N° B1/ENTR/06/054 – Sustainable Competitiveness of the Construction Sector.* [Online]. Available at: <https://citeseerx.ist.psu.edu/viewdoc/download?doi=10.1.1.357.3412&rep=rep1&type=pdf> [Accessed 5 March 2021].

Fischer, R. 2015. *Barriers to Effective Risk Management on Small Construction Projects in South Africa.* Thesis (MSc). Johannesburg: University of Witwatersrand.

Gbahabo, P. T. and Ajuwon, O. S. 2017. Effects of Project Cost Overruns and Schedule Delays in Sub-Saharan Africa. *European Journal of Interdisciplinary Studies*, 3(2), pp. 46–59. Available at: <http://journals.euser.org/index.php/ejis/article/view/1842> DOI: http://dx.doi.org/10.26417/ejis.v3i2.p46-59. [Accessed 15 April 2021].

Gul, M., Guven, B. and Guneri, A.F. 2018. A New Fine-Kinney-based Risk Assessment Framework Using FAHP-FVIKOR Incorporation. *Journal of Loss Prevention in the Process Industries* [online], 53. Available at: <https://www.sciencedirect.com/science/article/abs/pii/S0950423017307489> [Accessed 11 September 2019].

Hanna, A. S., Thomas, G., & Swanson, J. R. 2013. Construction Risk Identification and Allocation: Cooperative Approach. *Journal of Construction Engineering and Management* [online], 139(9). Available at: <https://ascelibrary.org/doi/10.1061/%28ASCE%29CO.1943-7862.0000703> [Accessed 15 March 2021].

Jongo, J.S., Tesha, D.N.G.A.K., Kasonga, R., Teyanga, J.J. & Lyimo, K.S. 2019. Mitigation Measures in Dealing with Delays and Cost Overrun in Public Building Projects in Dar-Es-Salaam, Tanzania. *International Journal of Construction Engineering and Management* [online], 8(3). Available at: <http://article.sapub.org/10.5923.j.ijcem.20190803.01.html#Sec2.5> [Accessed 17 April 2021].

Krejcie, R.V. & Morgan, D.W. 1970. Determining Sample Size for Research Activities. *Educational and Psychological Measurement*. [Online]. Available at: <http://www.kenpro.org/sample-size-determination-using-krejcie-and-morgan-table/> [Accessed 3 August 2021].

Losby, J. and Wetmore, A. 2012. *CDC Coffee Break: Using Likert Scales in Evaluation Survey Work*. [pdf]. Atlanta. Available at: <https://www.cdc.gov/dhdsp/pubs/docs/cb_february_14_2012.pdf> [Accessed 5 August 2021].

Makombo, H.M. 2011. *The Risk Framework for the Organisations Dealing With Construction Project Management in South Africa*. Treatise (Masters). Pretoria: University of Pretoria.

Morphy, T. 2020. *Risk Management, Risk Analysis, Templates and Advice*. [Online]. Available at: <https://www.stakeholdermap.com/risk/risk-management-construction.html> [Accessed 04 March 2021].

Olaru, M., Sandru, M. and Pirnea, I.C. 2014. *Monte Carlo Method Application for Environmental Risks Impact Assessment in Investment Projects*. [Online]. Available at: <https://www.researchgate.net/figure/Risk-matrix-according-to-risk-occurrence-probability-and-impact_tbl2_275537561> [Accessed 5 March 2021].

Osipova, E. 2008. *Risk Management in Construction Projects: A Comparative Study of the Different Procurement Options in Sweden*. [Online]. Available at: <https://www.researchgate.net/publication/270163040_Risk_management_in_construction_projects_a_comparative_study_of_the_different_procurement_options_in_Sweden> [Accessed 11 March 2019].

Peckiene, A., Komarovska, A. & Ustinovicius, L. 2013. Overview of Risk Allocation between Construction Parties. [Online]. *Procedia Engineering*. Available from: <https://www.researchgate.net/publication/271638057_Overview_of_Risk_Allocation_between_Construction_Parties> [Accessed 10 September 2019].

Roseke, B. 2018. *Project Risk Management According to the PMBOK*. [Online]. Available at: <https://www.projectengineer.net/project-risk-management-according-to-the-pmbok/> [Accessed 4 March 2021].

Roslan, N.B. 2015. *Mitigation Measures for Controlling Time and Cost Overrun Factors*. [pdf]. Available at: <http://eprints.uthm.edu.my/id/eprint/7984/1/NADZIRAH_BINTI_ROSLAN.pdf> [Accessed 18 May 2020].

Rwelamila, E.K. 2017. *Understanding the Risk in South African Construction Projects – A Case of the Western Cape*. [Online]. Thesis (MSc). Cape Town: University of Cape Town. Available at: <https://open.uct.ac.za/bitstream/handle/11427/28107/thesis_ebe_2018_rwelamila_esther_kagemulo.pdf?sequence=1> [Accessed 20 May 2020].

Selepe, M.M. 2019. The Appropriateness of Project Management Mechanisms Within the South African Public Sector Environment. *The 4th Annual International Conference on Public Administration and Development Alternatives* [online], Johannesburg 3–5 July 2019. Pp. 697–705. Available at: <http://ulspace.ul.ac.za/bitstream/handle/10386/2702/selepe_appropriateness_2019.pdf?sequence=1&isAllowed=y> [Accessed 25 May 2020].

Shadish, W. R., Cook, T. D., & Campbell, D. T. 2002. *Experimental and Quasi-experimental Designs for Generalized Causal Inference*. [Online]. Boston: Cengage Learning. Available at: <https://iowareadingresearch.org/blog/technically-speaking-random-sampling#:~:text=Random%20sampling%20ensures%20that%20results,equal%20chance%20of%20being%20selected.> [Accessed 22 September 2020].

Sharma, S. and Goyal, P.K. 2015. *Cost Overrun Factors and Project Cost Risk Assessment in Construction Industry – A State of the Art Review*. [Online]. Available at: <http://www.academia.edu/7130439/COST_OVERRUN_FACTORS_AND_PROJECT_COST_RISK_ASSESSMENT_IN_CONSTRUCTION_INDUSTRY_-_A_STATE_OF_THE_ART_REVIEW> [Accessed 6 March 2019].

Shunmugan, S. and Rwelamila, P.D. 2014. An Evaluation of the Status of Risk Management in South African Construction Projects. Proceedings of the Project Management South Africa (PMSA) Conference, Johannesburg 29, 30 September and 1 October 2014. Pp. 2–16.

Simushi, S.J.S. 2017. *An Integrated Management Strategy to Reduce Time and Cost Overruns on Large Projects*. [Online]. Unpublished Thesis (PhD). Stellenbosch: Stellenbosch University. Available at: <https://scholar.sun.ac.za/bitstream/handle/10019.1/102691/simushi_integrated_2017.pdf?sequence=1&isAllowed=y> [Accessed 20 May 2020].

Tembo, C.K. 2017. *Risk Allocation on Building Projects in the Zambian Construction industry*. Thesis (PhD). Johannesburg: University of Witwatersrand.

Walonick, D.S. 2010. *Survival Statistics*. [pdf]. Bloomington: Statpac. Available at: <https://youthsextion.files.wordpress.com › 2011/04> [Accessed 03 August 2021].

Causes of job burnout amongst Female Quantity Surveyors (FQS) in the Nigerian Construction Industry (NCI)

F.C. Jayeola
Department of Civil and Environmental Engineering, University of Zambia, Lusaka, Zambia

A.J. Ogungbile
Department of Building and Real Estate, The Hong Kong Polytechnic University, Hong Kong

A.E. Oke
Department of Quantity Surveying, Federal University of Technology, Akure, Nigeria

E.M. Mwanaumo
Department of Civil and Environmental Engineering, University of Zambia, Lusaka, Zambia
Department of Civil Engineering, University of South Africa

ABSTRACT: Construction workers are more exposed to burnout than in any other industry but female construction workers are more likely to experience mental health issues such as burnout than their male counterparts. This study aims to examine the factors causing burnout among female quantity surveyors in the construction industry. To achieve this aim, a cross-sectional survey was used alongside a well-structured questionnaire asking the FQS the causes of burnout at their place of work. 86 responses were collected and were analyzed using mean item score (MIS), standard deviation, and factor analysis. The findings showed that environment of work, social support and interaction contribute more to FQS job burnout. Therefore, construction organizations should make provision for social interaction among workers and a benchmark for measuring their performances. There is a need to find the important roles played between job burnout and job demands to support future research.

Keywords: Job Burnout, Job Demand, Construction Industry, Factors, Female QS.

1 INTRODUCTION

The construction industry is prominent due to the work-connected stress of the construction workers. This is due to the nature of the industry's project-based and pressure from their activities and the long working hours they encounter daily (Jignyasu *et al.* 2019). Male participates more in obtaining white collar professionals and managers, construction workforce and as a matter of fact, most companies are male-dominated, while on the other hand for female, poor rates of participation together with poor rates of retention decline them from the increased enthusiasm they have for the industry, this, as a result, makes them quit construction industry faster than their male counterparts (Diane *et al.* 2013). Natalie *et al.* (2015) postulated that the reduction of female participation in the construction industry is not just a matter acknowledged by the government and industries to be justice and fairness but the important factor the sector needs to realize fully is the ability of productivity and innovation. Many professionals in the construction industry, including Architects, Project Managers, and QS in particular, experience stress as a result of the fast-paced, competitive nature of building projects (Naoum *et al.* 2018). Because it has already been established that workplace stress might result in a lasting adverse impact if ignored, research has examined the high-stress levels and job burnout of Female QS (Zhang et al. 2020).

Several features were revealed to be associated with burnout experienced in many organizations. Some of these factors are educational level, available resources, recovery status, and personal age. While (Salyers et al. 2015) proceeded by explaining how the research was linked with burnout which gives negative outcomes on employee health and can thereby reduce employee satisfaction. This burnout syndrome includes anxiety, headaches, chronic pain, loss of appetite, difficulty digesting food, depression, excessive sleeping, or deregulated body clock. Failure to create a balance between job demands and job control might result in occupational stress. This balance can ease the level to which job demands induce harmful effects on workers such as low productivity, high absenteeism, and poor job performance (Paul & Keith 2014).

2 FACTORS CAUSING BURNOUT

Yang et al. (2017) stated that heavy project task is one of the core stressors that causes job satisfaction. It is always good to put in mind that interaction with others at work as a result of emotional hard work also causes job burnout for FQS. The categories of stress associated with burnout are job role, lack of authority, employee conflicts, and organizational stress contribute to the factors causing burnout (Timotius & Octavius 2022).

2.1 *Organization factors*

2.2.1 *Job demands*
Overloaded burnout may likely arise in the situation of economizing, which frequently does not slim down the organization's aims, but needs lesser workers to achieve those aims (Maslach et al. 2001). Workload and lengthy working hours are factors that are subjective to construction workers' safety performance and this is because construction itself has natural traits of generating great job demands (Adnan & El-Rayyes 2015). The findings agreed with (Ibem et al. 2011) that the most significant work-related stress in the Nigerian CI is high work measurements. The major cause of burnout includes an environment that influences stress on a person who is not capable to survive fully. Unnecessary workload and insufficiency of development material flow as the three major environmental conditions which can influence burnout among FQS. Job burnout springs up due to high job demand, insufficient motivation, and low support from the organization (Tsai et al. 2020).

The JD-R model proposes that the features of a job can be job demands which are works that cause stress and its high level is related to the high level of exhaustion and a low level of job resources which is the second job feature (Scanlan & Still 2019). Job resources are some aspects that provide the employee the support, strength, and help needed to maintain their well-being. In addition to the JD-R model, (Bowen & Zhang 2020) added that stress is created when a worker's task challenges the ability to perform or fulfill the demands of another task. The greatest stress factor that increases FQS burnout, according to the measure of work exhaustion developed by (Yang et al. 2017) involves oversight of stakeholders.

2.2.2 *Lack of appropriate resources*
This means inadequate resources and a lack or less training to perform the job commendably. Human resources are considered one of the main resources in an organization and work carried out in an organization should be done as a division of labor among trained employees. Poor organizational communication, inadequate remuneration, job role, lack of authority, unspecified employment requirements, and lack of adequate facilities are resources needed to overcome stress at work (Timotius & Octavius 2022). Therefore, if the required number of workers is not available, the few workers available experience work overload which will ultimately lead to burnout.

2.2.3 *Role ambiguity*

It was ascertained from the findings by (Ibem *et al*. 2011) that the nature of the training received from work by construction employees leads to long hours of work and thereby exposed them to being burned out. Lack of proper communication might lead to a series of and this might ultimately lead to physiological and emotional stress. The main demand associated with a job that has taken an important part in job burnout is the ambiguity of role which is a result of personal duty created from impartial work situations, psychological nature (Schieman & Young 2013), and work pressure (Bowen & Zhang 2020). Individuals experiencing excessive impartial work situations, after-hours, and work pressure are most likely to feel that ambiguity of roles results in psychological distress, burnout, and sleeping complications (Bowen *et al*. 2018b) according to a study carried out on construction professionals in South Africa.

2.2 *Personal factors*

Decreased levels of endurance are always linked with complex burnout cuts such as a lack of relationship between workers and top management, poor support from the top management, and a high-quality workforce (Poon *et al*. 2013). Burned-out beings cope with hectic actions in a submissive and defensive way. These are the tense employees who have no further hope for their job. These factors can also be affected by age, maturity, (Nanda & Chawla 2010) lack of socialization, and also extends to lack of interaction with others.

2.2.1 *Workload mismatch*

It is identified that job burnout reacts to depersonalization, emotional exhaustion, and a low level of personal accomplishment which all emanated from the workplace (Elham *et al*. 2019). This burnout victim is overloaded with work or they are involved in the immoral work type. Female employees with workloads might be stressed due to a mismatch of work which supersede the stretch and drive available to accomplish tasks required by individuals and this leads to sleeping problems, impaired health, and exhaustion (Schieman & Young 2013).

2.2.2 *Loss of a positive connection with others*

Burnout can also affect an employee who is being revealed as ostracized. Such employees cannot share similar values with the group or create a good relationship between the employees and the employers especially the top management (Adnan & El-Rayyes 2015) which leads to burnout for workers that have gone rascal.

2.3 *Social factor*

This study aid in identifying the causes of burnout among FQS and the study revealed that the greatest cause of burnout for FQS is social contact or interaction at work (Zhang *et al*. 2020). Although earlier studies tended to concentrate mostly on the job stress that emerges from supposedly difficult duties. The biggest factor causing burnout in FQS is interpersonal interaction (Naoum *et al*. 2018). Consequently, they participate in psychological efforts to control what they are feeling. The workplace stress theory of Job Demands-Control-Support stated that jobs that are low control or decreased social support at the workplace, and high job demand is termed to be the most stressful and cause a great impact on workers' health (Paul *et al*. 2013). A dangerous working environment or bad site settings merged with great numbers of projects can lead to putting workers under pressure which most of the time leads to loss of concentration on-site and thereby cause or contribute to accidents (Ibem *et al*. 2011). This study further stated that inadequate temperature (too hot or cold) and poor lighting at site offices can contribute to workers being burned out. These points were further stressed (Adewa & Agboola 2020) that the performances of workers are encouraged by self-fulfillment in the environment in which the employee works, organizational backing, and motivations. The study (Qiao 2019) established that the high the level of social support of workers the weaker the feelings of the burnout experience. The support obtained from social life helps to reduce the experience of burnout at work. Social support could effectively reduce the job burnout of workers by enhancing their self-efficacy (Zhao et al. 2019). Xie et al. (2022) constructed a model of the relationship between social workers, social support, and job burnout, it thus says enhancing social support helps in supporting the sustainability and stability of the industry.

3 RESEARCH METHODS

The design adopted for this is a quantitative approach to quantify the behavior, attitude, and opinions of the selected population. Data was collected using a well-structured questionnaire that was administered to female QS in the contracting firms, consulting firms, academics, and government establishments in Lagos and Ondo States. Female QS were selected because they are most likely to experience burnout than their male colleagues. A non-probabilistic convenience sampling with 122 sample size was used to collect data from every member of the population who actively and conveniently participated in the study. A closed-end questionnaire was administered asking questions from the three divisions of factors causing burnout which are organizational, personal, and social factors causing burnout. a total of 86 responses were retrieved out of the 122 questionnaires that were sent out. The data was analyzed using Mean Item Score (MIS) and Standard deviation (SD) to rank the major factors of burnout among female QS in the construction industry. Factor analysis was conducted to subdivide the variables into smaller classifications and Cronbach's Alpha tested the reliability of the data set.

4 RESULTS

In examining the factors that cause job burnout amongst FQS, questions were asked based on three (3) main divisions which are Organization Factors, Personal Factors, and Social Factors. The results from the respondents showed that the organizational factors of burnout are caused by conflicts, role ambiguity, little participation in decision-making, and job demands with MIS of 3.51, 3.63, 3.80, and 3.95 respectively.

Table 1. Factors causing burnout among female quantity surveyors.

S/N	Factors	Variables	Mean	SD	Rank
1	**Organizational Factors**	Conflicts	3.51	1.00	4
		Role ambiguity	3.63	0.98	3
		Little participation in decision making	3.80	0.99	2
		Job demands	3.95	0.96	1
		Average		3.72	
2	Personal Factors	Age	3.23	1.12	10
		Maturity	3.48	1.07	9
		Conflict between values	3.59	1.03	8
		Unfairness	3.62	1.11	7
		Lack of interaction with others	3.62	1.04	6
		Socialization	3.63	1.04	5
		Level of Education	3.67	1.01	4
		Loss of a positive connection with others	3.67	0.99	3
		Lack of appropriate resources	3.80	1.07	2
		Workload mismatch	3.87	0.97	1
		Average		3.62	

(*continued*)

Table 1. Continued

S/N	Factors	Variables	Mean	SD	Rank
3	**Social Factors**	Counter-transference	3.60	0.90	6
		Lack of social interaction	3.64	1.03	5
		Staff turnover	3.72	1.00	4
		Nature of the social work	3.73	0.83	3
		Population and workload	3.85	0.95	2
		Environment of work	4.07	0.78	1
		Average		**3.77**	

Personal factors of burnout are age, maturity, a conflict between values, unfairness, lack of interaction with others socialization, level of education, loss of a positive connection with others, lack of appropriate awards, and workload mismatch with MIS of 3.25, 3.48. 3.59, 3.62, 3.62, 3.63, 3.67, 3.67, 3.80, 3.87 respectively. The last division is the social factors with MIS of 3.60, 3.64, 3.72, 3,73, 3.85, and 4.07 and these factors are counter-transference, lack of social interaction, staff turnover, nature of the social work, population or workload, and environment of work respectively. The overall factors causing burnout have 3.72 on average for organizational factors, 3.62 for personal factors, and 3.77 for social factors. This implies that generally, the factor causing burnout among female QS is social factors resulting from the environment of work, population or overload, nature of social work, staff turnover, lack of social interaction, and counter-transference. However, workload mismatch and lack of appropriate resources as well as job demands and little participation in decision-making also contribute to factors causing burnout because they were highly ranked in their categories.

4.1 *Factor analysis*

Kaiser-Mayer-Olkin Test (KMO) serves as the first step in recommending a data set for factor analysis. The KMO results showed a 78% level of sampling adequacy and 0.000 significance level which makes the data set suitable for factor analysis as shown in Table 2.

Table 2. KMO and Bartlett's test.

Kaiser-Meyer-Olkin Measure of Sampling Adequacy.		0.781
Bartlett's Test of Sphericity	Approx. Chi-Square	191.457
	df	45
	Sig.	0.000

The variables were subdivided into four smaller classifications to deal with smaller groups instead of larger numbers (Shrestha 2021) using the rotated component matrix which shows a significant level of satisfaction of 60.53% as represented in Table 3. Variables were grouped accordingly meeting all the requirements and groups were renamed based on the author's discretion.

Table 3. Rotated component matrix for factors of Burnout experienced by FQS.

Component	Initial Eigenvalues			Extraction Sums of Squared Loadings		
	Total	% of Variance	Cumulative %	Total	% of Variance	Cumulative %
1	5.214	32.586	32.586	5.214	32.586	32.586
2	2.040	12.749	45.335	2.040	12.749	45.335
3	1.309	8.179	53.515	1.309	8.179	53.515
4	1.122	7.011	60.526	1.122	7.011	60.526

Groups were renamed as Organizational Factors at 32.59%, Personal Factors at 12.75%, Emotional Factors at 8.18%, Overloaded Factors at 7.01%, and a total cumulative variance of 60.53%. Organizational factors are the first set of variables causing burnout as they are classified under group 1, followed by personal, and emotional factors. The last group is the overloaded factors which are job demands and workload mismatch with 0.828 and 0.504 values which imply a higher rate of factors causing burnout. The following factors: F2 (Conflicts), F3 (Role ambiguity), F15 (Nature of social work), F17 (Population and workload), and F20 (Lack of social interaction) have been deleted from the grouping because they have values <0.5 which makes them unsatisfactory and unable to meet the requirements for the classification. The internal consistency and reliability scale of the data set was tested using Cronbach's Alpha. The results showed a high level of correlation and suitability by alpha values of 0.769, 0.756, and 0.689 for organizational, personal, and emotional factors causing burnout respectively. But the overloaded factor category showed an alpha value of 0.413 which means that the reliability scale of the overloaded factors data set is not consistent.

Table 4. Factor analysis for factors causing Burnout amongst FQS.

Factors	Codes	Items	1	2	3	4	Alpha
Organizational Factors	F7	Loss of a positive connection with others	0.730				
	F6	Lack of appropriate resources	0.679				
	F9	Conflict between values	0.641				
	F4	Little participation in decision making	0.574				
	F10	Unfairness	0.573				0.769
Personal Factors	F12	Level of Education		0.855			
	F13	Socialization		0.825			
	F11	Maturity		0.640			
	F8	Age		0.586			
	F14	Lack of interaction with others		0.503			0.756
Emotional Factors	F16	Environment of work			0.801		
	F18	Staff turnover			0.730		
	F19	Counter-transference			0.671		0.689
Overloaded Factors	F1	Job demands				0.828	
	F5	Workload mismatch				0.504	0.413

5 DISCUSSION

5.1 *Organizational factors*

The results showed that factors (F7, F6, F9, F4, and F10) were regarded as organizational factors. These factors do not connote stress only at the workplace as a result of work overload but the probability of being stressed as a result of the deficiencies faced by an employee from the management. Loss of a positive connection with others had the highest factor loading and this indicates that a FQS who does not build a good relationship with her colleagues will experience burnout. This poor connection can be a result in being declared ostracized (Adnan & El-Rayyes 2015) or being banished from attending some workplace activities. Along the same line, conflicts between values also contribute to job burnout of FQS. An employee who fails to have similar values or cordial relationships with others can be burned out. This factor is detrimental to health and this can make such FQS go rascal because there is no room for letting out their minds on whatever they are going through. Moreover, lack of appropriate resources such as no or less training to perform a commendable job, and good and timely organizational communication could be reasons for FQS in dealing with job burnout. Unspecified job roles or requirements, lack of authority, and inadequate facilities could also contribute to job burnout. Furthermore, in a situation where there is a shortage of human resources which is considered to be a main resource for executing jobs in an organization (Timotius & Octavius 2022), there is always a shortage in the division of labor which lead to work overload. It is also possible that if the required trained workers are not provided for the available work, the available few workers tend to overwork themselves (unfairness and imbalanced evaluations) which untimely lead to job burnout. Little participation in decision making in addition makes it difficult for FQS to carry out their jobs without stress. Given authority on jobs to be done without proper description or knowledge turn out to be a challenge to deliver a successful job.

5.2 *Personal factors*

In this study, F12, F13, F11, and F8 denote personal factors causing job burnout amongst FQS in the NCI. These factors were level of education, socialization, maturity, age, and lack of interaction with others were dovetailed and react to workplace emotional exhaustion, depersonalization, and low level of personal accomplishment which is also supported by (Elham *et al*. 2019). According to (Nanda & Chawla 2010), any burnout victim who is overloaded with work or undergoes undying locus of regulation, and copes with hectic actions in a submissive and defensive way is affected by age, maturity, and lack of socialization, and this also extends to lack of interaction with others. These beings have no further hope for their jobs. The personal relationship between FQS with top management and a high-quality workforce (Poon *et al*. 2013) combined their support fight against burnout and enhance job satisfaction. Social or interpersonal interaction with others at work is said to be very important as said by (Paul *et al*. 2013) that there is low control or decreased social support at the workplace. Social support coupled with social interaction could effectively enhance self-efficacy and reduce the level of burnout and thereby improving the work performance of FQS.

5.3 *Emotional and overloaded factors*

The analysis revealed F16, F18, and F19 as the representation of emotional factors contributing to job burnout amongst FQS in the NCI. The emotional factors (environment of work, staff turnover, and counter-transference) give excessive impartial work situations and pressure known as role ambiguity. The consequences of ambiguity of roles are psychological or emotional distress, sleeping complications, burnout according to (Bowen *et al*. 2018b), and emotional entanglement which is known as a counter-transference. Emotional burnout could be a lack of proper communication which makes a worker repeat works or do a certain work for a long hour. It could also be a lack of proper training (Ibem *et al*. 2011) resulting in overthinking and taking longer time to provide solutions to tasks or problems. The emotional thought of colleagues leaving the organization could also add to the level of burnout experienced by FQS as well as work pressure or work overload which is an overloaded factor of burnout. In this setting, job demand and workload mismatch (F1 and F5) are causes of overloaded burnout amongst FQS. The impact of these factors contributes substantially to poor performance, untimely deliverables, employee dissatisfaction, and insufficient motivation (Tsai *et al*. 2020). It was recorded (Adnan & El-Rayyes 2015) that workload and long hours of work by the FQS subject them to naturally great job demands. Overloaded burnout also gives rise to system economization, but this does not reduce the aims of the organization but rather aids lesser workers to achieve the aims. To achieve successful job satisfaction, there is a need to get rid of unnecessary work overload, multitasking at a particular period, and insufficient development of material flow.

5.4 The implication of the study

The factor analysis carried out in this study showed that the causes of burnout amongst FQS in the NCI could be denoted by four factors, that is organizational factors, personal factors, emotional factors, and overloaded factors. The study revealed that organization factors were the first set of burnout affected by FQS and these were not only a result of work stress resulting from too much work done but also as a result of some difficulties caused by the management such as unfairness, lack of good relationships with the top management, lack of employees being involved in the decisions made before tasks are performed. The industry must create chances for the employees to make good relationships with their management Management should also provide enough trained staff and resources needed for a particular job to balance the division of labor to avoid overloaded work or unnecessary work pressure. Proper communication should be provided to avoid the repetition of work and unnecessary time consumption in performing tasks. It is very important to improve social support and interaction to enhance the sustainability and stability of the construction industry.

6 CONCLUSION

Results from the study showed how female FQS are burned out due to a lot of factors contributing to stress at work their various workplace. It was established that social interaction and support between management and employees contribute the larger percentage of burnout. As a result of this, there is need for improving the level social interaction by creating positive social relationship. They can also reach out to people who they can brainstorm together on a particular task or having social support in overpowering conservational tasks by keeping themselves motivated, improved managing abilities, and extensive good health. There is also need for the FQS to be involved in the organization's decision-making. This is to prepare them ahead for the task they need to carry out as an employee. It is thereby recommended that organizations should provide means for social interaction among the workers as well as provide a benchmark by which organization can measure their performance and minimize the key cause of burnout. Future research can find out how FQS or other construction stakeholders get satisfied with their jobs when they are burned out and how to strike a balance between job burnout and job satisfaction.

REFERENCES

Adewa, K. A. & Agboola, A. A., 2020. Effects of Job Burnout on Employees Satisafaction in Selected Health Service Sector in southwestern Nigeria. *Open Journal of Applied Science*, Volume 10, pp. 877–890.

Adnan, E. Y. & El-Rayyes, S. A., 2015. Job Stress, Job Burnout and Safety Performance in the Palestinian Construction Industry. *Journal of Financial Management of Property and Construction*, 20(2), pp. 170–187.

Bowen, P. A., Govender, R., Edwards, P. J. & Cattell, K. S., 2018b. "Work-related Contact, Work-family Conflict, Psycological Distress and Sleep Problems Experieced by Construction Professionals: An Integrated Explanatory Model". *Construction Management Economics*, 36(3), pp. 153–174.

Bowen, P. & Zhang, R. P., 2020. Cross-boundary Contact, Work-familty Conflict Antecedents, and Consequences: Testing an Integrated Model for Construction Professionals. *Journal of Construction Engineering and Management*, 146(3).

Diane, b., Pat, C., wazir, J. K. & Routlegde, 2013. *Gendered Fields: Women, Men and Ethnography*. s.l.:s.n.

Elham, A., Steven, W. B. & Cem, T., 2019. Workplace Bullying, Psycological Distress, Resilience, Mindfulness, and Emotional Exhaustion. *The service Industries Journal*, Volume 39, pp. 1–25.

Halkos, G. & Bousinakis, D., 2010. The Effect of Stress and Satisfaction on Productivity. *Productivity and Performance Management*, 59(5), pp. 415–431.

Ibem, E., Anosike, M. & Azuh, D. T., 2011. Work Stress Among Professionals in the Building Construction Industry in Nigeria. *Australasian Journal of Construction Economics and Building*, 11(3), pp. 45–57.

Jignyasu, P. J., Nabsiah, A. W., Hemalatha, S. & Lavinsaa, P., 2019. *Determinants of Work Stress for Construction Industry Employees in Malaysia.* s.l., Atlantis Press SARL.

Joe, G. W., Broome, K. W., Simpson, D. D. & Rowan-Szal, G. A., 2007. Counselor Perceptions of Organizational Factors and Innovations, Trainning and Experiences. *Journal of Substance Abuse Treatment*, 33(2), pp. 171–182.

Lee, H. S., Jin, F. J. & Park, M. S., 2012. A study on Factors Influencing Turnover Intention of New Employees in Construction Company. *Journal of Construction Engineering and Management*, Volume 13, pp. 137–146.

Maslach, C. et al., 2001. Job Burnout: Annual Review of Psycology. *Theory and Practice of Counselling and Psycotherapy*, Volume 52, pp. 397–422.

MCNulty, T. et al., 2007. Counselor Turnover in Substance Abuse Treatment Centers: An Organizational-level Analysis. *Journal of Sociological Enquiry*, 77(2), pp. 166–193.

Mostert, T. K., 2011. Job Characteristics, Work-home Interference and Burnout: Testing a Structural Model in the South Africa Context. *Human Resource Management*, Volume 22, pp. 1036–1053.

Nanda, P. & Chawla, A., 2010. Impact of Age and Maturity Type on Emotional Maturity of Urban Adolescent Girls. *Ejournal*, 19(107).

Naoum, S. G., Herrero, C., Egbu, C. & Fong, D., 2018. Integrated Model for the Stressors, Stress, Stress-coping Behaviour of Construction Project Managers in the UK. *International Journal of Managing Projects in Business*, 11(3).

Natalie, G., Abigail, P., Martin, L. & Louise, C., 2015. Gender Equality Among Workers in the Construction Industry. *Construction Management and Economics*, Volume 33, pp. 375–389.

Paul, B. & Keith, C., 2014. Occupational Stress and Job Demand, Control and Support Factors Among Construction Project Consultants. *International Journal of Project Management*, 32(7), pp. 1273–1284.

Paul, B., Keith, C. & Peter, E., 2013. *Workplace Stress Experienced by Quantity Surveyors.* 20(2).

Poon, S. W., Rowlison, S. M., Koh, T. & Deng, Y., 2013. Job Burnout and Safety Performance in the Hong Kong Construction Industry. *International Journal of Construction Management*, 13(1), pp. 69–78.

Qiao, Y., 2019. The Study on the Study of Job Burnout of Social Worker from the Perspective of Social Support-based on the Survey of Chenngdu City, Southwest. *Finance Ecnomics*, Volume 7, pp. 1–62.

Rajeshwari, R. R. & Mano Raj, S. J., 2017. Study on the Impart of Environmental Factors on Emotional Maturity. *IOSR. Journal of Business and Management*, 19(5), pp. 1–09.

Salyers, M. P. et al., 2015. Burnout and Self-reported Quality of Care in Community Mental Health. *Administration Policy Mental Health Mental Health Service Resources*, 42(1), pp. 61–90.

Scanlan, J. N. & Still, M., 2019. Relationships Between Burnout, Turnover Intention, Job Satisfaction, Job Demands and Job Resoursces for Mental Health Personnel in an Australian Mental Health Service. *Scalan and BMC Health Service Research*, January.

Schieman, S. & Young, M., 2013. "Are Communications About Worrk Outside Regular Working Hours Assocoiated with Work-to-family Conflict, Psychological Distress and Sleep Problems?" *Work Stress.* 27 (3), pp. 244–259.

Shrestha, N., 2021. Factor Analysis as a Tool for Survey Analysis. *American Journal of Applied Mathematics an Statistics*, 9(1), pp. 4–11.

Timotius, E. & Octavius, G. S., 2022. Stress at the Workplace and Its Impact on Productivity: A Systematic Review from Industrial Engineering, Management, and Medical Perspective. *Industrial Engineering & Management Systems*, 21(2), pp. 192–205.

Tsai, J., Jones, N., Klee, A. & Deegan, D., 2020. Job Burnout Among Mental Health Staff at a Veteran's Affairs Psycology Rehabilitation Centre.. *Community Mental Health Journal*, Volume 56, pp. 294–297.

Xie, X., Zhou, Y., Fang, J. & Ying, G., 2022. Social Suppot, Mindfulness, and Job Burnout of Social Workers in China. *Frontiers in Psycology*, Volume 13.

Yang, F. X., Li, Z., Zhu, Y. & Wu, C., 2017. "Job Burnout of Construction Project Managers in China: A Cross-sectional Analysis.". *International Journal Project Management*, 35(7), pp. 1272–1287.

Zhang, L., Yao, Y. & Yiu, T. W., 2020. Job Burnout of Construction Project Managers: Exploring the Consequences of Regulating Emotions in Workplace. *Journal of Construction Engineering Management*, 146(10).

Zhao, Y. G., Li, X. X. & Cui, Y., 2019. Analysis of the Impact of Social Support on Job Burnout Among Doctors and Nurses: Based on Mediation Effect of Self-efficacy. *Journal Chinese Hospital Management*, Volume 39, pp. 48–50.

Infrastructure: Economic, social / environmental sustainability

An exploratory and comparative assessment of cost estimates on infrastructure projects: A client's perspective

F. Muleya
Department of Quantity Surveying and Construction Management, University of the Free State, South Africa
School of the Built Environment, Copperbelt University, Kitwe, Zambia

M. Beene
Department of Engineering Services, Lusaka City Council, Lusaka, Zambia

A. Lungu & C.K. Tembo
School of the Built Environment, Copperbelt University, Kitwe, Zambia

K. Kajimo-Shakantu
Department of Quantity Surveying and Construction Management, University of the Free State, South Africa

ABSTRACT: The study assessed client's perspective of cost estimates on infrastructure projects prepared by quantity surveyors on public and private infrastructure works in Zambia. Convenience and purposive sampling were used to identify the respondents from two provinces. A total of 12 semi structured questionnaires were collected from clients in the public sector, private sector, parastatal, and quasi-institution. Further, 12 semi-structured questionnaires were collected from quantity surveying firms. Results indicated that quasi-government and public sector clients did not expect final costs to exceed 25% above the initial cost estimate. Private sector clients did not expect the cost estimate to exceed 10% of the initial estimate. However, selected projects recorded cost overruns more than 90% above the initial estimates for public works and 0 to 30% for private sector works. Results demonstrate the differences in cost estimate expectation between the public and private sectors.

Keywords: Clients, Cost Estimates, Satisfaction, Quantity Surveyors, infrastructure

1 INTRODUCTION AND BACKGROUND

Cost estimation is defined as a procedure of establishing the scope of work and the financial resources needed to meet the requirements of the client from inception to completion of the project, (Hatamleh *et al.* 2018). At each project inception, be it public or private, it is every client and quantity surveyor or cost engineer's concern to have a quality, reliable and accurate conceptual cost estimate. An estimate is qualified as a quality estimate if it meets the expectations of the client, usually expressed as a percentage of the amount. The accuracy and reliability of the cost estimate is of critical importance to the client for budgeting purposes and importantly to the cost engineers as a way building confidence and reputation (An *et al.* 2011; Ismail, *et al.* 2021; Cheng, *et al.* 2010; Doloi 2011). Despite cost estimation aiming to provide guidance on keeping project design and scope with the budget of the client, projects still suffer from cost overruns. Cost estimates in developed and developing countries still suffer from errors, omissions, and inaccuracies (Cheng *et al.* 2010; Heravi & Mohammadian 2017; Vaardini *et al.* 2016).

2 COST ESTIMATES IN PUBLIC AND PRIVATE PROJECTS

There has been a debate on the difference in accuracy of cost estimates in public and private projects. Studies reveal that large public infrastructure projects such as bridges and tunnels have had their final costs exceed the initial cost estimate by between 20% to over 100% (Akinradewo et al. 2020; Bakr 2019; Skamris et al. 1996). There seems to be more data on public projects than private projects. This can be attributed to the public nature of infrastructure projects and its subjectivity to scrutiny by the public (Agyekum-Mensah 2019). As a result, there is less private work that is documented for analysis between initial estimate and final cost.

3 RESEARCH MOTIVATION

Studies by Muya et al. (2013) and Kaliba et al. (2009) reveal that the Zambian construction industry has been experiencing cost overruns on projects despite the availability of qualified Quantity Surveyors operating as quantity surveying consulting firms, individually employed quantity surveyors and those employed as inhouse consultants in the public sector known as buildings department. Kaliba et al. (2009) related this to failure to carry out sufficient cost analysis at the initial stages of contracts. It must be mentioned that changes driven by clients will inevitably render inaccuracies in the initial cost estimates. Quantity surveyors and cost engineers have no control over such overruns. According to information obtained from the buildings department, cost overruns ranged from 16% to 87.5% for projects which included low-cost housing, office blocks, hospitals construction, government lodges and road works. Some questions stand out: Is it necessary to produce cost estimates if some overruns are close to 100% above the original estimate? Are the cost overruns because of quantity surveyors lack of competence? Or are these cost overruns triggered by client changes of scope or poor understanding of the scope. With significant cost overruns recorded, this study aimed to establish whether cost estimates produced by Quantity Surveyors are reliable and accurate by determining how satisfied public and private clients are with these cost estimates in Zambia. The underpinning motivation of the study was that there is no information that indicates differences in cost estimate satisfaction levels between public and private clients. Further, the extension of the study to the private sector was to establish if there is a difference in levels of satisfaction between public and private clients. To achieve this aim, the following objectives were formulated:

(a) Establish the difference between initial cost estimates and final cost estimates between public and private clients.
(b) Establish the main drivers of cost overruns between public and private client projects.
(c) Establish the understanding levels of clients on project scope at the project inception stage

4 LITERATURE REVIEW

Different countries experience dissatisfaction with cost estimates of construction projects because of cost overruns when compared to initial estimates. Some of the countries cited in literature for cost estimate dissatisfaction include United Arab Emirates, Nigeria, Iran and Vietnam (Fazeli et al. 2020; Johnson & Babu 2020; Ritz 1994). Percentage cost overruns in relation to cost estimates range from 10% to over 100% especially in developing countries. Reasons cited for cost overruns include price fluctuation, fraudulent practices, design changes, change of material and method specifications, rate fluctuation, complexity of work, poor project management and corruption (Doloi 2011; Heravi & Mohammadian 2017;

Johnson & Babu 2020; Vaardin et al. 2016). During the cost estimate discussion, efforts to improve cost estimation include the use of Building Information Modelling (Fazeli et al. 2020). Cost estimation is the process of projecting what the project will cost. It is also used as a financial planning tool at inception of a project by the client (Ritz 1994). The projected figure is a cost estimate. A detailed estimate or definitive estimate is finally prepared (final cost estimate) and becomes the basis for selecting bids submitted by contractors and as such, it is expected to be inclusive of every aspect of the design (Fazeli et al. 2020). Estimates are not only used by clients or financiers for financial planning or deciding between alternative projects. Architects as well use the estimates to ensure that designs are kept within the clients' budget (Fazeli et al. 2020). The process involves project briefing, visitation of proposed and acquired site. Buchan et al. (2003) indicated that it is at the preliminary stage that the client gives their briefing and basic designs are drawn up. Aibinu & Jagboro (2002) refers to this as the planning stage in which initial estimates are determined to make the client aware of the cost implications of some of the components they may want to be added to the design. At this stage, the client ascertains whether the project is worth undertaking.

Sepasgozar et al. (2022) asserts that referring to past similar projects can increase the possibility of creating a cost estimate that covers all possible costs. For complex civil engineering projects, the cost of the project may go above the final estimate by >10% or may be less than the estimate by 5%, however, clients, who become more informed of the project and what is expected demand for the project to be done within a stipulated budget and if anything, ask the architect to produce designs that will be within the given budget. Factors affecting accuracy of cost estimates may be prepared by qualified and registered Quantity Surveyors but there are different factors that result in different accuracies and these factors include experience, time available, detail of designs and the actual process of preparing the cost estimates (Al Juboori 2021; Dandan et al. 2019; Zhong et al. 2022). Cost estimates include unit rate estimating, operational estimating and manhour estimating. Further, Quantity Surveyors make use of computer software that aide their preparation of cost estimate rates. These types of estimates are considered time consuming. It is essential for the purpose of obtaining accurate work as it allows for the control of omissions and duplications which the person preparing can identify with ease. According to Yang & Peng (2008), client satisfaction explains that customer satisfaction has become a key to improvement of quality and works as a tool to measure and aide the development of the construction industry. Among other industries, there is more need for the construction industry to investigate client satisfaction because their marketability is solely dependent upon customer loyalty and word-of mouth reputation (Khairuddin & Rashid 2016; Tembo et al. 2021; Pishgar et al. 2013; Subram et al. 2018) Oyewobi (2013) further states that it is for this reason that Quantity Surveyors should be very concerned with the quality of the services they provide to the public, which can be done by finding out what clients think of the service received. Omonori & Lawal (2014) identified the following reasons as to why it is necessary for customer satisfaction to be assessed: Customer satisfaction acts as an indicator for the future financial success of a company or firm. A satisfied customer will always return to the client under what is referred to as customer loyalty or a customer base Creates a trend of monitoring and taking note of achievements and loses by the firm and accounting them to their causes. Xiao & Proverbs (2002). states that seeking feedback from clients improves communication and collaboration of client. This makes the client participate and feel like the works are being executed as they sought. Literature has demonstrated that client sati*sfaction*, inaccurate estimate, cost overruns exist in other countries as well with *statistics* attached as indicted in the literature review section. Further literature hardly contains information that compares satisfaction levels between *public* and private clients. Results from this study would *benefit* other readers and scholars in understanding the difference between public and private clients. This study would also be used as a guide in *carrying* out *similar works* in *different* countries in order to *understand* the difference between pu*blic* and private clients in *terms* of *clients'* satisfaction.

5 RESEARCH METHODOLOGY

A mixed method approach was used to obtain both quantitative and qualitative data in order to gather as much information as possible. As a result, a semi structured questionnaire was adopted as data collection instrument administered to 12 clients and 12 quantity surveying firms. The population for quantity surveying firms at the time of data collection was 18. This translated to a response rate of 66%. The number of clients willing to participate in the study was low due to the sensitive nature of the subject at the time. This is one of the reasons that led to embedding a qualitative component in the approach to create the opportunity to obtain in-depth information that would be analysed qualitatively. With the limited number of targeted respondents, qualitative data was used to provide data that would add value to the quantitative data. Data was collected from public clients, private clients, parastatal firms and quasi-government clients.

5.1 Sampling

Sampling is the process of selecting a target group from a population. The non-random sampling method was adopted for this research. Purposive sampling was used to allow the selection of clients based on information required for the study as shown in Table 1. This method has advantages of control over the information that is received. Data was provided by clients' representatives for both public and private institutions. Table 1 gives a summary of the target groups, the sampling methods used, and justification used to select participants within those groups

Table 1. Sample size selection criteria.

Target Group for in-depth structured questionnaire	Justification	Sampling Method	Population	Sample Size
Quantity Surveying Consultancy Firm	• Zambian Quantity Surveying Firm. • Registered with SIZ and QSRB. • Worked on notable projects. • Practiced for over 2 years.	Purposive;	18	12
Private Sector Clients	• Privately funded a project. • Project underwent cost overrun. • Hired Quantity Surveyor at inception of project.	Purposive	Unknown	4
Public Sector Clients	• Funded a partly or fully funded Government Project. • Project underwent cost overrun. • Hired Quantity Surveyor at inception of project.	Purposive	Unknown	8

6 DATA ANALYSIS

This section discusses the methods that were used for analysing the collected data through semi structured questionnaires. Microsoft Excel 2019 was used to analyse data to generate frequencies and graphs to have an appreciation of outcome statistics to support the qualitative data. Thematic analysis was used to analyse additional data provided by respondents.

7 FINDINGS

Results shown in Table 2 indicate that the public sector clients allowed cost overruns exceeding 10% of the contract sum as acceptable considering that that prices of goods and services were constantly changing while the private sector preferred to stay without any cost overrun. Private clients however indicated that a maximum of 10% would be allowed with strong justification that is beyond their control. Table 2 shows an extract from thematic analysis showing analysis and outcome of the themes based on additional data provided by the respondents in the semi structured questionnaires. The table provides a summary of the practice variations surrounding cost estimates between the public sector and private sector.

7.1 Satisfaction with cost estimates

The first indicator on cost efficiency was deemed to be acceptable if it falls within the expected range of 5% to 10%. Measuring this range against that expected in the public sector, it was revealed that cost overruns that range between 0% and 25% over the agreed contract sum were regarded as a cost-efficient project. The findings revealed that for government projects, a strict position is given to the construction team to keep the costs within this range, and this applies for all public projects. The quasi-government and parastatal sectors, which are partly run by the government, apply the same range (0% to 25%) as their measure of cost efficiency in terms of maximum cost overrun allowance. The private sector, however, considers a project to be cost efficient if there are absolutely no cost overruns above the initially agreed contract sum.

Table 2. Thematic analysis measuring client satisfaction against noted indicators.

THEME/ INDICATOR	Standard	Public Sector	Private Sector	Quasi-Government	Parastatal
Cost Efficiency	−5% to 10% under or over the contract sum, respectively.	0–25% over contract sum.	0% increase; stay within the budget.	0–25% over the contract sum.	0–25% over the contract sum.
Involvement of Professionals	Involves Quantity Surveyor during planning stage. Receive cost reports	Involve in-house Quantity Surveyors and external Quantity Surveyors in different projects. Receive cost reports approximately 50% of the time.	Involves external Quantity Surveyors in projects after briefing stage. Receives cost reports weekly, monthly, quarterly, depending on agreement or contract stage.	Involves project manager after briefing stage. Receives cost reports 50% of the time.	Involves Quantity Surveyor after briefing stage. Receives cost reports.
Customer Relations	Willingness of client to rehire quantity surveyor. Clients understanding of cost reports.	Prefers using in-house staff.	Willing to rehire and utilise consulting Quantity Surveyors but do not fully understand the cost reports.	Willing to rehired consulting Quantity Surveyors as they help control costs.	Prefers using in-house staff for projects.

7.2 Factors for re engaging quantity surveyors

With the public clients accepting a final figure of 0–25% over the contract sum at completion of the works, it was necessary to establish under what conditions clients would re-engage Quantity Surveying firms for other construction projects. This interest in this inquiry was triggered by the difficulties that Quantity Surveyors face because of the restrictions to advertising their services. Clients advanced several reasons for re-engaging Quantity Surveyors as seen in Figure 1. Out of 12 clients. 7 experienced cost overruns in the range of 0% to 30% as shown in Figure 2. This is also the range that public sector, quasi-government, and parastatal clients indicated to be acceptable. Nevertheless, some clients indicated experiencing cost overruns of 31% to 60% (3/10), and 91% to 100% (8/10). These results indicate cost overruns up to 100% above the initial estimates still exist. On the other hand, private sector and quasi-government clients mostly experienced cost overruns ranging from 0%–30%.

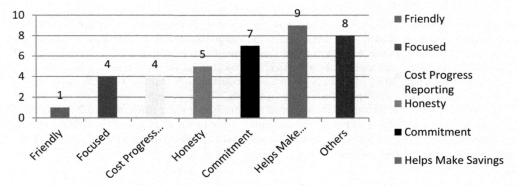

Figure 1. Factors influencing re-engagement of quantity surveyor firms.

7.3 Levels of cost overruns

Figure 2 shows the percentage cost overruns in ranges as provided by clients.

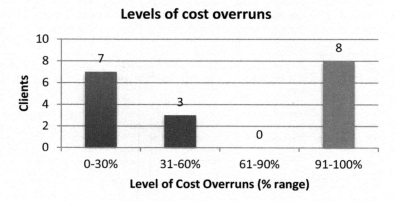

Figure 2. Percentage ranges of cost overruns experienced.

7.4 *Causes of cost overruns*

Results from the study indicate that the causes for cost overruns as observed by clients included indecision on the part of clients, delayed drawings, unforeseen circumstances and other reasons that were found to be common from clients as indicated in Figure 3.

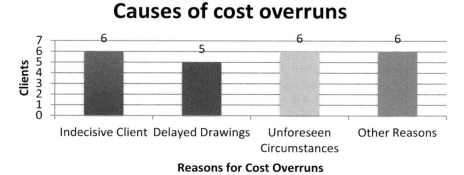

Figure 3. Causes for cost overruns.

Causes of cost overruns are triggered with numerous variations and as a result, prolonged project duration, which increases the costs of the contractor being on site. An overrun, however, is only an overrun if the client is aware of the changes to cost and approves of it. Some clients however do not seem to be aware of the cost implications of the changes they have made to the scope. Further interrogation of the data revealed that delayed drawings and client indecision were the most prominent cause of overruns in the public sector, and this was higher than in other organisations. This is not surprising because public and related clients are dominated by bureaucracy and long approval procedures compared to private clients

7.5 *Clients understanding of scope of project*

The study reveals that all the types of clients generally had acceptable levels of understanding the project scope with the highest being 88% as average in the public sector as seen in Figure 4. Figure 6 shows combined data from clients and quantity surveyors on levels of

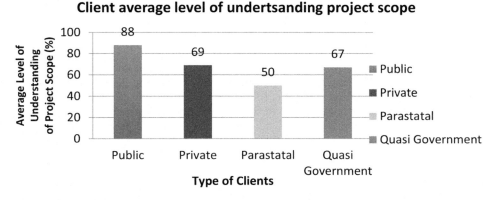

Figure 4. Client average level of understanding project scope.

understanding the project scope. Both respondents indicated that majority of the clients understood up to 75% of the project scope. While this result may be above average, there is still room for improvement in understanding projects scope and reducing cost overruns above the initial budget.

8 DISCUSSION

Findings indicate that the private sector clients handled cost overrun with the maximum of 30%, better than the public sector with worst case scenario being 90% cost overruns when the documented threshold is 25%. While the public clients are subjected to bureaucracy and a large spectrum of stakeholder consultation, they can learn how to manage cost overruns from the private sector by revising procurement approval systems. This provided answers to the first objective of the study that the final estimates between private sector clients was different with public sector clients. Honesty, commitment, and support to make savings were three main factors that were found to influence quantity surveying firm re-engagement by the contractor. This means that these are the areas the quantity surveying firm would need to strengthen to improve satisfaction levels in clients. Significant causes of cost overrun as cited by clients and consultant quantity surveyors were indecision, delayed drawings and unforeseen circumstances. This finding provided answers to the second objective which was to determine what challenges Quantity Surveyors face when preparing cost estimates This implies that there is need to improve risk management in order to reduce the risk of unforeseen circumstances. Drawing handling has been circumvented by many clients through the ISO certification under the care of document- drawing handling and control. The study further reviewed that the public sector and the quasi-government were struggling with prompt decision making resulting in cost overruns driven by inflation due the volatile nature of the local currency. The study also established that the level of understanding the project scope in the initial stage was >50% with most respondents from the clients and quantity surveying firms reporting 75% understanding. These findings provided answers to objective number three on levels of scope understanding by clients. The findings based on the clients and QS data according to Figure 6, 25% of the clients did not understanding the full scope of work as obtained from the clients and QS data could results in very high-cost overruns. This implies that there is great need by clients to fully understand the scope of works before making commitments. While similar studies have been undertaken in other regions (Al Hosani *et al.* 2020; Samarghandi *et al.* 2016; Tobechi *et al.* 2022), this study considered the comparative study between public and private sector clients which presents gaps in allowance for cost overrun thresholds that could be considered for closing up based on the results from the study.

9 LIMITATIONS AND ASSUMPTIONS

The sample size for the clients was small. Despite the small number, many clients, private and public belong to the same categories so the study assumes that large sample would not yield different results. The study could also benefit from a different clientele which includes the mining industry. A project-by-project study approach could also present interesting results that could unlock information of the dynamic subject of cost estimates.

10 CONCLUSION

It is every client's desire and expectation to have the initial cost estimate as the final cost. This study has however indicated that it is not the case most of the times. The study has also revealed that those public and private clients are separated by bureaucracy, indecision, delayed drawing among other factors. Understanding of the scope was also established as a factor that could lead to overruns thereby pushing the final cost estimate away from the initial cost estimate. The quantity surveyor or estimator has the mandate of ensuring that the final account sum and the initial sum is zero or minimal. The study concludes that there are still gaps that need to be closed in keeping cost estimates under control in the public and private sectors. The study recommends that the public sector focusses on improving identified areas of weaknesses such as indecision, delayed drawings excessive cost percentage of overruns.

REFERENCES

Agyekum-Mensah, G., 2019. The Degree of Accuracy and Factors that Influence the Uncertainty of SME Cost Estimates. *International Journal of Construction Management*, 19(5), pp.413–426. https://doi.org/10.1080/15623599.2018.1452094

Aibinu, A. A., & Jagboro, G. O. (2002). The Effects of Construction Delays on Project Delivery in Nigerian Construction Industry. *International Journal of Project Management*, 20(8), 593–599.

Akinradewo, O.I., Aigbavboa, C.O. and Oke, A.E. (2020), "Improving Accuracy of Road Projects' Estimates in the Ghanaian Construction Industry", *Journal of Financial Management of Property and Construction*, Vol. 25 No. 3, pp. 407–421. https://doi.org/10.1108/JFMPC-11-2019-0087

Al Hosani, I. I. A., Dweiri, F. T., & Ojiako, U. (2020). A Study of Cost Overruns in Complex Multi-stakeholder Road Projects in the United Arab Emirates. *International Journal of System Assurance Engineering and Management*, 11, 1250–1259.

Al Juboori, A. A. (2021). Practices That Increase the Budget Estimate's Accuracy During the Initial Phase of the Construction Project's Life Cycle. *International Journal of Construction Project Management*, 13(2), 129–159.

An, S.H., Cho, H. and Lee, U.K., 2011. Reliabilty Assessment of Conceptual Cost Estimates for Building Construction Projects. *International Journal of Civil Engineering*, 9(1), pp.9–16. http://ijce.iust.ac.ir/article-1-463-en.pdf

Bakr, G.A., 2019. Identifying Crucial Factors Affecting Accuracy of Cost Estimates at the Tendering Phase of Public Construction Projects in Jordan. *International Journal of Civil Engineering and Technology*, 10(1), pp.1335–1348. ISSN Print: 0976-6308 and ISSN Online: 0976-6316

Buchan, R. D., Fleming, F. W. & Grant, E. K., 2003. *Estimating for Builders and Surveyors*. 2nd ed. Great Britain: Butterworth-Heinemann.

Cheng, M.Y., Tsai, H.C. and Sudjono, E., 2010. Conceptual Cost estimates Using Evolutionary Fuzzy Hybrid Neural Network for Projects in Construction Industry. *Expert Systems with Applications*, 37(6), pp.4224–4231. https://doi.org/10.1016/j.eswa.2009.11.080

Dandan, T. H., Sweis, G., Sukkari, L. S., & Sweis, R. J. (2019). Factors Affecting the Accuracy of Cost Estimate During Various Design Stages. *Journal of Engineering, Design and Technology*, 18(4), 787–819.

Doloi, H.K., 2011. Understanding Stakeholders' Perspective of Cost Estimation in Project Management. *International Journal of Project Management*, 29(5), pp.622–636. doi.org/10.1016/j.ijproman.2010.06.001

Fazeli, A., Dashti, M.S., Jalaei, F. and Khanzadi, M. (2020), "An Integrated BIM-based Approach for Cost Estimation in Construction Projects", *Engineering, Construction and Architectural Management*, Vol. ahead-of-print No. ahead-of-print. https://doi.org/10.1108/ECAM-01-2020-0027

Hatamleh, M.T., Hiyassat, M., Sweis, G.J. and Sweis, R.J. 2018. Factors Affecting the Accuracy of Cost Estimate: Case of Jordan", *Engineering, Construction and Architectural Management*, Vol. 25 No. 1, pp. 113–131. https://doi.org/10.1108/ECAM-10-2016-0232

Heravi, G. and Mohammadian, M., 2017. Cost Overruns and Delay in Municipal Construction Projects in Developing Countries. *AUT Journal of Civil Engineering*, 1(1), pp.31–38.

Ismail, N.A.A., Rooshdi, R.R.R.M., Sahamir, S.R. and Ramli, H., 2021. Assessing BIM Adoption Towards Reliability in QS Cost Estimates. *Engineering Journal*, 25(1), pp.155–164. DOI: https://doi.org/10.4186/ej.2021.25.1.155

Johnson, R.M. and Babu, R.I.I., 2020. Time and Cost Overruns in the UAE Construction Industry: A Critical Analysis. *International Journal of Construction Management*, 20(5), pp.402–411. doi.org/10.1080/15623599.2018.1484864

Kaliba, C., Muya, M. and Mumba, K., 2009. Cost Escalation and Schedule Delays in Road Construction Projects in Zambia. *International Journal of Project Management*, 27(5), pp.522–531. https://doi.org/10.1016/j.ijproman.2008.07.003

Khairuddin, S. S., & Rashid, K. A. (2016). Quantity Surveyors and Their Competencies in the Provision of PFI Services. *Malaysian Construction Research Journal* (MCRJ), 153.

Muya, M., Kaliba, C., Sichombo, B. and Shakantu, W., 2013. Cost Escalation, Schedule Overruns and Quality Shortfalls on Construction Projects: The Case of Zambia. *International Journal of Construction Management*, 13(1), pp.53–68.

Omonori, A., & Lawal, A. (2014). Understanding Customers' Satisfaction in Construction Industry in Nigeria. *Journal of Economics and Sustainable Development*, 5(25), 115–120.

Oyewobi, L. O. (2013). Influence of Public Service Motivation on Job Satisfaction and Organisational Commitment of Quantity Surveyors in Nigerian Public Service. *Social and Management Research Journal (SMRJ)*, 10(1), 41–76.

Pishgar, F., Dezhkam, S., Ghanbarpoor, F., Shabani, N., & Ashoori, M. (2013). The Impact of Product Innovation on Customer Satisfaction and Customer Loyalty. *Oman Chapter of Arabian Journal of Business and Management Review*, 34(976), 1–8.

Ritz, G. J. (1994). Total Construction Project Management.

Samarghandi, H., Mousavi, S., Taabayan, P., Mir Hashemi, A., & Willoughby, K. (2016). *Studying the Reasons for Delay and Cost Overrun in Construction Projects: The Case of Iran*

Sepasgozar, S. M., Costin, A. M., Karimi, R., Shirowzhan, S., Abbasian, E., & Li, J. (2022). BIM and Digital Tools for State-of-the-Art Construction Cost Management. *Buildings*, 12(4), 396.

Skamris, M.K. and Flyvbjerg, B., 1996. Accuracy of Traffic Forecasts and Cost Estimates on Large Transportation Projects. *Transportation Research Record*, 1518(1), pp.65–69. https://doi.org/10.1177/0361198196151800112

Subram, K. S., Khan, M. N., & Srivastava, C. (2018). The Impact of Marketing Mix Elements on Brand Loyalty: A Case Study of Construction Industry. *Sumedha Journal of Management*, 7(3), 77–98.

Tembo, C. K., Muleya, F., & Bulaya, G. K. (2021). Developing a Social Media Marketing Framework for Small-Scale Contractors in the Construction Industry. *Open Journal of Business and Management*, 10(1), 77–100.

Tobechi, E. B., Chioma, E. P., & Ezinne, O. J. (2022). Evaluation of the Effects and Risks of Cost Estimation on Project Costs and Claims: A Study in Imo State, Nigeria. *International Journal of Environment, Engineering and Education*, 4(2), 36–47

Vaardini, S., Karthiyayini, S. and Ezhilmathi, P., 2016. Study on Cost Overruns in Construction Projects: A Review. *International Journal of Applied Engineering Research*, ISSN 0973-4562 11(3), pp.356–363.

Xiao, H., & Proverbs, D. (2002). The Performance of Contractors in Japan, the UK and the USA: An Evaluation of Construction Quality. *International Journal of Quality & Reliability Management*.

Yang, J. B., & Peng, S. C. (2008). Development of a Customer Satisfaction Evaluation Model for Construction Project Management. *Building and Environment*, 43(4), 458–468.

Zhong, Q., Tang, H., & Chen, C. (2022). A Framework for Selecting Construction Project Delivery Method Using Design Structure Matrix. *Buildings*, 12(4), 443.

Investigation of cost management factors influencing poor cost performance on large to medium sized projects in the construction industry

C.K. Tembo & A. Kanyembo
Department of Construction Economics and Management, Copperbelt University, Zambia

F. Muleya
Department of Quantity Surveying and Construction Management, University of the Free State, South Africa
Department of Construction Economics and Management, Copperbelt University, Zambia

K. Kajimo-Shakantu
Department of Quantity Surveying and Construction Management, University of the Free State, South Africa

ABSTRACT: Various factors contribute to poor cost performance on projects. A study was conducted to investigate the factors contributing to poor cost management in the Zambian Construction industry. A positivist approach was used a cross-sectional manner using self-administered questionnaires for data collection. Respondents were consultants (random sample) and contractors (stratified random sample) carrying out the cost management function in construction firms and consulting firms. A total of 118 self-administered questionnaires were distributed. The response rate of 79% and the reliability test using Cronbach alpha of 0.779 was achieved. The data was analyzed using means and Spearman's rho. Findings show that the top factors contributing to cost management were inflation, variations, extra work, unclear scope, and poor cost control. The afforemetioned are indications of the need for improved forecasting and detailed scoping in the planning and estimation stages of a project.

Keywords: cost management, cost performance, projects, construction

1 INTRODUCTION

There are many factors that influence the cost of construction throughout the life cycle of a project (Konior & Szóstak 2020). The desired position is being able to conduct the project within the budget established. One of the most common indicators for poor project delivery is cost overruns that would imply that the project has exceeded its budget line. There is vast evidence of undesirable cost performance of construction projects worldwide. This is mainly in form of cost overruns (Aljohani *et al.* 2017; Okerereke *et al.* 2022; Vaardini *et al.* 2016). The resultant effects of poor cost performance are reduction in contractors profit, resulting in loses and other ills associated with poor cash flow (Vaardini *et al.* 2016). Studies have been conducted to identify the causes of the cost overruns which is the major indicator of poor cost performance in different sectors of the construction industry in both developed and developing countries (Aljohani *et al.* 2017; Vaardini *et al.* 2016). In the Zambian context it is unclear the factors influencing cost performance of projects in medium to large projects as this account for over 90% by value of projects done in the construction industry (NCC 2019). Since 2003 to date the Auditor general's reports on public projects; who is the major client in the construction industry have revealed that the industry has been under performing in terms of cost management of projects. This finding is reinforced by Muya *et al.* (2013) and Tembo (2018); it is therefore important to identify the factors contributing to this poor cost performance so as to improve the industry's cost performance.

The next section reviews the literature on cost factors contributing to poor cost performance in the construction industry. Thereafter, the methodology for the study is presented. Subsequently, the study results are presented, analyzed and discussed before conclusions are given. Indications for future research are also given.

1.1 Factors influencing cost management on construction projects

Cost management is one of the basic criteria of success measurement in construction projects. This therefore implies that the cost should be managed and controlled throughout the lifecycle of the project. Some of these costs may be variable others fixed; direct or indirect in nature (Kerzner 2018). While plans can be made on how to manage and control for cost; the plans may not materialise due to several factors that may affect the cost of a project. Various studies have been conducted globally to ascertain the factors that affect cost of a project as shown in Table 1. The causes have various origins and are researched in different parts of the world. These are outlined below in a brief manner. Nevertheless the focus of this research is on factors closely tied to cost management. In India, Bhargava *et al.* (2010) identified the significant factors that result in poor cost performance to include: conflict among project participants, ignorance and lack of knowledge, presence of poor project-specific attributes, nonexistence of cooperation, hostile socio-economic, climatic conditions, reluctance in timely decision, aggressive competition at tender stage and short bid preparation time. A study done in Zambia showed change in scope of the project, schedule delays, inflation, technical challenges, inclement climate and environmental issues as causual factors for cost overruns on construction projects (Muya *et al.* 2013).

Sawalhi & Enshassi (2014), researched on cost overruns on projects in the Gaza Strip and revealed the major causes being, material shortage, late procurement of materials and equipment, fluctuation of material prices, delay in progress payment, differentiation of currency prices, rework, poor control of cost, poor site management, poor communication and coordination between stakeholders, inadequate planning and scheduling, mistakes in design documents, project complexity and delay in design approval. Similarly, Polat *et al.* (2014) stated that factors that cause cost overruns on construction projects which are cost-related consist of design problems and mistakes in cost estimates which originate in the planning stage of the projects, bribes, extra works not included in the contract, penalties resulting from safety incidents, obstacles from other contractors, unclear project scope, waste of materials and lack of control over payment on site.

Jedhav *et al.* (2016) uncovered the critical factors of cost overrun to consist of sudden changes made by the client in the specification of materials, design changes during construction, delayed payments by the client, variation in cost of materials, and rework at site due to mistakes. A study by Kaming *et al.* (1997) on the 31 construction projects in Indonesia, revealed that that from a contractor's point of view, cost overruns were mainly caused by inaccuracy of material take-off, increase in material costs and cost increase due to environmental restrictions. In another study by Le-Hoai *et al.* (2008), it was concluded that cost overruns on construction projects in Vietnam were caused by, poor site management, material cost increase due to inflation, inaccurate quantity take-off, poor supervision, financial issues faced by client and contractor, many changes in design and labour cost increase due to environment restriction. Another study carried out in Pakistan revealed that major factors of cost overrun are a shortage of experienced contractors, project site location, security problems, low productivity, and mistakes in the estimation of the project costs (Zaafar *et al.* 2016).

Sohu *et al.* (2018) researched in theBahrain construction projects regarding causative factors cost overrun. A questionnaire was designed and distributed among experts and engineers to find out the causes of cost overruns. Results of study unveiled that factors that cause of cost overruns include: frequently design changes, mistakes during construction, schedule delay, inadequate supervision and site management, mistakes in time and cost estimates, delay in making and approval of different design and drawings, and poor design. Furthermore, 173 causes of cost overrun on construction projects were found in 17 contexts with the main potential causes being: frequent design change, contractors' financing, payment delay for completed work, lack of contractor experience, poor cost estimation, poor tendering documentation and poor material management (Aljohani et al. 2017). In the UAE Johnson & Badu (2020) through a questionnaire survey found design variation, poor cost estimation, delay in clients descion making process, fiancial constraints of client and inappropriate procurement method as top causes of cost ovverruns result in poor cost management of projects. Further in Nigeria Okereke et al. (2022) found that major factors influencing construction cost management are; experience and competence of the project managers, weak management support and control, poor project communications, external economic environment, and lack of use of project management tools (technology).

Table 1. Factors influencing cost performance.

Cause	Cost Management Stage	Author
• Inaccurate cost estimate • Due to changes in scope or material specifications, • Absence of national database for prices to rely on and lack of estimators' experience. • Lack of cost data	Pre-contract stage	Polat et al. (2014), Adjohani et al. (2017), Zaafar et al. (2016), Niazi & Painting (2017), Sohu et al. (2017), Leo-Hoai et al. (2008).
• Unclear project scope due to poor communication poor estimation practices (wrong/inaccurate)		Polat et al. (2014) Ikediashi et al. (2014), Peeters & Madauss (2008)
• Inadequate supervision due to incompetency and unreliable supervisors	Contract stage	Sohu et al. (2017)
• Inflation (changes in material and labour prices)		Muya et al. (2013), Jedhave et al. (2016) Chimwaso (2017)
• Inclement weather		Muya et al. (2013), Amusan et al. (2018)
• Variations		Chimweso (2017) Annor-Asubonteng (2018)
• Design changes due to poor designs		Jedhave et al. (2016), Leo-Hoai et al. (2016), Sawalhi & Enshassi (2014), Sohu et al. (2017)
• Bribes • Poor contract management • Schedule delays due to technical constraints and change in the size of the project poor financial discipline		Polat et al. (2014) by Amusan et al. (2018), Annor-Asubonteng (2018) Muya et al. (2013)
• Poor budgetary and resource management		John & Michael (1994) Meeampol & Ogunlan 2006

Table 1 shows the various factors with the indicative stage in the cost management process of where it is likely to occur. The cost management process in pre-contract has planning, estimating and budgeting. In the contract phase there is controlling. From this brief analysis using the literature it apears that the most causal factors are more evident in the construction phase were the actual management of cost is implemented. The next section gives the methodology used in the study.

2 METHODOLOGY

The research sought to determine the causes of poor cost management in the Zambian construction industry. A positivist approach was used in cross-sectional manner. The population of this research comprised of -contractors and consultants as shown in Table 2. The focus was on construction companies engaged in the building category the National Council for Construction registration as this entity that registers contractors. The Quantity surveyors targeted registered under the Quantity Surveying Registration Board (QSRB) and had at least been involved on a medium to large sized project. The grades of contractors considered were grade 1 to 4 of the National Council for Construction registers. Of these grades 1 and 2 are large scale contractors while 3 and 4 are medium scale. Stratified probability sampling was used to select the sample size for the contractors using the formula shown below. The confidence level of 95% was used and a precision of ±5% was adopted

$$nh = [Nh/N] \tag{1}$$

Where; nh = Sample size, Nh = population size, N = Total population size and n = Total sample size.

Table 2. Cost management tools and techniques.

POPULATION	RESPONDENT DETAILS	POPULATION	SAMPLE	DISTRIBUTED	RECEIVED	RESPONSE RATE %
Contractors	Grade One	8	3	3	3	100.00%
	Grade Two	15	8	8	3	37.50%
	Grade Three	13	6	6	5	83.33%
	Grade Four	52	56	26	22	84.62%
Consultants	Quantity surveyors	30	30	30	27	90%
	Total	118	73	73	64	79.09%

The questionnaire had open and closed questions with three sections. The sections included were demographic information, factors contributing to cost performance and cost management tools and techniques. The cost management techniques are not reported in this paper. The questionnaire used a four point (A = always, B = often, C = sometimes, D = never) and a five point (1 = Applied to a very great extent, 2 = Applied to a great extent, 3 = Applied to some extent, 4 = Applied to a small extent, 5 = Never applied) to determine frequency and application respectively; and six point scale the mean of 1 = Applied to a very great extent, 2 = Applied to a great extent, 3 = Applied to some extent, 4 = Applied to a small extent, 5 = Never applied, 6 = Not indicated. In the analysis closed questions were analyzed quantitatively using descriptive statistics (percentages, means etc.) and inferential statistics (one sample Wilcoxon test, Spearman rho, Shapiro Wilk and Cronbach Alpha). For The Wilxcon test the hypothesis test for each stage in terms of methods used was *H1: The Median of cost factors for cost management stages equals 2*. The hypothesis test for the Spearman rho was *(H2: There is no significant relationship between cost performance factors* (Table 4). To test for normally a hypothesis of Ha The sample comes from a normal distribution was used using the Shapiro Wilk test.

3 RESULTS PRESENTATION

3.1 Data characteristics and demographic information

A reliability test of 0.779 for 65 items was obtained using the Cronbach Alpha test. The value demonstrated reliability of the instrument (Kothari 2014). A 79% response rate was achieved. While many contractors were registered, few were found at the registered addressed or were reachable by phone or email using the registration details resulting in a low response for contractors as shown in Table 2. To determine the distribution of the data the Shapiro Wilk test used due to the small sample size. Shapiro-Wilk test ranged from .781 to 884 (df = 64, p = 0.000 to 0.002 < 0.05). It therefore provided evidence that the population from which the sample was drawn was not normally distributed hence the use of non-parametric statistics.

Of the respondents 29% were females and 71% males for Consultants while 30% were female and 70% were male for contractors. In terms of sector of origin 27% of the consultants are from the public sector while 73% are from the private sector. 46% of the contracting firms' respondents were quantity surveyors, 19% of them were contracts managers, 16% were directors of the firm, 11% were construction managers, 5% were project managers and 3% were site managers. All the respondents from the quantity surveying consultancy firms were quantity surveyors by profession.

3.2 Projects handled

The projects were building projects in nature ranging from commercial, residential and industrial buildings in nature. From the findings, 62% of the contracting firms' respondents had 0 to 2 projects with overruns, 24% of them had 3 to 5 projects with cost overruns and the remaining 14% had over 5 projects with cost overruns for the past 5 years. On the other hand, 41% of the quantity surveying consultancy firms' respondents had 3 to 5 projects with cost overruns, 37% had 0 to 2 projects with cost overruns and the remaining 22% had over 5 projects with cost overruns for the past 5 years.

On average quantity surveying consultancy firms' respondents had projects as follows with an exchange rate of 1US$ = K17.98; 37% handled projects of about 9 to 13 million Kwacha, 26% of about 13 to 25 million Kwacha, 15% of 25 to 55 million Kwacha and 4% had projects of about 55 million kwacha and above. 11% of the respondents were unsure of the value of the projects they handled while the remaining 7% stated that the projects' value was not applicable to them because they were non-profitable organizations. For contracting firms, 60% of the respondents were registered in grade 4), 13% were in grade 3, 19% were in grade 2, and only 8% were in grade 1.

3.3 Causes of poor cost performance

Several factors as indicated in the literature influence cost performance. Table 3 has ranked the factors from the perception of the consultant and contractor individually and the two groups combined. Extra work, variations and change in scope are the top three causes from quantity surveyors perspective while inflation, variations and unclear scope are top three causes for contractors. The implication of these differences implies that in managing costs contractors and quantity surveyors have difference focuses for managing these factors. Overall inflation, extra work and variations are the top contributors. Extra work and change in scope are outside the control of the QS as these normally come at a time when the project is being executed.

A one sample Wilcoxon Test was carried with a hypothesized median of 2 (contributing to a great extent) test was computed. A Wilcoxon signed rank test showed that there was a significant difference for factors 9–11 (Z = 4.780 to −0.875, $p < 0.005$). The hypothesized median score showed that factors 1 to 8, 13 and 22 in Table 4 contributed to a great extent to cost management in the Zambian building sector. While the factors 16 to 21 contributed to a small extent and factors 12 and 14 contributed to some extent to cost management.

Another inferential statistic of interest was to determine relationships or associations that describe cost management factors therefore a non-parametric correlation analysis was conducted using spearman's' rank correlation. The hypothesis being tested was H_1: *There is no significant relationship between cost factors*. In view of the findings shown in Table 5, this hypothesis is rejected as there are several moderate to high correlations. For interpretation Rovai et al. (2013) suggest the following guide to describe strength of statistically significant relationships follows: between 0 and ±0.20 very weak; between ±0.20 and ±0.40 weak; between ±0.40 and ±0.60 moderate; between ±0.60 and ±0.80 strong and between ±0.80 and ±1.00 very strong. Correlations were computed among 16 cost factors for a sample of 64 respondents as shown in Table 4. The results suggest that 26 out of 256 correlations were statistically significant

Table 3. Causes of cost poor cost management.

		Consultant				Contractors				Overall		
NO	Factor	N	Mean	Std. Deviation	Rank	N	Mean	Std. Deviation	Rank	Mean	Std. Deviation	Rank
1	Extra work	27	1.93	.730	1	37	2.08	.894	7	2.02	.682	3
2	Variations	27	1.96	.808	2	37	1.86	.887	3	1.98	.721	2
3	Change in scope	27	2.04	.854	3	37	2.22	1.004	11	2.14	.941	6
4	Poor cost control	27	2.19	.786	4	37	1.95	.911	5	2.05	.744	5
5	Unclear scope	27	2.19	.921	6	37	1.92	.829	4	2.03	.761	4
6	inflation's	27	2.22	.801	7	37	1.81	.811	2	1.91	.682	1
7	Material cost	27	2.26	.813	87	37	2.16	.928	10	2.20	.876	8
8	Inaccurate cost estimates	27	2.41	.971	9	37	1.97	.866	6	2.16	.930	7
9	Poor project specifications	27	2.52	.935	10	37	2.14	1.058	8	2.23	1.019	9
10	Short bid preparation	27	2.81	1.039	13	37	2.59	.832	12	2.69	.924	11
11	Poor tender documentation	27	2.85	.864	14	37	1.97	.866	7	2.34	.963	10
12	Poor communication within project team	27	3.19	0.962	115	37	3.81	1.371	12	3.50	1.167	13
13	Delayed payment	27	2.19	0.962	5	37	3.97	1.554	13	3.08	1.258	12
14	*Estimates prepared by an inexperienced person	27	3.19	0.921	16							
15	*Inadequate client knowledge	27	2.74	1.059	11							
16	*Change in client requirements	27	2.78	0.892	12							
16	*Poor workmanship					37	4.11	1.430	15			
17	*Inappropriate procurement process					37	4.41	1.092	18			
18	Inadequate design					37	4.35	1.086	17			
19	*Poor contract administration					37	4.41	1.093	19			
20	*Poor planning					37	4.16	1.280	16			
21	*Errors in estimates					37	4.08	1.422	14			
22	*Poor supervision					37	1.51	0.731	1			

• risk indicated by that category only

There was a negatively significant moderate relationship between inflation and extra work, change in clients' requirement and estimate prepared by inexperienced personnel. There was a negative significant weak relationship between clients' knowledge and short bidding period. The negative relations ships meant that as one factor increased the other decreased and vice-versa. Positively moderate relationships were observed for inflation and material cost, inaccurate. Estimate and short bidding period, extra work and change in scope, extra work and variations, change in scope and variations, inaccurate estimate and poor project specifications. A significantly weak relationship was found for poor cost control and poor tender documentation. Positive relationships basically mean that as one variable increases the other also increases and vice-versa.

Table 4. Correlations between cost overrun factors.

Factor		1	2	3	4	5	6	7	8	9	10	11	12	13	14	15	16
1. Inflation's extent contribution to cost overruns on building projects	Correlation Coefficient	1.000	.228	.071	-.018	-.234	-.073	-.421*	.559**	-.009	.226	.133	.095	-.320	.212	.109	.120
	Sig. (2-tailed)		.253	.726	.931	.239	.718	.029	.002	.965	.257	.509	.636	.103	.290	.588	.552
2. Poor cost control's extent contribution to cost overruns on building projects	Correlation Coefficient	.228	1.000	.374	-.236	-.378	.291	-.173	.303	.318	.137	.396*	-.096	-.048	-.020	.080	-.079
	Sig. (2-tailed)	.253		.055	.236	.052	.140	.387	.124	.106	.496	.041	.634	.812	.923	.692	.695
3. Inaccurate cost estimates extent contribution to cost overruns on building projects	Correlation Coefficient	.071	.374	1.000	-.042	-.034	.173	-.071	.049	.442*	.488**	.610**	-.141	-.205	.008	.023	.039
	Sig. (2-tailed)	.726	.055		.837	.866	.389	.726	.808	.021	.010	.001	.483	.306	.970	.911	.847
4. Change in scope's contribution extent to cost overruns on building projects	Correlation	-.018	-.236	-.042	1.000	.552**	.183	.479*	-.043	.176	.358	.216	-.181	-.233	.134	.122	.128
	Sig. (2-tailed)	.931	.236	.837		.003	.360	.011	.831	.380	.067	.278	.367	.242	.506	.544	.523
5. Variations contribution extent to cost overruns on building projects	Correlation Coefficient	-.234	-.378	-.034	.552**	1.000	.341	.481*	-.195	-.009	.054	-.016	.002	.074	.098	.133	-.263
	Sig. (2-tailed)	.239	.052	.866	.003		.081	.011	.329	.965	.788	.938	.994	.714	.626	.508	.185
6. Unclear scope contribution extent to cost overruns on building projects	Correlation Coefficient	-.073	.291	.173	.183	.341	1.000	.369	-.253	.092	.201	.284	-.269	-.239	.186	.078	-.152
	Sig. (2-tailed)	.718	.140	.389	.360	.081		.058	.203	.649	.315	.151	.175	.230	.353	.697	.450
7. Extra work contribution extent to cost overruns on building projects	Correlation Coefficient	-.421*	-.173	-.071	.479*	.481*	.369	1.000	-.381	.098	.076	-.072	-.223	-.059	-.091	.016	-.074
	Sig. (2-tailed)	.029	.387	.726	.011	.011	.058		.050	.626	.708	.720	.263	.772	.651	.938	.712
8. Material cost due to inflation contribution extent to cost overruns on building projects	Correlation Coefficient	.559**	.303	.049	-.043	-.195	-.253	-.381	1.000	.289	.278	.284	.341	-.093	-.053	-.031	.245
	Sig. (2-tailed)	.002	.124	.808	.831	.329	.203	.050		.143	.161	.151	.082	.644	.792	.877	.218
9. Poor project specifications contribution extent to cost overruns on building projects	Correlation Coefficient	-.009	.318	.442*	.176	-.009	.092	.098	.289	1.000	.542**	.611**	-.226	-.240	.098	-.035	.366
	Sig. (2-tailed)	.965	.106	.021	.380	.965	.649	.626	.143		.004	.001	.258	.227	.626	.862	.060
10. Short bid preparation contribution extent to cost overruns on building projects	Correlation Coefficient	.226	.137	.488**	.358	.054	.201	.076	.278	.542**	1.000	.836**	-.155	-.397*	.100	.021	.361
	Sig. (2-tailed)	.257	.496	.010	.067	.788	.315	.708	.161	.004		.000	.441	.040	.619	.919	.064
11. Poor tender documentation contribution extent to cost overruns on building projects	Correlation Coefficient	.133	.396*	.610**	.216	-.016	.284	-.072	.284	.611**	.836**	1.000	-.113	-.264	-.075	.003	.224
	Sig. (2-tailed)	.509	.041	.001	.278	.938	.151	.720	.151	.001	.000		.576	.184	.708	.987	.262
12. Poor Communication between the client, consultants and contractors	Correlation Coefficient	.095	-.096	-.141	-.181	.002	-.269	-.223	.341	-.226	-.155	-.113	1.000	.301	.158	-.302	-.004
	Sig. (2-tailed)	.636	.634	.483	.367	.994	.175	.263	.082	.258	.441	.576		.127	.432	.125	.985
13. Clients having little or no knowledge of construction	Correlation Coefficient	-.320	-.048	-.205	-.233	.074	-.239	-.059	-.093	-.240	-.397*	-.264	.301	1.000	-.350	.073	-.193
	Sig. (2-tailed)	.103	.812	.306	.242	.714	.230	.772	.644	.227	.040	.184	.127		.074	.717	.334
14. Delayed payments with interest	Correlation Coefficient	.212	-.020	.008	.134	.098	.186	-.091	-.053	.098	.100	-.075	.158	-.350	1.000	-.313	.316
	Sig. (2-tailed)	.290	.923	.970	.506	.626	.353	.651	.792	.626	.619	.708	.432	.074		.112	.108
15. Estimates prepared by unexperienced personel	Correlation Coefficient	.109	.080	.023	.122	.133	.078	.016	-.031	-.035	.021	.003	-.302	.073	-.313	1.000	-.403*
	Sig. (2-tailed)	.588	.692	.911	.544	.508	.697	.938	.877	.862	.919	.987	.125	.717	.112		.037
16. Change in clients requirements	Correlation	.120	-.079	.039	.128	-.263	-.152	-.074	.245	.366	.361	.224	-.004	-.193	.316	-.403*	1.000
	Sig. (2-tailed)	.552	.695	.847	.523	.185	.450	.712	.218	.060	.064	.262	.985	.334	.108	.037	

*. Correlation is significant at the 0.05 level (2-tailed).
**. Correlation is significant at the 0.01 level (2-tailed).

4 DISCUSSION

The study uncovered the causes of poor cost performance using cost overruns as a proxy with specific emphasis to factors contributing to poor cost management' additionally the relationships that exist within the cost factors was unearthedThe top five causes of cost overruns common between the contractors and consultants on building projects are: inflation, variations, extra work, unclear scope, poor cost control. Scholars like Chimwaso (2017), Kaminng et al. (2017), Le-Hoai et al. (2016) had similar results on the factors that caused cost overruns on building construction projects. Variations in Botswana was found to be highly significant in the causation of cost overruns on building projects (Chimwaso 2017). Similarly, in Vietnam inaccurate quantity take off and material cost increase due to inflation were highly significant. Scoping inadequencies could be contributing to variations, extra works and unclear scope. Time should be taken by designers to undertstand what their clients really want in turn clients should give professionals enough time for cost management (Lings et al. 2021). Additionally, consultants should use the correct pricing mechanism and should be actively monitoring projects to mitigate inflation and poor cost control respectively. In the application of all tool and technique what stands out is the experience held by the person applying the methods as one of the major contributing factors to effective cost management (Obi et al. 2021). Consultants also perceived a noice client and one that is constanting making changes as one prone to contribute to poor cost management while for contractors owed most of their poor cost performance to poor supervision. The study reveal various relations that exist among the factors contributing to cost overruuns.

Many of the studies conducted in the existent literature focus on just identifying factors contributing to cost overruns and subsequently cost performance as shown in Table 1. This study after identifying the factors sought to determine any relationships between factors. Several moderate relations were found from the hypothesized tests set out for the study on establishing relations between cost performance factors. There was a negative significantly weak relationship between clients' knowledge and short bidding period. Positively moderate relationships were observed for inflation and material cost, inaccurate estimate and short bidding period, extra work and change in scope, extra work and variations, change in scope and variations, inaccurate estimate and poor project specifications. A significantly weak relationship was found for poor cost control and poor tender documentation. The implication for these relationships is that when one factor has eventuated then the factor in which it is in a relationship with should be closely monitored. Though this could be done, best practice would be to control and/or mitigate both factors in a relationship so as not to have any effects from either. In practice clients ought to give conclusive briefs so that extra works and variations can be avoidance at the guidance of design consultants who should advise on the best procurement method depending on design information available. Contractors should closely supervise work on site to help keep costs to a minimum.

5 CONCLUSION

Cost is one of the most important aspects in deciding the success or failure of a project. The top cost influencing factors include extra work; variations and change in scope are the top three causes from quantity surveyors perspective while inflation, variations and unclear scope are top three causes for contractors. This study provides guidance for better cost management in the ZCI so that factors do not impact project negatively. For consultants, an experienced estimator who understands what the client wants to avoid having a lot of extra work is needed while for the contractor good supervision is needed for good cost performance. This in turn will posively influence cost management in the construction phase. In addition to chosing the appropriate procurememnt route based on information available. Prudent supervision by both consultanst and contractors would help minimise costs.

Construction firms in grades 1 to 4 of the NCC register were considered in these studies which are medium to large scale in nature. However, there is need to study left out grades 5 to 6 which constitutes the largest number of the contractor population. Additionally, future studies can take a multiple case study approach which would incorporate qualitative methods to understand the extent to which the factors affect cost management. The study can also be extended to other sectors of the construction industry.

REFERENCES

Aljohani, A., Ahiaga-dagbui, D. and Moore, D. 2017. 'Construction Projects Cost Overrun: What Does the Literature Tell Us?', *International Journal of Innovation, Management and Technology*, 8(2), pp. 1–143. doi: 10.18178/ijimt.2017.8.2.717.

Amusan, L. M. et al. 2018. 'Data in Brief Data Exploration on Factors that in Fl uences Construction Cost and Time Performance on Construction Project Sites', *Data in Brief*. Elsevier Inc., 17, pp. 1320–1325. doi: 10.1016/j.dib.2018.02.035.

Annor-Asubonteng, J, Callistus. T, Tom Mboya Asigri, & Napoleon Kuebutornye D. K. 2018. Investigating the Cost Management Practices of Indigenous Firms in the Ghanaian Construction Industry, *Journal of Economics and Behavioral Studies* (ISSN: 2220 -6140) Vol. 10, No. 5, pp. 179–186.

Arif, M., Awuzie, B., Islam, R., Gupta, A.D. and Walton, R. 2021. "Critical Success Factors for Cost management in Public-housing Projects", *Construction Innovation*, Vol. 21 No. 4, pp. 625–647. https://doi.org/10.1108/CI-10-2020-0166

Bhargava, A. et al. 2010. 'Three-Stage Least-Squares Analysis of Time and Cost Overruns in Construction Contracts', (November), pp. 1207–1218.

Ikediashi, D. I., Ogunlana, S. O. & Alotaibi, A. 2014. Analysis of Project Failure Factors for Infrastructure Projects in Saudi Arabia: A Multivariate Approach. *Journal of Construction in Developing Countries*, 19, 35–52.

Kaliba, C., Muya, M. and Sichombo, B. 2013. Causal Factors of Cost Escalation, Schedule Overruns and Quality Shortfalls in Construction Projects in Zambia. *International Journal of Construction Management*, vol. 13, no. 1, pp 53–68.

Kaming P.F, Olomolaiye P.O, Holt G.D, Harris F.C. 1997. Factors Influencing Construction Time and Cost Overruns on High-rise Projects in Indonesia. *Construction Management and Economics* 15(1):83–94.

Kerzner, H. 2018. *Project Management: Best Practices: Achieving Global excellence*, 4th Edition, USA Wiley

Konior, J. and Mariusz, S. 2020. "Methodology of Planning the Course of the Cumulative Cost Curve in Construction Projects" *Sustainability* 12, no. 6: 2347. https://doi.org/10.3390/su12062347

Le-Hoai, L., Lee, Y.D. and Lee, J.Y., 2008. Delay and Cost Overruns in Vietnam Large Construction Projects: A Comparison with Other Selected Countries. *KSCE Journal of Civil Engineering*, 12, pp. 367–377.

Ling, S.C.A., Jun, L.W., Hashim, N., Kamarazaly, M.A.H., Yaakob, A.M. and King, L.S., 2021. Factors and Impacts on the Accuracy of Cost Planning: Pre-contract Stage. Factors and Impacts on the Accuracy of Cost Planning: Pre-contract Stage, *Malaysian Construction Research Journal- special issue* vol. 12 no 1, p. 159

Malkanthi, S. N., Premalal, A. G. D. and Mudalige, R. K. P. C. B. 2017. 'Impact of Cost Control Techniques on Cost Overruns in Construction Projects', 1(04), pp. 53–60.

Memon, A. H. 2011. 'Preliminary Study on Causative Factors Leading to Construction Cost Overrun', *International Journal of Sustainable Construction Engineering & Technology*, 2(1), pp. 57–71.

Okereke, R.A., Zakariyau, M. and Emmanuel, E.Z.E., 2022. The Role of Construction Cost Management Practices on Construction Organisations' Strategic Performance. *Journal Of Project Management Practice (JPMP)*, 2(1), pp. 20–39.

Peeters, W. & Madauss, B. (2008). A Proposed Strategy Against Cost Overruns in the Space Sector: The 5C Approach. *Space Policy*, 24, 80–89.

Polat, G., Okay, F. and Eray, E. 2014. 'Factors Affecting Cost Overruns in Micro-Scaled Construction Companies', 85, pp. 428–435. doi: 10.1016/j.proeng.2014.10.569.

Project Management Institute. 2017. *A Guide to the Project MAnAGeMent Body of KnowledGe*.

Sawalhi, N. El and Enshassi, A. 2014. 'Cost Management Practices by Public Owners and Constractors in the Gaza Strip', *International Journal of Construction Management*, 1(January 2004), pp. 17–28.

Tembo-Silungwe, C. and Khatleli, N. 2017. 'Deciphering Priority Areas for Improving Project Risk Management Through Critical Analysis of Pertinent Risks in the Zambian Construction Industry', *Acta Structilia*, 24(2), pp. 1–43.

Sohu, S. et al. 2018. 'Causative Factors of Cost Overrun in Building Projects of Pakistan', *International Journal of Integrated Engineering*, 10(9), pp. 23–27.

The Auditor General. 2020. *Report of the Auditor General on the Accounts for the Financial Year Ended 31st December 2020.*

Vaardini, U.S, Karthiyayini, S. and Ezhilmathi, P. 2016. Study on Cost Overruns Inconstruction Projects – A Review. *International Journal of Applied Engineering Research*, 11(3), pp. 356–363.

Vasista, T. G. K. 2017. 'Strategic C ost m anagement f or C onstruction P roject S uccess: A s ystematic S tudy', *Civil Engineering and Urban Planning: An International Journal*, 4(1), pp. 41–52. doi: 10.5121/civej.2017.4105.

Zafar I, Yousaf T, Ahmed S. 2016. Evaluation of Risk Factors Causing Cost Overrun in Road Projects in Terrorism Affected Areas Pakistan–a Case Study. *KSCE J Civ Eng*. 20(5):1613–1620.

Gender equity, social justice & social inequality in construction

Adoption of microgrids as an energy solution to uplift rural communities in the Eastern Cape, South Africa

S. Xulaba & C.J. Allen
Nelson Mandela University, Gqeberha, Eastern Cape, South Africa

ABSTRACT: Rural communities in South Africa face still experience the challenges of not having access to electricity. The use of generators has been common; however, alternative sources of electricity are required due to cost and environmental concerns. The study aims to provide strategies for successfully implementing renewable energy microgrids to promote economic development in rural areas of the Eastern Cape. The data was collected from small businesses operating in the rural areas of the Eastern Cape. The study used a qualitative method design and structured interviews supported by a few closed-ended questions. The study revealed that businesses spend most of their earnings on fuel for diesel-powered generators, making their operations non-profitable. Further studies can focus on the life cycle assessment of the microgrids for the various businesses in rural areas.

Keywords: Renewable energy, Microgrids, Rural economic development, Primary economic activities, District Municipalities

1 INTRODUCTION

Rural areas in the Eastern Cape are underdeveloped and need more access to electricity. Cook (2011) states that rural electrification has the potential to act as a catalyst for rural development, so the idea of not only providing electricity for these areas but also looking at how electrification can boost their economy and promote growth is what people in rural areas need to ensure a more promising future. According to the Organisation for Economic Co-operation and Development (2018) report, rural areas possess more potential for global opportunities than what meets the eye. The options identified in the report include developing new energy sources that meet climate change challenges, innovation in food production for growing populations and providing natural resources that will enable the next production revolution. Rural areas consist of diversified rural economies beyond agriculture and other natural resources-based sectors. According to Kanagawa (2008), poverty has been one of the significant obstacles for sustainable development. In addition to economic deprivation, Kanagawa (2008) describes poverty as the low attainment of social services such as education, health, and nutrition, and one of the ways to cope with the multi-dimensional aspects of poverty is to promote opportunities such as access to modern energy. Therefore, renewable energy microgrids are considered an optimal solution for rural areas in the Eastern Cape. Microgrids are considered a potential source for electrification in rural areas because generation is geographically close to the loads, reducing transmission construction requirements and costs. Furthermore, they have a flexible utilisation of distributed generation, increased reliability of critical loads, and provide better reactive power management and optimisation of operations at the distribution level (Kermani *et al.* 2015). Having access to affordable and clean energy is one of the sustainable development goals. Promoting electricity in rural areas is essential as it enables these areas to improve their standard of living and enhance production income expenditure and educational outcomes. The first section of the paper provides a brief introduction of the study, then followed by the review of the relevant literature. The third section focuses on the methodology adopted by the study. The fourth section discusses the findings, and the last sections discusses the conclusion and recommendations of the study.

2 RURAL ECONOMIC ACTIVITIES FOR DEVELOPMENT

According to the European Network for Rural Development rural areas can be classified based on population density and distance to the nearest city and makes mentions of four different types of typologies, namely, agricultural regions, consumption countryside region, diversified region with strong manufacturing base and diversified region with strong market service. For an agricultural region, the economy is mainly dependent on primary activities. For a consumption rural region, the economy comprises of small-scale farming and has tourism and recreation as key activities. The diversified region with strong manufacturing base can be a combination of agricultural region and consumption region with a strong manufacturing base. And lastly, the diversified region with a strong market service consists of an agricultural region with a strong market service. The rural non-farm activities include manufacturing or setting up and studies have shown that non-agricultural activities are starting to emerge as substantial contributors towards the rural economy and have influenced rural economic growth, creation of employment, livelihood diversification and reduced poverty (Gibson & Olivia 2010). The lack of electricity is seen as the most important obstacle in the investment climate and the difficulties of extending the national grid to rural areas consists of factors such as high capital costs, the scattered nature of the area, low load factor, distribution and electricity loses and a lack of policy and legal framework (Sahu et al. 2014). Studies have shown that utilising renewable energy technologies to generate electricity can help bridge the development gap in rural areas. Thurner (2019) states that renewable energy technology can assist rural areas in South Africa to convert sunshine into income, a flexible grid created by a decentralized energy system enables operators of the electricity system to balance demand and supply. The most effective use of energy generated from renewable energy sources is to create goods or services that have an impact of income or create value, and a high degree of correlation between energy use, economic growth, and level of development. For rural development, the systematic view of productive use of energy is mainly connected with the provision of power for agricultural and industrial or commercial purposes. In terms of the concept of rural development, the energy component was used for two distinct purposes, namely for residential use and for productive use (Cabraal et al. 2005). The use of electricity for domestic purposes can improve and have a positive effect on the quality of life and standard of living in rural areas. The industrious use of energy in rural areas, on the other hand, can increase rural productivity and economic growth, which can reduce the high rate of unemployment (Cabraal et al. 2005)

3 MICROGRIDS

Cook (2011) argues that electricity infrastructure as a consumption and an intermediate good is linked to growth in income and therefore the connection between income and infrastructure may be in both directions. The author continues to state that changes in income can lead to changes in the demand for electricity and electricity generation. According to Cook (2011), the slowness to extend electricity to rural areas in developing countries has resulted in a large proportion of the world's population to still not have access to electricity and that the disappointing progress towards providing sufficient rural electricity has partly contributed towards the failure to raise the income of rural households and effectively design tariffs and adapt regulatory systems that can make electricity more affordable to poorer communities. Renewable energy solutions as an energy provider for rural areas not only meet the criteria of being cost effective and socially beneficial to the community but they also offer significant environmental conservation methods that will help these areas to retain their cultural sentimental value. Kirubi (2008) states that the failure to extend electricity to rural areas means a loss for development potential. Research has shown that one of the major causes for energy loss is technical losses which occurs through transmission and distribution and at least 18% of the generated electricity is lost through distribution (Khonjelwayo & Nthakheni 2021). With the significant loss of electricity through transmission, microgrids can be considered as a viable option due to their ability of being a power system that provides clear economic and environmental benefits as compared to the expansion of modern power system (Hatziargyrious 2005). Microgrids also prove the option of being installed as hybrid power supply and the advantages of hybrid microgrids, includes being able to provide sustainable solutions to supply power in case of an emergency and power shortage during interruptions in the primary grid (Wang *et al.* 2011). They possess a plug-and-play functionality which has a feature for switching to a suitable mode of operation to be either grid connected, or island operated and each system and be structured differently for each applicable scenario for that specific area. The access to electricity is one of the decisive components for business success in developing countries (Akpan *et al.* 2013). From a study conducted by (Neelsen & Peters 2011), it was found that access to electricity has minimal direct impact on business profits or work remuneration however there is a significant indirect effect mainly due to the increase in demand for goods and services prompted by migration from non-electrified communities to electrified communities. An additional study done by Arnold (2008) also found that the reliability problems of the electricity grid had a statistically significant negative impact on the firm's total factor production.

4 METHODOLOGY

The philosophy adopted by the study was interpretivism. The interpretivist research philosophy seeks to understand the ways through which people experience the world (Zukauskas 2018). Saunders *et al.* (2009), describes the inductive approach as an approach that focuses on formulating a theory, it is used when a researcher wants to have a better understanding of the nature of the problem therefore the research approach adopted by the study was the inductive approach. The study adopted a qualitative research methodology in a form of structured interviews. The study used a non-probability sampling method and the technique that was used was the quota sampling technique (Unit of Statistical Consultation (USC) 2019). The population group of the study was business owners in the rural areas of the Eastern Cape, the study targeted all six district municipalities and a sample of 10 participants was set to be interviewed from each district municipality. The structured interviews consisted of closed ended and open- ended questions and it was divided into four sections. The first section the demographic section, the second section was used to determine the skills possessed by the participants in their respective areas and what kind of businesses were being conducted in these areas, section three covered the availability of resources and the last section covered information relating to renewable energy knowledge. Before data was collected, the study went through the ethical clearance process, and this is because the participants from the rural areas are considered as a vulnerable group. There was an enrolment session were the participants briefed about the purpose of the study and what was required to them as the target population group. Participants were given to make decisions regarding the willingness to participant in the study and after that process, participants were than given consent forms to complete and some of the participants indicated that they would rather withdraw from the study. The open-ended questions of the structured interviews were analysed using thematic analysis. The data was collected and organised in order and then analysed to create codes. After the coding was done, data was then analysed again to create themes. The closed-ended questions were analysed using descriptive statistics, the study mostly used measures of central tendencies with included the mode, median and graphical representation of data such as tables, bar graphs and pie charts.

5 FINDINGS AND DISCUSSIONS

The study was conducted in the Eastern Cape province, in South Africa. The province has six district municipalities, and the data was collected from three distinct municipalities. These three district municipalities are very distinct from each other. The differences are also used to distinguish the various factors that affect rural areas. Majority of the were from the Sarah Baartman District Municipality (SBDM), participants were asked to indicate their current working role, the data showed that most of the participants were farmers and they either specialized in crop, livestock or poultry farming. The study interview 20 participants and the profile of the participants is below in Table 1:

Table 1. Demographic details of the participants.

	Participants	Gender	Current Work roles	Skills availability
Sarah Baartman District	12	6 Males 6 Females	10 Farmers 1 Entrepreneur 1 Farm Manager	Electricians Farming Mechanic
Nelson Mandela Bay Municipality	3	3 Females	3 Farmers	Brick Laying Sewing Catering Dress making Plumbing Driving Firefighting
O.R. Tambo District	5	3 Females 2 Males	2 Entrepreneurs 1 Assistant Farm Manager 2 Farmers	Electrician Carpenter Farming

5.1 Challenges affecting businesses in rural areas

The data showed that 83% of the participants are farmers and have no other profession or jobs besides being involved in agriculture. This shows that rural areas in the Eastern Cape are still focused largely on primary activities to boost their economy. Figure 1 shows that 42% of these businesses are not making profit and even though 58% of the businesses are profitable, participants have indicated that there is no business growth and expansion because majority of the business income is not being reinvested back into the business, but it used for personal and household expenses.

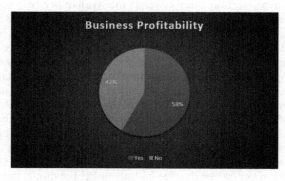

Figure 1. Business profitability in the SBDM.

Figure 2. Business profitability in the NMBMM.

There is no distinction between business revenue for re-investment and personal use. The participants were also asked to identify their daily operating challenges – electricity and business finances were the highest ranked. Other challenges that are being faced by these businesses in the rural areas include water, proper infrastructure, finances, transport and customers. The participant indicated that not having electricity or limited access to electricity was very detrimental to the operating of their business. Cook (2011) argues that electrification enables livelihoods by stimulating employment and income generating activities where people build assets and achieve better cash flows. Electrification enables people in rural areas to use surplus resources made possible through their entrepreneurship that contribute to the emergence of credit and saving schemes based on the newly available cash.

Figure 3. Business operational challenges in the SBDM.

Figure 4. Business operational challenges in the ORTDM.

At least 33% of the participants stated they do not have any access to the electricity and have to make use of alternative energy sources like a diesel-powered generator to operate their business. And the other 67% of the participants indicated they do have access to the grid however they were experiencing high electricity costs due to increasing tariffs and they also stated that electricity from the grid was not stable. In addition to ongoing loadshedding in the country, the participants stated that they would not have electricity for weeks without any valid reasons and due to such issues, are forced to alternative measures like the use of diesel-powered generators to operate their businesses. Cook (2011) states that cost continues to be a barrier to accessing off-grid electricity for poor households in rural areas. All participants indicated that in the absence of electricity, they make use of diesel-powered generators, and they were asked about renewable energy sources and their level of knowledge for these energy sources and at least 80% of the participants indicated they had minimal to no knowledge about renewable energy sources. The various district municipalities are situated in different geographical locations and therefore as much as the participants experience similar problems, the extent into which these challenges are being experienced is different. Participants situated in municipalities closer to the metro such as the Sarah Baartman District Municipality have access to the main grid and their major business challenges is unstable electricity, finances, and water, and 65% of the businesses situated in this municipality have profitable businesses. The data shows the businesses situated in the rural outskirts of the metro had different challenges such as customers, transportation, and water. And for participants that were situated in municipalities that are further from the cities and the metro indicated that electricity and water was their biggest challenge and 60% of these participants indicated that they need not have access to electricity. Most participants indicated that they are primarily involved in agricultural businesses, the participants in the SBDM and NMBM are most invested in livestock farming and the participants in the ORTDM are invested in crop farming. The unavailability of sustainable electricity causes high mortality rate in the livestock especially for poultry farmers and this has a negative impact in the business's revenue. Participants also indicated that without electricity, it is difficult to irrigate their crops because most of the farms operate on boreholes and with loadshedding and unsustainable electricity, they are unable to meet their production targets. Participants also indicated that having the option of diesel-powered generators is also an inconvenience to them because of the increasing petrol costs. Participants have stated that with the constant increasing diesel costs, it is practically impossible to make a profit.

5.2 Theme 1: Availability of inefficient electricity

According to (Onu *et al.* 2023) microgrid investments in developing countries are mainly encouraged by the increasing need for energy access for economic developments and over one billion people in developing and undeveloped countries cannot access electricity. In most cases, rural dwellers are victims of energy poverty and deprivation, and the rural population constitutes a vital sector of any developing economy due to their role in agriculture for food and raw materials. Participants indicated that they have load shedding and other electricity problems that affects their businesses significantly.

P3: "Our everyday business challenge is electricity. The community does not have electricity and in our farm, we use a generator."

The productive use of energy is vital to microgrid sustainability in rural communities. The productive use of energy in rural microgrids has economic and social dimensions and the economic aspect of the design considers an optimal system to accommodate the energy requirements of productive energy users like small businesses (Onu *et al.* 2023). The participants involved in poultry farming indicated that without electricity they are unable to maintain the suitable thermal comfort temperatures in their structures in order to grow their stock. They also indicated that they experience a huge mortality rate of their stock during winter and therefore they need to use alternative measures such as wood or paraffin to warm up the structures.

P2: "We do not have access to electricity, our community is not connected to Eskom. For our poultry farming business we use gas to warm up our poultry structures. We also bought a generator that we use for lighting and when we cannot afford to buy diesel for the generator, we buy paraffin and use a paraffin heater for cooking and keeping our poultry structures warm."

According to Onu *et al.* (2023), microgrids are potential technologies for electricity infrastructure expansion and they are flexible enough to fit into urban and rural electricity infrastructure expansion. Apart from load shedding issues, participants especially in the ORTDM indicated they sometimes go for days without electricity, and they might be different causes as to why they spend days without electricity and during this time they need to use alternative measures for cooking and lighting.

P5: "Our electricity is not reliable, in a case of a storm, strong winds or heavy rains then the electricity will go out for 2–3days.

P8: "Our electricity infrastructure is not in a good condition, one day we have it and then something gets broken and then we don't have it."

5.3 Theme 2: Electricity expenditure

Participants stated that the increase in electricity prices was negatively affecting their businesses and the use of paraffin and diesel-powered generators was not cheap either due to increasing fuel prices. The alternative options available to the participants is not cost effective and does not assist with the growth of their businesses.

P10: "Electricity is very expensive for us farmers because we use electricity for our irrigation systems 7 days a week and that affects our financial proceedings. The only downfall with having access to have electricity is that it is very expensive.

P11: "Our business is not making profit because of the problems we are experiencing such as electricity, the cost of purchasing electricity is very high."

OECD (2018) states that renewable energy microgrids provides remote rural regions with the opportunity to produce their own energy rather than importing conventional energy from external sources moreover being able to generate reliable and cheap energy can trigger economic development.

6 CONCLUSIONS

With challenges such as climate change and energy loss through distributions, renewable energy micro grids can provide a sustainable energy supply to rural areas. The need for stable electricity supply in rural areas cannot be overly stated. The study has shown that electricity is a significant challenge in rural areas. If most of the residents without electricity were to have access to electricity using renewable energy microgrids businesses in rural areas can have increased productivity. The productivity of these businesses in the rural areas has been hugely affected by lack of electrification. The study has also identified that the lack of electrification in rural areas has not motivated the participants to diversify their products and services and this has resulted in a limited opportunities for economic growth. If participants are struggling to mitigate the existing operational challenges in their businesses, it would be difficult to invest in other businesses because they might also be prone to the same challenges. The study has shown that the even though these businesses are situated in different locals, but the main problems are the same. It cannot be disregarded that rural areas have great business potential, assisting these areas with solutions to overcome their operational challenges could create a healthy and vibrant rural economy.

It is therefore recommended that rural development policy address the possibility of using hybrid micro grids solutions as an alternative to the expansion of the Eskom. In addition, the municipalities should incentivize the use of alternative energy solutions, alongside existing funding opportunities available to small business enterprises. Municipalities can also promote the diversification of businesses in rural areas. An increase in the various businesses which require electricity can the help balance the demand and supply of electricity in rural areas. Primary activities use less electricity than secondary and tertiary economic activities and if more businesses were operating in the latter economic sectors, a greater demand of electricity can assist municipalities to increase supply, which can attract investors because the magnitude of the investments can generate a huge return investment. Microgrids can act as agents for development if adequately managed and are designed to meet the economic needs of each specific area in rural areas. Some of the innovative ways that electricity generated from alternative renewable energy microgrids can have on rural areas include: the use of solar powered boreholes and irrigation systems to provide water, creating renewable energy substations to supply electricity. Accessibility to some parts of these rural areas was one of the study's limitations, and this was due to poor road conditions. Further studies can focus on the life cycle assessment of the microgrids for the various businesses in rural areas.

REFERENCES

Akpan U, Essien M, & Isihak. 2013. The Impact of Rural Electrification on Rural Micro-enterprises in Niger Delta, Nigeria. *Energy for Sustainable Development*. 17:504–509.

Arnold JM, Mattoo A, & Narciso G. 2008. Services Inputs and Firm Productivity in Sub-Saharan Africa: Evidence from Firm-level Data. *Journal of African Economics* 17(4):578–99.

Cabraal, Anil, Barnes, Douglas & Agarwal, Sachin. 2005. Productive Uses of Energy for Rural Development. *Annual Review of Environment and Resources*. 30: 117–144.

Cook, P. 2011. Infrastructure, Rural Electrification, and Development. *Energy for Sustainable Development*.15 (3): 304–313.

Gibson, J. & Olivia, S., 2010. The Effect of Infrastructure Access and Quality on Non-Farm Enterprises in Rural Indonesia. *World Development*. 38(5): 717–726.

Kanagawa, M. & Nakata, T. 2008. Assessment of Access to Electricity and the Socioeconomic Impacts in Rural Areas of Developing Countries. *Energy Policy*. 36(6):2016–2029.

Khonjelwayo, B., & Nthakheni, T. 2021. Determining the Causes of Electricity Losses and the Role of Management in Curbing Them: A Case Study of City of Tshwane Metropolitan Municipality, South Africa. *Journal of Energy in Southern Africa*. 32(4), 45–57.

Kirubi C. 2006. *How Important is Modern Energy for Micro-enterprises? Evidence from Rural Kenya*. Berkeley: University of California.

Neelsen S., & Peters J. 2011. Electricity Usage in Micro-enterprises — Evidence from Lake Victoria, Uganda. *Energy for Sustainable Development*.15:21–31.

Organisation for Economic Co-operation and Development (OECD) 2011, *Regional Outlook 2011*. Paris

Onu, U.G., Zambroni de Souza, A.C. & Bonatto, B. D. 2023. Drivers of Microgrid Projects in Developed and Developing Economies. *Utilities Policy*. 40: 101–487.

Papathanassiou, S., Hatziargyriou, N. & Strunz, K. (2005) A Benchmark Low Voltage Microgrid Network, Power Systems with Dispersed Generation: Technologies, Impacts on Development, Operation and Performances. *CIGRE Symposium*, Athens, 1–8.

Sahu, A., K., Shandilya, A., M., & Bhardwaj, S.K. 2014. Rural Electrification: Issues and Challenges of Sustainable Development. *International Journal of Emerging Technology and Advanced Engineering*. 4 (8):1–7

Saunders, M., Lewis, P. & Thornhill, A. 2009. *Research Methods for Business Students*. 5th ed. Essex: Pearson Education. https://theconversation.com/how-technology-could-help-rural-south-africa-turn-sunshine-into-income-117348

Unit for Statistical Consultation. 2019. *Introduction to Quantitative Research Methodology*. Nelson Mandela University

Wang, J., Yang, Li, X., Yang & H., Kong, S. 2011. Design and Realization of Microgrid Composing of Photovoltaic and Energy Storage System. *Energy Procedia*. 12: 1008–1014

Zukauskas, P. & Vveinhardt, J. & Andriukaitienė, R. 2018. *Philosophy and Paradigm of Scientific Research*. 10.5772/intechopen.70628.

A framework for sustainable human settlement building methods for low-cost housing in Northern Cape province

T. Bremer & S.C. Monoametsi
Department of Quantity Surveying and Construction Management, University of the Free State, Bloemfontein, South Africa

ABSTRACT: The aim of the study was to evaluate a framework for sustainable human settlement building methods for low-cost housing in the Northern Cape Province. The objectives were to assess drivers and barriers and propose recommendations for a suitable framework. The study involved 15 participants from the Department of Cooperative Governance, Human Settlement, and Traditional Affairs (COGHSTA) in the City of Kimberly specifically the Sol Plaatje Municipality. Qualitative research methods were employed, including semi-structured interviews and thematic analysis. Nine key themes emerged, addressing the research objectives. Recommendations included a buyer-supplier code, well-structured policies, and procedures for housing projects, and the establishment of project management offices. Other suggestions included conducting feasibility studies, providing training for managers and officers, and adopting innovative approaches to affordable housing structures. Implementing the recommendations could lay the foundation for a new sustainable framework for low-cost housing in the Northern Cape Province.

Keywords: Low-cost housing, sustainable, human settlement

1 INTRODUCTION

A framework for sustainable Human Settlement building methods was the focus of the research. The research has been conducted at the Department of Co-Operative Governance, Human Settlement and Traditional Affairs (COGHSTA), and Sol Plaatje Municipality in Northern Cape Province. As part of service delivery, the department provides housing to impoverished communities.

The goal of the department and the municipal government is aimed at advancing the standard of living for its people, and to stimulate, collaborate, and observe organisations and structures which meet the livelihood of citizens and how the service is being delivered to them in Northern Cape. The major housing projects were within the Human Settlements unit (construction of houses), which was half the budget allocation of the entire department under the grant allocation, making it a key focus for this study. It took the government 21 years for the improvement and execution of human settlements housing processes, legal structure, guidelines, and programs, which caused innovative thoughtful strategies to handle the most recent developing difficulties while also meeting forthcoming demands (Bredenoord 2017). Notwithstanding advances in the supply of houses, human settlement arrangements continue to be problematic nationally. The market for housing is deemed to be broken and unbalanced, and lacking accessibility to its mechanisms and advantages, which causes tenacious housing affordability problem across multiple sub-markets, notably the gap submarket (Department of Human settlement 2016).

Many investigations, studies, reviews, and evaluations of human settlements policies and programs have been carried out primarily by the National Planning Commission, National

Treasury, Financial Commission, Department of Performance Monitoring and Evaluation, Department of Cooperative Governance, and Statistics South Africa (Ganiyu 2016). Further the Department of Human Settlements, as well as conducted studies research organisations, revealed a slew of flaws because of policy gaps and inconsistencies in program execution. Thorough examination was therefore necessary in managing these challenges effectively (Bredenoord 2017).

Human settlements in South Africa are still being built without the essential pre-planning for township development, as well as the essential infrastructure and services to make them effective and sustainable. Communities remain fragmented, with impoverished households moved in distant locations of Northern Cape Kimberley (Ganiyu 2016). Homes for the low-income earners are frequently constructed with no or little attention for infrastructure, services, formal planning, or conformity to construction requirements which hampers the sustainability of the low-cost houses in Northern Cape, specifically in Kimberley, which leads to the communities being unable to engage actively and effectively in all aspects of human settlements and its development, especially in assisting with building their own dwellings (Department of Human Settlements 2016). Even though the NHBRC increased the feature of government-subsidized housing, deficiency of capability in the building project, management remains a tremendous concern for the industry. Statistics South Africa discovered in 2014 that 15.9 percent of families with subsidized housing had unsustainable structures and 15.3 percent had unstable rooftops (Ganiyu 2016). Poor construction quality has led to a considerable requirement for cleanup or reconstruction, both of which have come at a significant cost to the state (Department of Human Settlements: 2016).

The Minister of Housing implemented the norms and standards for the National Construction of the Stand-Alone Houses, on April 1, 1999. All houses built under the National Housing Programs must at the very least meet these norms and standards (Bredenoord 2017). These National Norms and Standards should be consistent with the National Builders Regulation and the National Home Builders Registration Council's house building guidelines (NHBRC). The mandated contractors to be registered with the Council for builders for them to be eligible to build houses and for the five-year structural failure warranty to apply (Department of Human Settlements 2016)

2 LITERATURE REVIEW

According to Golubchikov & Badyina (2012); Lin *et al.* (2015) Sustainability is commonly defined as development that meets the demands of the present generation without jeopardizing future generations. Li *et al.* (2016) explain sustainability as the creation of low-cost housing so that meeting the housing requirements of current medium-low-income groups does not jeopardize the ability to satisfy the housing requirements of future groups. The goals were to achieve total quality in terms of environmental, social, and economic performance Roufechaei *et al.* (2014) argued that while tackling the housing deficit, particularly in developing nations, sustainability considerations are frequently disregarded Ross *et al.* (2010), the economic viability of low-cost housing schemes has drawn significant attention (Pullen *et al.* (2009); Winston & Montserrat 2007).

Pullen *et al.* (2010) provided a different view by adding environmental sustainability as not always compatible with housing affordability, which was measured by the cost of affordable housing expected to rise if the environmental sustainability was taken into considerations. Such misunderstanding showed that people rarely used sustainability to achieve affordability (Golubchikov & Badyina 2012; Nubi & Afe 2014). As a basic social necessity, governments in developing nations are committed to providing suitable, sustainable low-cost housing of satisfactory features to the low-income group. Many places globally have been dominated in recent decades by the development of a new stock of dwellings. The practice of modifying existing houses is becoming increasingly essential. This has given a chance to improve the properties and communities' social, economic, and environmental performance, as needed by many sustainable development principles.

The study concerned not just with the building components, but also with the interaction between the dwellings, the entire surroundings, and the inhabitants (Isnin *et al.* 2018). According to Bredenoord (2017) high-quality building materials must always be used for the construction of the low-cost housing, to ensure its long-term viability. High-quality materials, for example, are long-lasting, have attractive qualities, and require little care, because of its durability. A house must give safety and security from the influences of the local environmental climate, like as cold and wind, rain, and heat of the Kalahari Desert, which can be extreme. The home should provide people with security and should never collapse. Natural disasters such as earthquakes, tropical storms, and floods place special quality requirements on housing material and its construction (Ganiyu 2016).

Sufficient planning laws must restrict construction of low-cost houses in places that are significantly endangered, particularly in places that technological solutions cannot safeguard such as dikes in flood zones. Natural occurrences are getting more violent because of climate change, causing a greater focus on the criteria and material used for constructing durable houses for the marginalized poor (Ganiyu 2016). Bredenoord (2017) further continues to say that Local, communities and municipalities usually have some understanding about local building materials and construction procedures. Metropolitan building cultures, and natural building materials which are frequently substituted by cement, concrete, and steel, affect many of the modern building methods used in rural regions.

As a result, one looks for chances to incorporate modern advancements into conventional construction materials and procedures. National and international agencies have set up and are monitoring pilot initiatives (Isnin *et al.* 2018). Using new technology to local construction materials can help to enhance traditional approaches in local communities. Individual homes and local communities should have access to professional support to:

- Elude disinvestment and use environmentally friendly products and practices.
- Get finance for the building of a house or sections of a house.
- Make use of safety precautions for places endangered by large natural events.
- In most cases, these methods are expensive, thus external funding from government's human settlement, humanitarian groups, private firms, banks, and so on is required (Isnin *et al.* 2018).

High-quality and technological structure mechanism must be supplied constantly to develop durable and sustainable human settlement for the Low-cost Housing. Disinvestment (shortcuts) must be avoided. Technical non-governmental organisations (NGOs) and municipalities together with governments each have their own set of responsibilities. The apex of building control would be municipal construction monitoring. The most efficient strategy to avert construction failures and disinvestments would be to provide technical support in advance. It could have been a perfect arrangement if the local government building oversight was available (Bredenoord 2017).

2.1 *The key drivers for sustainable human settlement building methods*

Bredenoord (2017) has found that it has taken the Government too long, over two decades to realise and attempt to get to a workable working solution to improve the execution of human settlements housing processes, legal structure, guidelines, and programs, which caused strategies to fail in meeting the demands of the people most in need for affordable houses. Odoyi & Riekkinen (2022) further found that the key driver for sustainable low costs housing was the need for the "masses" to have safety and security. People need service delivery and the desire to have access to a place they can refer to as "home".

2.2 *The key barriers that affect the successful sustainability of human settlement building methods for low-cost housing*

Ganiyu (2016) found that over 20% of families who had subsidised housing had unsustainable structures and 16% had unstable rooftops. Poor construction quality has led to a considerable requirement for cleanup or reconstruction, both of which have come at a significant cost to the state. Ndlangamandla & Combrick (2020) found that the construction methods for the development of housing and updating of the informal settlement that was considered having undesirable effects on the natural settlement, to determine the environmental sustainability of construction practices. Researchers found that the absence of ecological sustainability in the identified construction practices used, therefore, determined that the current construction practices used within informal settlements lead to negative environmental effects. Krishna et al. (2020) found that the lack of municipal funding, corruption, lack of management and building skills and lack of leadership support has created the most significant barriers to low-cost sustainable building projects in many communities most in need of housing.

2.3 The key recommended enablers for successful sustainability of human settlement building methods for low-cost housing

Celiker (2017) found and thereafter recommended that better and more advanced architecture be required for low-cost sustainable housing methods. There needs to be adequate research and development done to ensure that architecture is both of the highest quality and sustainable for mass housing projects and this requires the highest levels of innovation and creativity for cost management. Noorzai & Golabchi (2020) found that structures must be highly tested for durability and harsh weather before they are mass produced. Pilot structures should first be built in order to test durability and steel or mortar strength. Safety measures are key in the testing methods. Vananvati (2018) recommended as an enabler that proper and detailed drawings and building plans need to be first done, tested, and assessed, before taking on such massive housing units and building. Architecture approval and engineering experts need to test these structures first in a few pilots before mass production and subsequent sales to people.

3 DATA COLLECTION

In applying thematic analysis for this study, the differences and similarities in opinions and views were identified, the data was then organised in terms of codes and the emergence of the different themes noted (Creswell & Creswell 2018). After the data was coded, recurring topics were uncovered that provided answers to the study's questions. The target population in the study are the managers, supervisor and engineers employed at COGHSTA, in Kimberly, in the Northern Cape Province. The target population was therefore 120 managers at employees in that district. This study used frequency analysis, which shows how often the same 15 participants used particular words, phrases, and sentences. A non-probability sampling technique, judgmental sampling creates samples solely based on the researcher's expertise and experience. The judgment sampling method, also known as purposive sampling, involves selecting participants consciously based on their characteristics, as was done in this instance. They were chosen from various departments and units in order to get a diverse view and opinions on sustainable housing in the area. The departments included the supply chain and procurement division, Finance, human settlements, and environmental units directly involved in the housing projects in the district

The primary objective of the study was to develop an appropriate framework for sustainable Human Settlement Building Methods for Low-Cost Housing for Northern Cape Province. The sub-objectives looked at were:

- To identify the drivers for the successful sustainable human settlement building methods for low-cost housing in the Northern Cape.
- To identify the barriers that affect the successful sustainability of human settlement building methods for low-cost housing in the Northern Cape.
- To identify the enablers for the successful sustainability settlement building methods for low-cost housing in the Northern Cape.

This investigation focused on the Northern Cape Province's COGHSTA department in South Africa. The research study did not use the other departments (DENC and DPW) listed above. Due to time limits and potential volume issues, municipalities and national departments were to be excluded. Further and future studies may focus on more department and environmental units to attain a larger sample size, which could yield wider results, findings and recommendations that would add value in getting to a solution in the country to sustainable and affordable low-cost housing for most of the population in lower income levels.

4 RESULTS AND FINDINGS

A total of 15 participants were interviewed to collect the primary data required to establish findings and make the recommendations. The key emergent themes, which were derived, and such themes, would set forward the findings.

Table 1. Demographics of the participants.

No	Gender	Key Role	Function at Department
P 1	Male	Supply chain at COGHSTA	Manage the supply chain
P 2	Male	COGHSTA Management	Financial Reporting
P 3	Female	COGHSTA Management	Facilities Management
P 4	Male	COGHSTA Management	Town Planner
P 5	Female	Corporate Governance	Environmentalist
P 6	Male	COGHSTA Management	Civil Engineer
P 7	Male	COGHSTA Management	Compliance Reporting
P 8	Male	COGHSTA Management	Programme Manager
P 9	Female	COGHSTA Management	Financial Management
P 10	Male	COGHSTA Management	Supply chain management
P 11	Male	COGHSTA Management	Supply chain management
P 12	Male	COGHSTA Management	Management of assets
P 13	Female	COGHSTA Management	Analyst
P 14	Female	COGHSTA Management	Tenders Management
P 15	Female	COGHSTA Management	Policy Monitoring

Table 1 above, depicts the demographics and profiles of the 15 participants. The second column expressed whether they were male or female. The third column showed where they worked, which was COGHSTA in Kimberly in the Northern Cape Province, South Africa. The last column was a sign of their key functions at the Department. Participants had worked in Finance, human settlement, the environmental unit, others worked in engineering, compliance, and the tender / procurement offices. The results indicate that Male participants were at 53% of the total. The females were at 47% of the total. There is gender parity in the Department is commensurate with the EAP (economically active persons in South Africa (STATS SA 2020). The results do, however, indicate a skewness in favour of males to female employees in the Department, which may have originated from the Department employing more males at one stage. However, transformation is moving closer to the equal distribution of male and female employees.

4.1 Objective one: Participant discussion

Participants felt that the increasing of the sustainability of operations within the property sector is essential if the government wants to lessen the detrimental influence that is on the shoulders of the society, by changing people's habits is an important part of moving the

Table 2. Key emergent themes and frequencies.

No	Description	Key Emergent Themes
Objective one	The key **drivers** for the successful sustainable human settlement building methods for low-cost housing.	1. Need for shelter and dignity. 2. Affordable quality homes. 3. Framework to facilitate home purchases.
Objective two	The **barriers** that affect the successful sustainability of human settlement building methods for low-cost housing.	1. Lack of Funding. 2. Lack of Management accountability. 3. Lack of affordable quality material.
Objective three	The key **enablers** for the successful sustainability settlement building methods for low-cost housing.	1. Buyer-supplier code to facilitate house purchases. 2. Proper financial allocation and budgeting. 3. Well-structured and easy to follow policy guidelines.

government toward more environmentally friendly policies. Other participants felt that full and proper consultation with the beneficiaries or communities should be encouraged, in order for the human housing and settlements department to have access to data and then put in efforts to get the projects implemented well. This view as expressed by the participants was supported Ganiya (2016) who argued for strong infrastructure in low-cost housing to stand harsh weather and ensure that people in the lower LSM (living standard measures) had the safety and security of dignified housing structures. Communities in the Northern Cape Region are looking for what the rest of other communities need as well. These needs include the access to basic amenities including water, sanitation, garbage removal, and electricity, as well as suitable, appropriately located, cheap, and financially sustainable housing. Reduction of building operating cost, protection biodiversity, stakeholders' management are important in driving successful implementation. In terms of inspection and design stage, using locally sourced building materials and labour, community participation and health and safety are also drivers. Krishna *et al.* (2020) supported the above notion, by arguing that whilst housing in poorer communities needs to be relatively "cheap" they must still afford human beings' dignity in proper living conditions in whatever type of community they live in. These views were also supported by Odoyi & Riekkinen (2022).

4.2 *Objective two: Participant discussion*

Participants outlines the key barriers to sustainable provision of low-cost housing as: delays in decision making are caused by factors such as the lack of legislation that requires sustainability policies and the production of multiple policy documents to enforce sustainability in all elements of government developed housing. They further stated that in order for innovative ideas to be successfully implemented, the management and leadership of the construction sector and individual organisations need to work hand in hand with the government. There could be several challenges in putting the housing concept into action if it isn't backed by strong management and driven by new ideas. They were a lack of drive and accountability from senior officials and management. Other participants were negative in their responses and stated that there was no adequate capital or proper budgeting availability. Many township and rural communities had no "agreement," for good quality and low-cost sustainable housing. There was also a shortage or inadequate supply of good quality yet affordable hardware / and building material. Other participants further argued that there was a lack of skills in constructing houses and the infrastructure required to create sustainable communities. The participants also argued that the South African economy was historically and currently restrictive in ensuring there is an enough budget for the country's ambition to unlock alternative building. There was adequate skill for building the traditional way, but training and tertiary education might have to be considered to re-skill the current high potential and high-performance innovative builders. The government's policy makers and implementers should buy into it and make it a regulation. The lack of demand from what is presented, is creating some serious challenges according to other participants of the study. Participants from the environmental units stated that the climate of the Northern Cape was excessive in all conditions, which was hampering the success of the developments in the Northern Cape Province. The province has five regions with different cultural backgrounds, where many citizens and local community leaders have different views on housing requirements. Some of the local communities still are believing in mud houses, where we can make the fire inside them and cook from a black pot, whilst others have progressed and want larger brick and mortar type houses with modern infrastructure, water, electricity, and proper sanitation.

4.3 *Objective three: Participant discussion*

Participants (4 out 15) expressed the view that action is required from the South African National Government or Provincial Government to provide viable fiscal policies that make sustainable housing building cost-competitive with conventional construction. They felt that the demand would increase, stimulating economies of scale that would eventually favour more environmentally friendly construction methods and technology. Local municipal authorities must realise that their central role in the planning process would provide a good chance to advance sustainable development. Each local government should implement supplementary planning guidance to educate the public on the logistical challenges of sustainable building, and they should take this chance to coordinate the policies that affect their various departments, including those dealing with transportation, housing, waste management, energy efficiency, and environmental health.

Officials at the regional level should be able to advise regional developers and contractors on how they may aid the national government in accomplishing sustainable development's core social, economic, and environmental objectives. Further, there could be the establishment of a buyer-supplier code with obligations for both parties, which was mutually agreed upon. The stakeholders should define the responsibility of sustainable procurement efforts, complete with distinct roles and duties. Proper analysis of the requirements and benchmarks with other countries must be studied, where it has been successfully implemented must be done. Additional to the studies, is to do an environmental analysis of the country, to check all available resources, then refocus skills by educating people on how to build these houses. Other participants directly involved in human settlements from a building perspective argued for customised ABT/IBT specific buildings to the conditions of the province. The participants stated that there should be well researched, tested, and durable ABT/IBT building methods. Further to this, there should be meaningful community engagement and support and lastly, political agreement. The building contractors should be held accountable if they are not using quality materials in order for the houses to be sustainable, and to ensure effective administration and valuation of projects.

5 RECOMMENDATIONS & CONCLUSION

Emanating from both the literature review and the primary data analysis, the following recommendations are made to attain a suitable framework for the sustainable human settlement building methods for low-cost housing in the Northern Cape Province: A policy code or policy framework should be drafted to facilitate the purchase of homes to create a new framework for affordable housing methods. The code is recommended to guide and regulate the purchase of homes in an ethical and fair manner to prevent abuse or fraud.

The Human Settlements Department should appoint a proper and fully functional business unit which would be responsible for budget allocation, funding, and project management of low-cost housing in the areas. These Business Units would handle the Finance, raise funding from different platforms and then manage the costing and project timelines within a new framework for affordable housing schemes in the Northern Cape Province. It is recommended that before any start of a new housing project, that specialist practitioners be involved in feasibility studies, to assess the purchase of land, assess the need for infrastructure and plan accordingly. The feasibility studies should be done way in advance before any decisions are taken on where and how to build and for whom. Within the human settlements department, a proper project management office be established purely to manage housing projects for low-cost environments and communities.

This project management offices needs to employ individuals with the most suitable skills to manage such infrastructure and housing developments. All managers and staff be upskilled with the latest knowledge and business acumen in handling low-cost housing projects. Embark on proper research and be fully trained on competencies required to manage projects within a new framework of sustainable housing for certain lower LSM communities. Upskilling should be undertaken whilst in the job to make these projects sustainable and successful. Taking the recommendations and key findings of the study undertaken, new policies should be drafted. These Policies should form a framework to govern sustainable housing projects from inception all the way to monitoring and evaluation of housing structures both in the short and long terms.

5.1 Future research

The study was conducted at the COGHSTA Department in Kimberly, in the province of the Northern Cape. This was a limitation. Further studies may be conducted in other regions and provinces as well. Housing is a significant challenge in South Africa. Conducting research in other regions, communities, rural areas in the wider South Africa, would yield wider and more diverse findings. These findings may provide workable solutions to the housing shortages in South Africa. Second, the study used only 15 participants in a qualitative study, which is small and relatively restrictive. Conducting a quantitative study using a large sample size may be warranted in the future to attain other suitable alternatives for the housing challenges in South Africa. Global studies may also be conducted in similar emerging economies like South Africa, whereby other countries may have come closer to viable alternatives in solving their respective low-cost housing challenges.

5.2 Conclusion

The study was conducted in the Province of the Northern Cape, using 15 participants as a sample. The views, opinions and recommendations from the participants using qualitative research method were coded and thematic analysis was employed to derive key themes. There were nine key themes which emerged, and these themes formed the basis upon which the key findings were discussed, and key recommendations were derived from both the key themes and the literature review. The key recommendations, therefore, would form the basis upon which the new framework for sustainable human settlement building methods for low-cost housing is based. Should the Department invoke these recommendations into their current practices, then it is envisaged that a more sustainable and successful building method would emerge in due course.

REFERENCES

Bredenoord, J. (2017). Sustainable Building Materials for Low-cost Housing and the Challenges facing their Technological Developments: Examples and Lessons Regarding Bamboo, Earth-Block Technologies, Building Blocks of Recycled Materials, and Improved Concrete Panels. *Journal of Architectural Engineering Technology*, 6: 187.

Çeliker, A. (2017). Sustainable Housing: A Conceptual Approach. *Open House International*, 42 (2): 49–57.

Creswell, J. and Creswell, D. (2018) *Research Design: Qualitative, Quantitative Approach and Mixed Methods Approach*. 5th edition Sage, London.

Creswell, J. W. (2002). *Educational Research: Planning, Conducting, and Evaluating Quantitative and Qualitative Research*. Prentice Hall, New Jersey.

Department of Human Settlements (2016). *Programmes and Subsidies*: 1–28.

Ganiyu B. (2016): *Strategy to Enhance Sustainability in Affordable Housing Construction in South Africa*.

Ganiyu, B.O. Fapohunda, J.A. and Haldenwang, R. (2017). Sustainable Housing Financing Model to Reduce South Africa Housing Deficit. *International Journal of Housing Markets and Analysis*, 10(3): 410–430.

Golubchikov, O., Badyina, A. (2012). In: French, M. (Ed.), *Sustainable Housing for Sustainable Cities: Policy Framework for Developing Countries*. United Nations Human Settlements Programme, Nairobi.

Isnin, Z., Ramli, R., Hashim, A.E., Ali, I. (2018). Are House Alterations Sustainable? *Journal of ASIAN Behavioral Studies*, 3 (6):29.

Li, D., Chen, Y., Chen, H., Guo, K., Eddie, C.-M.H., Jay, Y. (2016). Assessing the Integrated Sustainability of a Public Rental Housing Project from the Perspective of Complex Eco-system. *Habitat International*, 53: 546–555.

Ndlangamandla, M.G and Combrick, C.C. (2020). Environmental Sustainability of Construction Practices in Informal Settlements. *Smart and Sustainable Environment*, 9 (4): 523–538.

Noorzai, E. and Golabchi, M. (2020). Selecting a Proper Construction System in Small and Medium Mass Housing Projects, Considering Success Criteria and Construction Volume and Height. *Journal of Engineering, Design and Technology*, 18(4):883–903.

Pullen, S., Arman, M., Zillante, G., Zuo, J., Chileshe, N., Wilson, L. (2010). Developing an Assessment Framework for Affordable and Sustainable Housing. *Australasian Journal of Construction Economics and Building*, 10 (1/2): 60.

Pullen, S., Zillante, G., Arman, M., Wilson, L., Zuo, J., Chileshe, N. (2009). *Ecocents Living*: Affordable and Sustainable Housing for South Australia University of South Australia.

Ross, N., Bowen, P.A., Lincoln, D. (2010). Sustainable Housing for Low-income Communities: Lessons for South Africa in Local and Other Developing World Cases. *Construction Management and Economics*. 28 (5):433–449.

Roufechaei, K.M., Bakar, A.H.A., Tabassi, A.A. (2014). Energy-efficient Design for Sustainable Housing Development. *Journal of Cleaner Production*, 65 (15): 380–388.

Vahanvati, M. (2018). A Novel Framework for Owner Driven Reconstruction Projects to enhance Disaster Resilience in the Long Term. *Disaster Prevention and Management: An International Journal*, 27 (4): 421–446.

Pathways to meaningful upgrading of urban informal settlements: Towards adequate housing infrastructure in South Africa

M.G. Mndzebele & T. Gumbo
Department of Urban and Regional Planning, University of Johannesburg, Johannesburg, South Africa

ABSTRACT: The challenges of how housing infrastructure impacts social equity have recently come to the forefront of social commentary across South Africa. The prioritization of equal access to conventional human settlements has emerged as a key element. The prime argument is the absence of clear pathways to dealing with the continual creation of urban informal settlements. The study adopted a secondary approach to gather relevant data on social justice in the redevelopment of urban informal settlements. The paper reveals that innovative practices such as community engagement, services and infrastructure, land tenure, and planning can proffer an effective upgrading of urban informal settlements in South Africa. The paper recommends that the National government play a leading role in enabling an environment that includes ensuring appropriate and flexible relevant national legislation for upgrading informal settlements.

Keywords: Informal Settlements, Social Justice, Social Inequality, Conventional Housing, In Situ Upgrading

1 INTRODUCTION

UN-Habitat (2016) slums are the most underprivileged and omitted form of informal settlements, categorized by poverty and large collections of dilapidated housing frequently situated in the utmost perilous city land. Over the past years, informality has been a central problem in many cities of the developing world (Guevara 2014). The absence of provision of acceptable planned surveyed and serviced plots for housing growth has been a key difficulty to secured land tenure for many developers in city areas. The only opportunity available to these people is to create their homes in the existing informal settlements, which are "affordable and accessible" (Jones 2017; Marutlulle 2017; Nazire & Kita 2016). As a consequence of population growth and rapid urbanization, informal settlements form a major part of the urban fabric globally. Despite the global propagation of informal settlements, UN-Habitat recognizes that the aspects of informal settlements remain inadequately addressed (Routledge 2013). Across sub-Saharan Africa, majestic statistics of 238 million people dwell in informal settlements (UN Stats 2020). Over half of the population in South Africa lives in urban centres, where more than 2,700 informal settlements are in existence and account for approximately 20 per cent of total households (SERI 2018). This has resulted in issues related to poor living conditions, poverty, inadequate infrastructure and housing. The ambition of urban informal settlements upgrading interventions as voiced in the Breaking New Ground is to transform informal settlements into integrated and sustainable human settlements that are inclusive and exposed to municipal services and opportunities. Elementary to the success of effective upgrading of urban informal settlements rely on community participation with their skills, finance and knowledge. In order to realise the research aim, the research is premised on the following objectives:

- Historical background of informal settlements in South Africa
- Providing precedent approaches to upgrading informal settlement
- Providing pathways enhancement to the current approach of upgrading informal settlements in South Africa

This paper recognizes the past approaches in upgrading informal settlements and the deficiencies in implementing the UISP in urban areas. Moreover, it presents pathways to meaning upgrading of informal settlements with an emphasis on the reform of implementation policies, achieving adequate housing infrastructure and finally government proffering an enabling environment for competent upgrading.

2 INFORMAL SETTLEMENTS

In the South African context, informal settlements are defined by legal, social and physical characteristics; hence it becomes difficult to explain the term 'adequate' housing (Housing Development Agency 2013). According to the HAD (2015a), many scholars highlight the issue of land tenure while others refer to the dwelling type (slums with poor-performing building materials). However, Roy (2011) recommends a progressive explanation of informal settlements as areas of livelihood, habitation, politics and self-organisation. Huchzermeyer (2011), stresses that informal settlements are spontaneous, popular and complex neighbourhoods offering an immediate response to housing needs and with their positioning crucial for the socio-economic activities of the involved settlement. This narrative shifts away from the genesis and pathology of informal settlements, envisioning instead their potential as dynamic areas of habitat.

2.1 *Precedent responses to upgrading informal settlements*

For many years, low-income and middle-income countries have reacted to informal settlements with various approaches and techniques, comprising of eviction, denying their existence, and demolishing settlements in part or whole (Vahapoğlu 2019). Recently it has been understood that eviction and demolishing does not address the realities that drive the creation of informal settlement. Ironically, there are still countries utilizing eviction and demolishing as an approach to deal with urban informal settlements. UN-Habitat (2018), millions are threatened with forced evictions, while at least 2 million people globally are forcibly evicted every year. These issues continue to persist despite the fact that the right to adequate housing is definite to all and a requirement for sustainable and inclusive urban centres.

2.1.1 *Demolishing and evictions*
The response of eviction and demolishing is evident in African cities and this confirms that African governments have no respect for the rights of their citizens. In Nigeria's capital, Abuja, over 75,000 houses were demolished (Harris 2008; Ogun 2009), while 5,000 homes were destroyed in Angola (Croese 2010). Forced evictions are widespread in Kenyan cities. In March 2021, the United Nations Special Rapporteur addressed the Kenyan government on incapacitating information of how residents of Kidos informal settlements were viciously evicted and their homes demolished without notice. It was further gathered that the forced evictions resulted in 3500 people being homeless during the rainy season. Reactions of evictions and demolishing have aggravated the growing interest in finding pathways towards effective upgrading of informal settlements without disturbance of socio-economic and livelihood of the affected communities.

2.1.2 *In situ upgrading*

The favoured approach by the growing body of literature is the *in situ* upgrading. This approach involves the formalization of informal settlements in their original location(Del Mistro & Hensher 2009; Huchzermeyer 2006; Massey 2014). It has been considered best practice in 'participatory slum improvement' and it is intensely supported by establishments such as the World Bank (D'Cruz *et al.* 2009). *In situ* approaches can be wide-ranging, from the provision of site-and-services associated with formal settlements to simply dealing with land tenure to incremental housing improvement (Georgiadou & Loggia 2021). *In situ* upgrading aims at avoidance and minimizing inconvenience and disruption of resettlement.

2.2 *Informal settlements in South Africa*

The phenomenon of informal settlements is not new in South Africa. They developed in the 1940's and 50's in the key cities (Huchzermeyer 2004, p. 95). With the formation of the National Party in 1948, the concentration was on forced removal of informal settlements and racial separation to support the apartheid model of white cities. During the process of the removal of informal settlements, the government initiated a mass low-income housing project situated in zones that became known as black townships, which were equally located far from white cities. Nevertheless, the provision of these houses did not meet the increasing demand for accommodation in urban zones and the National Party enforced 'influx control' through various regulations such as the Prevention of Illegal Squatting Act 52 of 1951 (PISA) to exterminate informal settlements (The Republic of South Africa 1951).

2.2.1 *Development of sites and services*

Huchzermeyer (2004), in the 1980s, South Africa was marked by fierce rejection of the apartheid system and urban unrest. The civic movements strengthened during this time and informal settlement communities found backing in NGOs such as the Development Action Group (DAG) and the Built Environment Support Group (BESG) (Huchzermeyer 2004, p. 118). The President's Council developed a new strategy for urbanization named 'Orderly Urbanization' in response to the growing unrest, and the PISA was amended for a new class of 'designated zones' to be declared permanent zones where urban planning laws, Group Areas Act and Slum Act did not apply, proffering a more pliable way to manage informal settlements at the time (Muller 2013). This showed a shift towards begrudging acceptance of informal settlements under controlled circumstances.

A crucial part of this acceptance was the building of 'sites and services schemes', in part borrowed from the World Bank Approach of the 1970s, and the Urban Foundation (UF), which was supportive in the design of this approach. The UF's approach to site and services was to deliver a layout plan of individual sites, with basic services (typically water standpipes and a pit latrine), and attain land tenure (Forster & Gardner 2014, p.125). The UF had a strong effect on the informal settlement upgrading approach that was later implemented by the Independent Development Trust (IDT), established in 1990 (Huchzermeyer 2004, p. 199). Forster & Gardner (2014), the IDT housing model was to deliver sites and services through a capital instrument. Huchzermeyer (2001), the capital subsidy did not contain a top structure but a pit latrine, allowing criticism of toilet towns by the civic movement. The developmental NGOs played a role in implementing numerous initiated huge projects that tackled *in situ* upgrading, the largest being Bester's Camp in Durban and Soweto-on-Sea in Port Elizabeth (Huchzermeyer 2004, p. 156).

2.2.2 *New Era of reformulated housing policy in South Africa*

The National Housing Forum (NHF) was formulated in 1992, comprising political groupings, development organizations and financial institutions in order to negotiate a new housing future for the new dispensation (Huchzermeyer 2004, p. 133). Significantly parallel to the IDT model (except that it included funds for a basic top structure), the government pledged funding to the NHF for a housing capital subsidy scheme. The 1994 democratic government's quantitative goal was to deliver 1 million houses in five years (Khan & Thurman 2001, p. 3). The dominant approach to housing delivery to the poor was proving single sites and services, land ownership and a top structure.

The Reconstruction and Development Programme (RDP) which is now defunct, became a central focus of the state-subsidized housing delivery programme. The period saw the involvement of community development organizations such as the Homeless People's Federation, but their influence was mostly in being able to have an impact on the top structure building rather than changing the approach of the RDP housing delivery programme (Huchzermeyer 2004, p. 30). The approach focused on supply (product-driven) and provided little development support as encouraged by the international community (Basset et al. 2003).

2.2.3 *Implementation of Breaking New Ground 2004*

The Breaking New Ground (BNG) expresses a restructured version of the RDP housing programme. BNG subsidy builds on the existing housing policy uttered in the *White Paper on Housing (1994)*, but changes the tactical attention from merely warranting the delivery of affordable housing to making sure that housing is brought in settlements that are sustainable and habitable. It comprises of a new housing subsidy programme, specifically the Informal Settlement Upgrading Subsidy Programme (UISP) which is a phased *in situ* upgrading approach to informal settlements with international best practices (Huchzermeyer 2006, p. 45). South Africa, for the first time, through the BNG and the UISP, had a human settlements policy that explicitly addressed *in situ* and zone-based upgrading (Huchzermeyer 2006, p. 59).

3 METHODOLOGY

This paper adopted secondary data approaches in the arrangement of journals, articles, research papers, books and government documents and websites which were presented in the form of a literature review to act as precedents for the data collected on upgrading of informal settlements in South Africa. Literature analysis was used to comprehend the gaps and opportunities in the utilization of the UISP. The data was analyzed in terms of each identified deficiency in the process of upgrading and then pathways to meaningful upgrading of informal settlements were presented with the support of different case studies and relevant literature.

4 FINDINGS AND DISCUSSION

This section discusses the key findings of the study and highlights the deficits that inform the need for pathways to a meaningful upgrading of urban informal settlements to achieve adequate housing infrastructure in South Africa. The UISP is the current guiding policy instrument for informal settlement upgrading and facilitates a phased approach to upgrading. The upgrading for each settlement is structured into four phases: Phase 1-Application, Phase 2-Initiation, Phase 3-Implementation, and Phase 4-Housing and Consolidation.

4.1 *Pathways to meaningful upgrading of urban informal settlements*

There is a high need for innovative practices such as community engagement, infrastructure, services and land tenure to substantiate an effective approach in the upgrading of urban informal settlements.

Communities are best placed to recognize local priorities and needs, and therefore to determine suitable development trajectories for their settlements. The Integrated Urban

Development Framework (IUDF), National Development Plan (NDP), Spatial Planning and Land Use Management Act (SPLUMA) and UN Sustainable Development Goals (SDGs) all highlight the importance of informal settlement upgrading. Isandla (2019:9), community involvement is a crucial element of our relevant policies and legislation in South Africa.

4.1.1 Community participation, spatial and social justice

Social justice in the redevelopment of urban informal settlements is critical. The delay and confusion of effective implementation of the UISP is a result of the government's failure to substantively and seriously involve the people whose lives are affected by the upgrading process (SJC). Isandla (2014), spatial justice submits that there is a link between the use of space and justice. If city dwellers do not have access to adequate basic services, shelter and land and to platforms for meaningful involvement in the construction of their living environments, then justice remains an abstract idea with no material outcome.

The findings of the UISP baseline assessment evaluation that active participation and community empowerment are fundamental beliefs of the UISP, however, that is currently poorly implemented (Rebel Group 2017). A central approach to informal settlement upgrading is an emphasis on societies as active agents with critical roles to play in the implementation, planning and maintenance of development interventions (Cirolia et al. 2016; Fieuw & Hendler 2018; Isandla Institute 2019; NUSP 2015; SERI 2019). The Department of Human Settlements (2015), acknowledges that there has been an inability for civil society and communities to effectively and meaningfully participate in the various dimensions of human settlements development, including urban management, project planning, and urban planning, and there is an inadequate ability to efficiently leverage the social capital and potential self-help, inherent in societies. SERI (2019), in informal upgrading projects, it is essential that current systems, procedures, patterns and arrangements that make up practices, norms and agency in the settlements that are being upgraded are understood and recognized, and that the lucidity of an intervention approach begins with what already exists.

Throughout all the phases of the UISP, the communities must be transparently and effectively involved. Isandla Institute (2019:76), co-production requires a change in mindset as it perceives communities as strategic partners with a deep-rooted understanding and knowledge of their settlements that must be attached. Essentially, co-production methods allow for more contextually-sensitive incremental settlement upgrading plans that answer well to the desires of communities, and these methods improve confidence in the municipality, reinforce community commitment to the upgrading progression, and empower communities.

In the upgrading of Slovo informal settlement in South Africa, different opinions of the community members were invited but not necessarily considered because the objectives of the meetings were to share information on the plans derived for the upgrading and proffer progress reports (Maganadisa et al. 2021). An elementary concern for community participation remains the legitimacy of engagement with the dwellers, and the ability to be involved beyond consultation to the empowerment of individuals (Theron & Mchunu 2016:115–147). The UISP also sets out principles of engagement and states that "it is of the utmost importance that the dwellers are involved in all aspects of the upgrading process of the settlements" (National Housing Code 2009). Community participation is critical for the success of any project delivered to people. Maganadisa et al. (2021) further emphasized that beneficiary education of prescripts of the UISP policy is essential so that the settlement dwellers can buy into a project they holistically understand.

4.1.2 Re-blocking

Basson (2019:35), though the UISP is part of incremental housing interventions and in theory prefers in-situ upgrading over moving dwellers, it does not fully support in-situ upgrading in practice. In the City of Ekurhuleni re-blocking does not serve as the first step in the application of in-situ upgrading through UISP, as the informal settlements are usually too dense to comply with traditional standards and norms, and can therefore not be wholly upgraded in-situ (Basson 2019:56). The delays and confusion within the UISP regarding re-blocking are a call for proffering and investing in allocating resources toward making informal settlements livable today as an objective, as opposed to formalization in the long term.

The prominent challenges of re-blocking projects are dwellers of communities feeling manipulated into agreeing to the re-blocking initiatives. Reports from Abahlali baseMjondolo (2018b) and Sacks (2018) discovered in Estineni, community members were uninformed of the re-blocking project while contractors were already on-site and commencing the dismantling structures. Basson (2019), municipal-driven re-blocking appears to not certainly result in communities taking ownership of their settlements. The success of re-blocking lies in the settlement dwellers as it is regarded as a community initiative (Sokupa 2012).

4.1.3 *Infrastructure*

The delivery of housing fits impeccably into the scope of providing adequate services; it aids to meet the needs of people for housing on macro and micro levels and contributes an important element to social development (Patel 2015). Moreover, a lack of housing which is of poor quality will inevitably reflect hostilely on the degree of social development of a certain community and the economic environment within which its individuals' members live (Social Work Policy Institute 2006). A study conducted in the Eastern Cape Province depicted that the houses provided under UISP are not of good quality. The majority of the 91.2% interviewed respondents rated the quality of windows in their UISP houses as poor and an equally significant majority of the 91.2% respondents maintained that the roofs of the structures were of poor quality, while 8.8% rated the quality as fair (Manomano & Tanga 2018). 93.6% of the respondents complained about the quality of the doors. The findings confirmed that the utilized materials were defective and weak.

From the literature, there are crucial findings that are affecting the implementation of the Upgrading of Informal Settlement Programme (UISP) and there is a call for a need for innovative practices to make it effective.

In South Africa, communities and government have often experienced tensions, marked by mistrust, violence and hostility (SDI South African Alliance 2013). It is important that from the initial stage of the UISP, there is transparent engagement with communities and are given the platform for true input regarding the upgrading of their settlements. Cheung (2022), community involvement reduced violent executions. The communities must be invited to engagement through radio announcements. In an informal settlement in eMahlaheni, Planact did broadcasts on radio on important issues to empower and inform authorities and communities (Centre for Development and Enterprise 2013, p. 4). These approaches give the communities a sense of belonging and meaning in the upgrading of their own settlements.

World Bank (2016), in the upgrading of conventional projects, infrastructure is supposed to be delivered using building standards and norms set at the city scale. All departments involved have standards which can be applied. There must be close supervision of quality assurance to ensure that informal settlement dwellers receive reliable and tested infrastructure houses. To further complement quality infrastructure, the UISP approach must consider utilizing off-the-grid services to increase efficiency, which is the usage of localized infrastructure instead of not relying on networked systems such as electricity but instead using solar power.

To further improve the execution of the UISP, upgrading of security of tenure should be a priority in the first phase of upgrading through a moratorium on evictions and by issuing occupancy certificates and further documenting current rights (Yunda & Sletto 2017). The prioritization of security of tenure over land titles will spur housing improvement and investments. In the process of providing tenure, the local government must have a 'twin track' approach to avoid working backwards, in which urban upgrading initiatives to provide infrastructure and security of tenure in the existing communities are accompanied by a strategy and system to ensure that new informal settlements do not get established.

5 LESSONS LEARNT, POLICY IMPLICATIONS AND CONCLUSION

South Africa has taken a positive approach to accepting the realities of urban informal settlements and has put the necessary avenues and approaches to counter the challenges. The gap in the current approach to upgrading informal settlements is a call to rethink and amend where possible for its effectiveness. Community engagement is a crucial element that needs the programme to centre its approach around it. NUSP (2015), participation of vulnerable groups in the phases of upgrading can promote capacity building and community empowerment, and develop a sense of ownership in the project rather than being passive beneficiaries. Delivering quality houses also reduces violent protests that cost the government more. It is equally essential that security of tenure is made a priority for beneficiaries and that the government continually monitor and assess its upgrading policies to make them relevant and inclusive to all different groups.

Indego (2018), informal settlement upgrading is principally a function of local government, as it is at the local radius where the complexities of human settlements development must be piloted. This directive vests within local government, which should deliver the coordination of urban development functions, and that provincial and national departments return to a housing emphasis and focus. Therefore, municipalities are not purely execution mediators of national human settlements programmes; they need to gather the mandatory processes and partnerships to efficiently manage the trade-offs, challenges, and contestation inherent to human settlement development and must do so in a transparent, accountable and engaged manner. Hendler & Fiew (2018:14), informal settlement upgrading projects are an essential means to draw in policymakers and politicians in order to transform and challenge policies and institutional arrangements. The deficiency in informal settlements needs to be recognized and standards need to be put in place for important services. Rebel Group (2017), municipalities need to be held accountable for ensuring that all informal settlements meet these standards regardless of whether the settlement is to be relocated or upgraded.

The disjuncture between practice and policy may be the result of a number of problems including unchanged attitudes concerning the function of informal settlements, a frail desire on the part of the state to partner with non-state players, stress on quantitative rather than qualitative results, a deficiency of capacity amongst local stakeholders (together with communities, NGOs and local government) and discrepancy between current instruments for developments and policy objectives. It is critical for the White paper to explicitly encourage and enable progressive practices as the underpinning for human settlements policy moving forward, which equally means that the new policy needs to be strongly founded in practitioner perspectives on the constraints and possibilities of partnership-based, participatory, incremental and in situ settlement development.

REFERENCES

Avis, W. R. (2016). *Urban Governance (Topic Guide)*. Birmingham, UK: GSDRC, University of Birmingham.

Cheung, Ck. Preventing Violence through Participation in Community Building in Youth. *Applied Research Quality Life* **17**, 1725–1743 (2022). https://doi.org/10.1007/s11482-021-09982-y

Cities Alliance (2021) "Informal Land Markets – City government interventions for enhancing land access and tenure security", Cities Alliance, Brussels.

D'Cruz, C., McGrahanan, G & Sumithre, U. (2009). The Efforts of a Federation of Slum and Shanty Dwellers to Secure Land and Improve Housing Moratuwa: From Savings Groups to Citywide Strategies. *Environment and Urbanization*, 21(2),367–388.

Del Mistro, R. and Hensher, D. (2009). 'Upgrading Informal Settlements in South Africa: Policy, Rhetoric and What Residents Really Value'. *Housing Studies*, 24(3): 333–54

Effectiveness of In Situ Upgrading in Improving the Quality of Life of Beneficiaries Living in Informal Settlements in South Africa. 2021 Kedibone Maganadisa, Vuyiswa Letsoko, Ockert Pretorius

Gardner, D. and Forster, C. (2014). *Financing Housing Consolidation in In-Situ Informal Settlement Upgrading Programmes in South Africa*. Pretoria: City Support Programme, World Bank and NUSP.

Georgiadou, M. C., & Loggia, C. (2021). "Beyond self-help". In *African Cities and Collaborative Futures*. Manchester, England: Manchester University Press. Retrieved Apr 16, 2023, from https://doi.org/10.7765/9781526155351.00010

Housing Development Agency (2013). *Kwazulu-Natal: Informal Settlements Status*. Johannesburg: HAD.

Huchzermeyer, M. (2011). *Cities With 'Slums': From Informal Settlement Eradication to a Right to the City in Africa*. Claremont: UCT Press.

Mandate of the Special Rapporteur on Adequate Housing as a Component of the Right to an Adequate Standard of Living, and on the Right to Non-discrimination in this Context (2021)

Massey R. (2014). 'Exploring Counter-conduct in Upgraded Informal Settlements: The Case of Women Residents in Makhaza and New Rest (Cape Town), South Africa'. *Habitat International*, 44: 290–6.

Muller, G. (2013). The Legal-historical Context of Forced Evictions in South Africa. *Fundaminabvol 19 No.2*.

Patel, L. 2015. *Social Welfare and Social Development in South Africa* (2nd ed.). Cape Town: Oxford University Press.

Roy, A. (2011). 'Slumdog Cities: Rethinking Subaltern Urbanism'. *International Journal of Urban and Regional Research*, 35(2): 223–38.

Satterthwaite, D., Archer, D., Colenbrander, S., Dodman, D., Hardoy, J., Mitlin, D. & Patel, S. (2020). Building Resilience to Climate Change in Informal Settlements. *One Earth*, Vol. 2(2): 143–156.

SERI (Socio-Economic Rights Institute) (2018). *Informal Settlements and Human Rights in South Africa*. Johannesburg: SERI

Theron, F. & Mchunu, N. (eds.). 2016. *Development, Change and the Change Agent: Facilitation at Grassroots*. 2nd edition. Pretoria: Van Schaik

Yunda, J. G., and B. Sletto. 'Property Rights, Urban Land Markets and the Contradictions of Redevelopment in Centrally Located Informal Settlements in Bogotá, Colombia, and Buenos Aires, Argentina.' *Planning Perspectives*, Vol.32, No.4, 2017, pp. 601–621

Failures of Small Medium Enterprise (SME) in the Kwazulu-Natal construction industry: Management, technical, and economic factors

A.O. Aiyetan
Department of Construction Management and QS, Faculty of Engineering and the Built Environment, Durban University of Technology, Durban, South Africa

A.B. David
Discipline of Civil Engineering, University of KwaZulu-Natal, Durban, South Africa

ABSTRACT: Small and medium-sized enterprises (SMEs) construction companies are vital for economic growth in many countries. The success or failure of these businesses has significant implications for owners, employees, and communities. This study explores the factors (Management, Technical, and Economic) that influence SME success and failure, providing insights for policymakers and owners. A non-probabilistic convenient sampling method was used to select the sample due to the small population size in the KZN province of South Africa. The entire sample was surveyed electronically, with a response rate of 60.7%. Data analysis involved basic statistical techniques such as mean scores and standard deviations. The research instrument's reliability was confirmed through feedback from industry professionals and academic experts. Multiple studies reveal major causes of SME failure, including financial mismanagement, lack of skilled personnel, insufficient capital, and poor financial management. Key success factors include business and management skills, government intervention, access to capital, recruiting young professionals, and expediting client payments. Addressing these factors is crucial for policymakers and SME owners to support small business growth and sustainability.

Keywords: Failures, Small, Medium, Enterprises

1 INTRODUCTION

Small and medium-sized enterprises (SMEs) play a significant role in the economic development of many countries, including South Africa. The National Small Business Act of 1996 classifies SMEs into four categories: micro, which includes survivalist enterprises, very small, small, and medium-sized. The South African government has committed to economic growth and transformation, and historically disadvantaged individuals (HDIs) and SMEs are being encouraged to participate in the mainstream economy, particularly in the construction industry (Construction Industry Development Board 2004; Department of Public Works 1995).

To support SMEs, various government agencies offer non-financial services such as mentorship, access to technology, training of trainers, trade and investment development, and materials development. These agencies also provide counselling and skills training services. For example, Ntsika Enterprise Promotion, Khula, and the Small Enterprise Development Agency (SEDA) are some of the programs that aim to develop SMEs (Bureau for Economic Research 2016). However, SMEs in South Africa face several challenges, including a lack of market research and understanding of competitors, chronic skills shortages, and difficulty obtaining the necessary capital (Deloitte 2020; Kalema et al. 2021; Mkhabela 2020).

In recent years, the South African government has recognized the importance of SMEs and established a Ministry of Small Business Development in 2014. The construction industry, in particular, is large and critical to the country's economy but has witnessed an increasing number of financial failures. SMEs in the construction industry, like other SMEs, face challenges in terms of management and capacity-building (Jahanshahi et al. 2021). SMEs must understand the opportunities and threats associated with international business and develop their strengths relative to international activities when expanding into international markets.

SME development is seen as socially imperative for income generation, increasing the living standards of citizens, reducing poverty, and minimizing internal problems affecting progress. Although countries worldwide define SMEs differently, the definitions are similar. In South Africa, the definition of SMEs should provide a basis for recording statistical information and data on the sector to measure and monitor small business growth and development over time (National Small Business Act 1996; National Small Business Amendment Act 2003). Based on the aforementioned, the objectives of the study are the causes of failures of SMEs and the identification of their success factors.

1.1 A look at SMEs in the South African construction industry

The South African construction industry provides a fertile ground for small and medium-sized enterprises (SMEs) to thrive, contributing significantly to the country's economic development (Construction Industry Development Board 2004; Department of Public Works 1995). SMEs in this sector encompass micro, very small, small, and medium-sized enterprises (National Small Business Act 1996; National Small Business Amendment Act 2003). These SMEs encounter unique characteristics and challenges, including market research gaps, chronic skills shortages, and obstacles in accessing capital (Deloitte 2020; Kalema et al. 2021; Mkhabela 2020). Understanding the nature of SMEs in the construction industry is crucial for formulating effective policies and strategies that support their growth, sustainability, and integration into the mainstream economy.

The South African government, through initiatives like the Ministry of Small Business Development, recognizes the importance of fostering SMEs in the construction industry and addressing their specific needs (Jahanshahi et al. 2021). By leveraging opportunities and mitigating threats associated with international business, SMEs can expand their reach and capitalize on emerging trends in the global construction market. The growth and development of SMEs in the South African construction industry are critical for generating income, reducing poverty, and addressing internal socio-economic challenges (Bureau for Economic Research 2016). Accurate statistical recording and monitoring of SMEs are essential for measuring their progress over time and evaluating their impact on the overall sector.

2 LITERATURE REVIEW

2.1 Definition of SME

The definition of Small and Medium-sized Enterprises (SMEs) varies across different countries, organizations, and sectors. In South Africa, the Department of Trade, Industry and Competition (DTIC) provides the official definition of SMEs based on the number of employees and annual turnover. However, the definition has undergone a series of revisions to align it with the current economic conditions and business environment. The most recent update was proposed by the Minister of Small Business Development and gazetted on 12 October 2018 for public comment. The revised definition considers factors such as job creation, innovation, and market access, and aims to promote the growth and sustainability of SMEs in South Africa (DTIC 2021).

2.2 The causes of SME failure

SMEs face several challenges that may lead to their failure, including inadequate financial resources, insufficient planning, and a lack of business management skills. Here are some of the common causes of SME failure in the Construction Industry:

2.2.1 Starting for the wrong reason

One of the reasons SMEs fail is that they are started for the wrong reason. Many entrepreneurs start businesses without proper research and planning, driven by factors such as the desire for financial independence or dissatisfaction with their current employment. However, starting a business requires more than just a passion for the product or service. Business owners need to possess the skills and acumen to manage a successful enterprise. According to Forbes, more than 500,000 businesses are started each month, but only a fraction of them succeed. Therefore, entrepreneurs should take the time to learn the skills required to run a business successfully (Forbes 2021).

2.2.2 Insufficient capital

Inadequate funding is one of the primary reasons why SMEs fail. Starting a business without sufficient operating capital is almost certainly a death knell. The lack of funds often leads to cash flow problems and an inability to meet financial obligations. According to a report by Hiscox, 2% of US entrepreneurs have resorted to using their credit cards to fund businesses. Entrepreneurs should protect their capital before starting a business and ensure they have enough funds to cover startup costs and keep the business running for a year or two. Getting the right liability insurance for a business is also crucial to managing cash flow and protecting the business from unforeseen events (Hiscox 2021).

2.2.3 Improper planning

Lack of proper planning is another common reason small companies fail and go out of business. Entrepreneurs often focus on achieving their dreams of financial independence and fail to create a strategic business plan. A business plan should include factors such as workforce needs, analysis of competitors, sales and expense forecasts, and marketing budgets. One entrepreneur started her salon business without first conducting market research to see if the area could support such an endeavor. She was never able to build a customer base strong enough to keep the doors open. Creating an effective business plan can help entrepreneurs better ensure success (Kappel 2021).

2.2.4 Poor management

Effective management and leadership skills are essential to the success of small and medium-sized enterprises (SMEs), and a lack of either can lead to confusion and conflict within the ranks, poor morale, and reduced productivity. Poor management has been identified as one of the major causes of business failure in many studies (Izedonmi & Ogbeide 2020; Zhang *et al.* 2021). While the causes of SME failures may vary, good management is critical to avoiding them. SME owners or managers need to develop basic managerial skills and knowledge to improve their chances of success (Tuyên 2019). This includes acquiring adequate skills in planning, organizing, directing, and controlling organizational resources. Managers also need specific skills such as technical, human, and conceptual skills to maintain efficiency in the way employees complete their working tasks (Drew 2019).

Lack of managerial experience has been identified as a factor that contributes to the failure of SMEs. SMEs lack managerial experience because they are unable to tackle issues that require previous knowledge or experience. Similarly, SME performance is closely linked to the entrepreneurial skills of the proprietor. Business owners with relevant qualifications tend to survive 30% more than non-qualified proprietors (Panda & Mishra 2021).

2.2.5 *Financial management*

Poor financial management is another factor that contributes to the failure of SMEs. Financial management is a very broad term that encompasses planning, organizing, activating, and controlling financial resources. SMEs in the start-up phase face critical areas such as financial needs and control of funding as well as the accountability of start-up capital. SMEs within the social welfare context increase their chances of obtaining funding if they maintain their accountability status. Therefore, accurate record-keeping is vital as a task under the organization, which can assist in keeping and obtaining more funding to grow (Chandra & Bapna 2020).

2.2.6 *Expansion*

Over-expansion can drive a company to higher-risk investment with financial debt, hence increasing its chance of business failure. A change in the type of work and where that work is going to be done also contributes to the expansion factor, when a contractor goes to do work outside his normal territory it might bring some difficulties as he will have to adapt to the new geographical location (Huffington 2020).

2.2.7 *Economic*

Economic factors are perceived to be beyond the control of management. While economic factors may be external to a firm's operations, failure by firms to recognize that their efforts may lead to the termination of the firm's operations. The construction industry is particularly susceptible to failure due to distinct characteristics such as trading within a highly uncertain environment, competitive tendering, and low fixed capital requirements for entry into the market resulting in the market being over-capacitated (Siddiqui & Akhtar 2020). High-interest rates, stringent rules, and regulations set by the government have also been identified as prominent causes of business failure (Zhou *et al.* 2021).

2.2.8 *Training and education*

The relationship between training and SME performance is inconclusive. However, firms with sophisticated training systems and strong management support for training are more successful in delivering training. Neglecting training can hold back organizational performance. Xiang and colleagues (2022) found that government-funded training investments in SMEs have a significant influence on setting proactive strategies to combat the recession. Effective training can increase employees' skills level, customer satisfaction, quality of products or services, technology transfer and productivity.

2.2.9 *Poor crime management*

Crime may be from within the organization or those working with the organization. Crime from those working with the organization is the workers themselves, while those working for the organization may be suppliers. In any capacity, crime not managed can adversely affect the success of the organization. On the other hand, Internal crime in small businesses is caused by several factors such as hiring personnel without careful background checks or employment references, failure to enforce strict, uniform rules for even minor infractions, failure to establish a climate of trust, confidence, and respect for employees as well incentives for outstanding and honest performance, and failure to apply techniques that will thwart opportunities for employee theft and cost-cutting measures (Hodgetts & Kuratko 2014).

2.2.10 *Poor business location*

A wrong selection of the location of the Business due to the cost of renting may have an adverse effect on the success of the company. The location of a business also serves as an advert to the business and eventual patronage relative to procurement and construction. Recent studies have shown that factors such as accessibility, parking, and convenience can significantly influence the success of a small business (Yang & Hu 2021). Furthermore, the importance of location varies across different industries, and small business owners should carefully analyze the factors that are relevant to their specific industry (Oxenfeldt & Kelly 2019).

2.3 Remedies

The following are remedies for the failures of SMEs in the Kwazulu-Natal province of the South Africa's Construction Industry.

2.3.1 The role of government in providing access to finance

Government has a lot to do in the development of the finance of SMEs. Research indicates that introducing SME Credit Bureaus significantly reduces the cost of lending to SMEs and improves access to finance. Moreover, research shows that governments have continued to use credit guarantee schemes to increase SME lending to avoid SME failure.

2.3.2 Government intervention

To address problems of start-off and finance challenges faced by small business contractors in South Africa, Chadhliwa (2015) noted that the government must review policies concerning contractors' Development Programs (CDP) and Construction Education and Training Agencies (CETAs) to ensure that the government contributes to the success of small contractors. This could be in the form of soft loans, training and financing their bills.

2.3.3 Business skills

For SMEs to succeed, it is very important for them to have adequate and relevant business skills. It is an important driver of innovation and economic growth (Yoshino & Taghizadeh-Hesar 2016). Business skills assist the manager to set specific targets for their business, carry out market research, employ qualified personnel and accord them responsibilities according to their skills, knowledge, and experience.

2.3.4 Management skills

While business skills can be referred to as technical skills, management skills denote the management style of managers, the incentive given for higher productivity, assurance of job security, and assessment of training needs that afford the worker to put in his best in carrying out his duties. Further, a good knowledge of financial management is required (Thwala & Mofokeng 2012). Financial management and management incompetencies are among the attributes that lead to construction failure.

The other factors that could enhance the success of SMEs are access to capital, recruiting young and qualified professionals, family or domestic situations and expecting payment from clients.

3 RESEARCH METHODOLOGY

This study focuses on investigating the causes of SME failures in the construction industry of Kwazulu-Natal province, South Africa.

3.1 Data collection

To gather data, a questionnaire survey was conducted among the construction firms graded 1-4 according to the Construction Industry Development Board (CIDB) grading system, which consists of a total population of 112 firms. Given the relatively small size of the population, a convenient sampling method was employed to survey the entire sample (Leedy & Omrod 2013). The survey was administered electronically, and a response rate of 60.7% was achieved, with a total of 68 completed questionnaires received. The surveyed construction firms in the sample have an average organizational existence of 6 years and have handled an average of 5 projects. In terms of monetary value, their average annual turnover is R176M. The educational qualifications of the owners indicate that 70% hold ordinary national diplomas, 15% hold B.Tech degrees, 10% hold bachelor's degrees, and 5% hold matric certificates or other qualifications.

3.2 Data analyses

For the analysis of data in this study, simple inferential statistics were employed. For reliability and validity, the research instruments were sent to experts in the discipline both in the industry and in academics. It was confirmed that the instrument has the capacity to measure the intent of its design. Furthermore, the questionnaire items were carefully designed to provide a comprehensive description of each construct, enhancing the construct's validity.

4 RESULTS

Table 1 presents the causes of SME failure in terms of management and technical factors. The mean score (MS) is a measure of the average score given by the respondents for each factor, while the standard deviation (St. Dv) measures the degree of variation or spread of the scores around the mean. The ranking column shows the order in which the factors were rated based on their mean scores.

Financial mismanagement and managerial skills were ranked as the top cause of SME failure, with a mean score of 3.58, indicating that respondents strongly agreed that this factor is a major contributor to SME failure. The high standard deviation of 1.12 suggests that opinions on this factor were widely dispersed, indicating that respondents' experiences with financial mismanagement and managerial skills varied widely. Funds are required for the daily activities of a project. This could be for the purchase of materials, machines, and payment of labor. The lack of proper planning of resources and quantification of cost, when required on-site, is crucial and relative to financial management and managerial expertise.

Lack of skilled people was ranked second with a mean score of 3.56, indicating that respondents strongly agreed that this factor contributes significantly to SMEs' failure. However, the relatively low standard deviation of 1.08 suggests that there was less variation in responses compared to financial mismanagement. Skilled labor avoids reworks on projects and causes profit gain and acceptance of work. These, when lacking, lead to failure. Experience is both technical and administrative, the lack of these causes failures.

Lack of experience in the industry was ranked third with a mean score of 3.34. The low standard deviation of 0.64 suggests that there was relatively little variation in responses, indicating that respondents generally agreed on the importance of this factor in SMEs' failure.

Poor business location and poor crime management were ranked eighth and ninth, respectively, with mean scores of 2.00 and 1.88. The high standard deviations of 0.92 and 1.52, respectively, suggest that opinions on these factors were widely dispersed, indicating that respondents' experiences with these factors varied widely. These factors have a negligible contribution to SMEs' failures, a good advert will negate poor business location, while improving security on site will reduce crime on site relative to pilfering.

Table 1. Causes of SME's failure; management, and technical factors.

S/No	Factors	MS	Av Ms	St. Dv	Ranking
1	Financial mismanagement & managerial skills	3.58	3.40	1.12	1
2	Lack of skilled people	3.56	3.40	1.08	2
3	Lack of experience in the industry	3.34	3.40	0.64	3
4	Expansion	3.28	3.40	0.53	4
5	Business skills	3.20	3.40	0.38	5
6	Training and Education	3.08	3.40	0.15	6
7	competition	2.40	3.40	0.92	7
8	Poor business location	2.00	3.40	0.92	8
9	Poor crime management	1.88	3.40	1.52	9

Table 2 reveals the economic factors that contribute to SME failure, in the Kwazulu-Natal Construction industry. Insufficient capital was rated as the primary economic factor contributing to SME failure, with a mean score of 3.74 and a high standard deviation of 1.39. Insufficient capital availability to SMEs is the main reason for their failure. This may be a result of inadequate knowledge in accessing funds. The lack of funds indicates a lack of progress on construction and may lead to relocation by the client.

Not enough work was rated second, with a mean score of 3.50 and a moderate standard deviation of 0.91. This suggests that while many SMEs struggle with a lack of work, it is the volume of work done or the job procured that leads to success. A company without work would fold up. The likely reason for this may be the lack of adequate knowledge on how to source for jobs or bids failing.

Financial management ranked third, with a mean score of 3.26 and a low standard deviation of 0.49. This suggests that the issue of financial management is a relatively consistent challenge for SMEs, with less variation in experiences among businesses. The management of funds is crucial to the success of a project and in turn the organization.

The least factor contributing to the failure of SMEs is starting for the wrong reason ranked sixth with a mean score of 2.40 and a high standard deviation of 0.92. This suggests that some SMEs may fail because they were started for the wrong reasons, but the prevalence and impact of this factor vary widely among businesses.

Overall, the data in Table 2 highlights the economic challenges that SMEs face and underscores the importance of proper financial management, securing sufficient capital, and generating a steady stream of work to succeed. It also suggests that competition and external economic factors can be significant barriers to SME success.

Table 2. Causes of SME's failure; economic factors.

S/N	Factors	Ms	Av Ms	St. Dv	Ranking
1	Insufficient capital	3.74	3.40	1.39	1
2	Not enough work	3.50	3.40	0.91	2
3	Financial management	3.26	3.40	0.49	3
4	Competition	2.40	3.40	0.92	4
5	Starting for a wrong reason	2.40	3.40	0.92	5
6	Economical	2.34	3.40	0.49	6

Table 3 indicates that business skill is the most significant factor for the success of SMEs, with a mean score (MS) of 4.81. This implies that having proficient business skills is crucial for the success of SMEs. Business skills include the ability to develop and execute a sound business plan, identify target markets, develop marketing strategies, and monitor and control business operations.

Management skills (MS = 4.77) follow closely as the second most significant factor for SME success. This indicates that having effective management skills is critical for SMEs to achieve their objectives. Management skills include the ability to lead, motivate, and manage employees, as well as the ability to make informed business decisions.

Government intervention (MS = 4.50) is rated as the third most important factor for SME success. The success factor may refer to various forms of government support or intervention, which respondents perceive to have contributed positively to their business success.

Finally, expediting payment (MS = 3.05) from clients is rated as the least significant factor for SME success. This suggests that while timely payment from clients is important for the cash flow of SMEs, it is not as critical as having proficient business and management skills, as well as access to adequate capital.

Table 3. Success factors.

S/N	Factors	Ms	Av. Ms	St. Dv	Ranking
1	Business skills	4.81	3.95	0.759	1
2	Management skills	4.77	3.95	0.825	2
3	Government intervention	4.50	3.95	0.689	3
4	Access to Capital	4.23	3.95	0.288	4
5	Recruiting young, qualified professionals	3.20	3.95	0.832	5
6	Family/ domestic situation	3.11	3.95	0.885	6
7	Expediting payment from a client	3.05	3.95	0.918	7

For SMEs to succeed, there is a need for a sound knowledge of contract administration. The administration of a contract indicates the following: the start and finish of every event, the natural resources, the technical skills, and adequate tasks to labour requirements including finance. A good knowledge of this is important to the success of SMEs. This factor is corroborated by the study of Thwala & Mofokeng 2012.

Business skills which include knowing the technical aspects in relation to methods and time management is a key factor in the success of SMEs.

Government intervention regarding the development of SMEs is important. These SMEs are those to develop the country's infrastructure. Therefore, their intervention in terms of training, less stringent conditions to access capital of SMEs and ensuring their being in business will ensure their success. This view is supported by the study of Jahanshahi *et al.* 2021.

5 CONCLUSION

Based on the analysis of the data, it is recommended that SMEs acquire adequate knowledge of contract administration. Where lacking, they should endeavour to learn in short courses.

Technical skills are very important to the success of SMEs. This is about every knowledge of what to do on sites and the various methods of construction. SME owners must be knowledgeable in the area of sources of funds, and the requirement to access funds must be well-known and in place with SMEs to access funds for their projects.

REFERENCES

Bureau for Economic Research. (2016). *SMEs in South Africa: Survival, Success, and Growth Constraints.* University of Stellenbosch.

Chadhliwa, T. (2015). *An Evaluation of the Contractors' Development Programme in South Africa.* Unpublished master's thesis, University of Johannesburg.

Chandra, Y., & Bapna, S. (2020). Factors Affecting the Performance of Small and Medium Enterprises in India: An Empirical Study. *South Asian Journal of Management*, 27(4), 70–85.

Construction Industry Development Board. (2004). *Black Economic Empowerment.*

Deloitte. (2020). Small Business 2020: *Building Resilience in Challenging Times.* Deloitte.

Department of Public Works. (1995). *White Paper on National Strategy for the Development and Promotion of Small Business in South Africa.* South Africa.

Drew, M. J. (2019). Factors Affecting the Success and Failure of Small and Medium-Sized Enterprises (SMEs). *The Journal of Entrepreneurship*, 28(2), 241–265.

DTIC. (2021). *Small Business Support.* [online] Available at: https://www.thedtic.gov.za/sme-development/small-business-support

Forbes. (2021). *Why Most Small Businesses Don't Work and What to Do About It.* [online] Available at: https://www.forbes.com/sites/stevedenning/2021/04/10/why-most-small-businesses-dont-work-and-what-to-do-about-it/?sh=1a765de0459fHiscox. (2021). Why Small Business Insurance is Crucial to Financial Success. [online] Available at: https://www.hiscox.com/small-business-insurance/blog/why-small-business-insurance-is-crucial-to-financial-success

Hodgetts, R. M., & Kuratko, D. F. (2014). *Effective Small Business Management: An Entrepreneurial Approach*. Cengage Learning.

Hufflington, C. (2020). *The Top 10 Reasons Why Small Businesses Fail*. Forbes. Retrieved from https://www.forbes.com/sites/allbusiness/2019/02/28/top-10-reasons-why-small-businesses-fail/?sh=6da36a0231e8

Izedonmi, P. F., & Ogbeide, O. S. (2020). Small and Medium Enterprises (SMEs) Failure in Nigeria: Causes and Solution. *African Journal of Management*, 6(1), 21–32.

Jahanshahi, A. A., Hossain, A., Li, X., & Marwala, T. (2021). Barriers to Growth of Small and Medium Enterprises in South Africa. *Journal of Small Business and Entrepreneurship Development*, 9(1), 1–17.

Jahanshahi, A. A., Karimzadeh, M., Razmi, J., & Jahanshahi, H. (2021). Dynamic Capabilities, Organizational Innovation and SMEs' Performance in the Construction Industry: *A Conceptual Framework. Business Strategy and Development*, 4(1), 27–39.

Kalema, B. M., Mugomba, B., & Kalema, D. (2021). An Exploration of the Key Challenges Faced by Small and Medium-Sized Construction Companies in South Africa. *Journal of Economics and Finance*, 12(2), 157–176.

Kalema, B. M., Mukiibi, W. M., & Balunywa, W. (2021). Effect of Financial Management Practices on the Growth of Small and Medium Enterprises in Uganda. *Journal of Small Business and Enterprise Development*.

Kappel, T. (2021). *Top 10 Reasons Small Businesses Fail*. [online] The Balance Small Business. Available at: https://www.thebalancesmb.com/top-reasons-why-small-businesses-fail-2951539

Leedy, P. D. and Omrod, J. E. (2013) *Practical Research Planning and Design*. Tenth Edition, Pearson New International Edition.

Mkhabela, B. (2020). SME Challenges and Strategies in South Africa. *Journal of Economics and Behavioral Studies*, 12(4), 51–61.

Mkhabela, T. (2020). Challenges Faced by Small and Medium Enterprises in South Africa. *Journal of Economics and Behavioral Studies*, 12(1), 1–16.

National Small Business Act. (1996). *Act No. 102 of 1996*. South Africa.

National Small Business Amendment Act. (2003). *Act No. 29 of 2003*. South Africa.

Oxenfeldt, A. R., & Kelly, W. E. (1969). Will the New Competitive Environment Eliminate Small Businesses? *Harvard Business Review*, 47(4), 105–118.

Panda, A., & Mishra, R. K. (2021). Determinants of Small and Medium-Sized Enterprises (SMEs) Survival: Evidence from India. *International Journal of Emerging Markets*, ahead-of-print(ahead-of-print).

Siddiqui, J., & Akhtar, S. (2020). Factors Influencing the Failure of Construction Projects in Pakistan. *International Journal of Civil Engineering and Technology*, 11(3), 299–307.

Thwala, W. D., & Mofokeng, T. M. (2012). Management Factors Affecting Small and Medium-sized Enterprises in South Africa. *Journal of Economics and Behavioral Studies*, 4(5), 267–275.

Tuyên, T. V. (2019). The Impact of Management Skills on the Performance of Small and Medium-Sized Enterprises in Vietnam. *Journal of Economics and Development*, 21(1), 80–102.

Wang, X., & Cheng, M. (2021). Why Do SMEs Fail in China? A Systematic Review and Future Research Agenda. *Journal of Small Business Management*, 59(4), 585–607. doi:10.1080/00472778.2020.1740301

Xiang, D., Chen, J., & Su, J. (2022). Does Government-funded Training Investment in Small and Medium-sized Enterprises Matter? Evidence from China. *Journal of Small Business Management*, 60(2), 308–323.

Yang, X., & Hu, Y. (2021). Spatial Determinants of Small Business Success: Evidence from Street Vending in Shanghai. *Cities*, 109, 103044.

Yoshino, Y., & Taghizadeh-Hesary, F. (2020). Challenges Faced by SMEs in East Asia: How Can the Region Foster Innovation-driven Entrepreneurship? *Asian Development Bank Institute*.

Zolin, R., & Fernandes, A. M. (2020). Factors Influencing the Performance of Small and Medium-sized Enterprises: A Systematic Literature Review. *Journal of Innovation & Knowledge*, 5(2), 124–133.

Zou, H., Chen, J., & Hu, Y. (2020). Entrepreneurship Education, Entrepreneurial Intention, and Venture Creation: The Mediating Role of Entrepreneurial Self-efficacy. *Journal of Small Business Management*, 58(2), 498–523.

Equalizing opportunity of female: Kwazulu-Natal construction industry

A.O. Aiyetan*
Department of Construction Management and QS, Faculty of Engineering and the Built Environment, Durban University of Technology, Durban, South Africa

A.B. David
Discipline of Civil Engineering, University of KwaZulu-Natal, Durban, South Africa

ABSTRACT: This study examines the causes of discrimination against women in the KwaZulu-Natal construction industry and proposes strategies for promoting equal opportunities. A sample of 50 female professionals was surveyed, and data were analysed using discrete analysis methods. To verify the validity and reliability of the research instruments, experts from both the industry and academia reviewed the instruments, confirming their capability to measure the intended purpose. The study found evidence of discrimination against women in the construction industry. Economic prosperity was identified as a factor that could enhance female equality, and equitable rewards based on relative input to a project were found to positively impact the retention of women in the construction industry. It is recommended that female counterparts be provided with equitable pay and rewards and that appointments to top management positions should be made based on merit, without discrimination against females. Additionally, ensuring job security would help retain many female workers in the industry.

Keywords: Discrimination, Equal opportunity, Female in Construction

1 INTRODUCTION

According to Statistics South Africa, as of the second quarter of 2018, women accounted for only 9% of the construction industry workforce in South Africa (Stats SA 2018). Equalizing opportunity refers to applying the same conditions and rules to people in all aspects of life, including in the case of female workers in the construction industry (Abdullahi & Adeniran 2020). Despite recent advancements in female education and skills development, women in the KwaZulu-Natal construction industry have faced issues of discrimination in areas such as recruitment, promotion, and remuneration (Mkhize & Ngwenya 2021). Women have shown great talents and performance in various fields, including traditionally male-dominated trades such as block and bricklaying (Sikakana *et al.* 2021). However, they continue to face discrimination in the industry, even when working as supervisors in extreme weather conditions and operating heavy machinery (Mphahlele *et al.* 2020).

To address these issues, the South African government has implemented initiatives to encourage women's participation in the construction industry, such as appointing women to the construction committee and creating recruitment programs that target women (Govere *et al.* 2020). However, women's representation in the industry remains low (Khumalo *et al.* 2021). This study aims to assess the factors that contribute to equalizing opportunities for women in the KwaZulu-Natal construction industry and enhance women's recruitment and retention in the industry. By identifying and addressing these factors, this study will contribute to the advancement of gender equality in the Kwazulu-Natal construction industry and the South African workforce.

---

*Corresponding Author: ayodejia@dut.ac.za

2 LITERATURE REVIEW

2.1 Background of women's involvement in the KwaZulu-Natal construction industry

Women's involvement in the KwaZulu-Natal's construction industry reveals a multifaceted landscape characterized by diverse challenges and opportunities. Drawing on insightful research, including Eige's (2016) examination of the sector, it becomes evident that persistent gender disparities, discrimination, and a predominantly male culture hinder women's advancement in the Kwazulu-Natal's construction industry. Moreover, a more recent study conducted by Nkomo et al. (2021) reinforces these findings, emphasizing the challenges faced by women in breaking barriers and achieving equal opportunities. Additionally, McKinsey & Company's (2015) earlier investigation highlights the potential for economic prosperity and the development of women's competencies to drive gender equality within the industry. Nkomo et al.'s (2021) research further highlight the critical role of equitable pay and rewards in retaining women professionals. By addressing these dynamics, stakeholders can pave the way for a more inclusive and empowering construction sector. Therefore, the study objectives are:

1. To identify barriers facing women in the South African Construction Industry of KwaZulu-Natal province.
2. To identify the factors that will enhance women equality in the KwaZulu-Natal Construction Industry, and
3. Motivational factors for retaining female employees in the KwaZulu-Natal Construction Industry.

2.2 Barriers facing women in the KwaZulu-Natal construction industry

2.2.1 Image of the industry
The construction industry in South Africa has historically been male-dominated and stereotyped as being a physically demanding and dirty job, which has deterred many women from entering the field (Cilliers & Pretorius 2016; Kekana 2021).

2.2.2 Career knowledge
Many women are unaware of the various career paths available within the construction industry, such as project management, quantity surveying, and design, which can be less physically demanding and more appealing to women (Cilliers & Pretorius 2016).

2.2.3 Culture and working environment
The construction industry in South Africa has a macho culture, which can make it difficult for women to fit in and be accepted. Additionally, the lack of gender diversity in the industry has created a hostile working environment for women, with sexual harassment being a significant concern (Cilliers & Pretorius 2016; Kekana 2021).

2.2.4 Discrimination
Women in the South African construction industry face discrimination in hiring, promotion, and pay (Agumba & Kihumba 2021). They are often paid less than their male counterparts and have limited opportunities for career advancement.

2.2.5 Glass ceiling and the pay gap
The glass ceiling effect is prevalent in the South African construction industry, as women are often excluded from leadership positions due to gender bias (Kekana 2021). Women in the industry also experience a significant pay gap, with some studies reporting a gap of up to 40% (Agumba & Kihumba 2021).

2.2.6 Biology
The physical demands of some roles in the construction industry, such as bricklaying and heavy lifting, are often seen as more suited to men. However, this belief is a misconception, as studies have shown that women are just as capable of performing these tasks with the right training and support (Kekana 2021).

2.3 Factors enhancing females' equality in the KwaZulu-Natal construction industry

2.3.1 Economic prosperity

Women's employment in the construction industry has a direct impact on the GDP of a nation. McKinsey Global Institute Report states that businesses with at least 30% women leadership are 15% more profitable (McKinsey & Company 2015).

2.3.2 Growing competencies in women

The ageing population and shrinking pool of young professionals make it essential to target both genders concerning the growing competencies of the workforce. Addressing gender inequality in employment, promotion, equal salary, and benefits, could lead to an additional 10.5 million jobs in 2050 (Eige 2016).

2.3.3 Attracting more females to the industry

Women should be attracted to the construction industry for supervisory and administrative tasks at the top management level. The attraction of women into the industry could be with additional benefits (Chinomona *et al.* 2020).

2.3.4 Health and happiness

A clean and safe work environment results in high productivity, enables decision making and can result in innovation (Deeb & Alam 2021).

2.3.5 Addressing gender inequality

The Constitution of South Africa enforces equal human rights for men and women. The Commission for Gender Equality (CGE) enforces this by way of its mandate. These rights are relative to employment, education, land, finance, and other resources. The South African Women in Construction (SAWIC) is the voice and driver behind women's initiatives in the built environment. They are representatives for the promotion, advocacy, and support for the advancement of women in the construction industry (National Association of Women in Construction 2019).

2.3.6 Diversity in leadership roles

Companies with leadership in the top quartile for gender diversity were 15% more likely to have financial returns above the industry median. A diverse team produces better solutions to complex problems (McKinsey & Company 2015).

2.3.7 Gender imbalances in housework

Women face a conflict between work and family obligations. To be effective, women need to transfer household obligations to another (Madonsela & Matolweni 2019).

2.4 Motivational factors for retaining female employees in the KwaZulu-Natal construction industry

The construction industry in South Africa has historically been male dominated, making it challenging for women to progress in their careers. However, retaining female employees in the industry is crucial to promote diversity and equal opportunities. Here are some key motivational factors that can help to retain female employees in the South African construction industry.

2.4.1 Equitable rewards

The equitable reward for relative input to a project is essential in promoting an atmosphere of equal rights and high productivity relative to women. Both male and female workers should be given even rewards relative to their input to a project, which creates a sense of confidence, loyalty, and belonging. This eliminates anger or bitterness, frustration, low input to work, and a sense of inferiority that can result in low productivity (Nkomo *et al.* 2021).

2.4.2 *Exercise of power*

The freedom to take decisions and exercise power is an act of recognition of competence, control, and a sense of belonging. This is essential for women employees to feel empowered and motivated to contribute to the success of the project. A lack of power can lead to frustration, low motivation, and a high turnover rate (Muthathi et al. 2017).

2.4.3 *Recognition of contribution made*

Recognition of the contribution made is crucial in retaining female employees in the construction industry. This gives an impression of high performance, appreciation for efforts and knowledge of discipline. Recognition can be rewarded in cash or kind, which goes a long way in retaining and encouraging staff to do more (Kalema & Maina 2021).

2.4.4 *Job security*

Job security is essential for retaining female employees in the South African construction industry. Some people remain in their job for the sake of job security, as moving from one job to another may lead to losing a job in the future. Ensuring job security will encourage female employees to stay and contribute to the project (Ergen & Aslan 2020).

2.4.5 *Sense of belonging and identification with project*

A sense of belonging and identification with the project is essential in retaining female employees in the South African construction industry. Access to benefits such as access to a vehicle, housing, furniture, education, and medical benefits, among others, without any prerequisite conditions, reveals a sense of belonging. This fosters a sense of loyalty and commitment to the project (Muthathi et al. 2017).

2.4.6 *Achievement from meeting complex tasks*

Recognition of achievement, when complex tasks are accomplished, is vital in retaining female employees in the South African construction industry. It indicates recognition for hard work and belongingness to a team. Providing opportunities for employees to work on challenging tasks promotes the development of skills and encourages them to stay (Nkomo et al. 2021).

2.4.7 *Opportunity to extend skills and experience*

Providing opportunities for female employees to extend their skills and experience is critical in retaining them in the South African construction industry. Increasing their competencies will enhance their value and make them more attractive to the industry. This will encourage them to stay and contribute to the project's success (Ergen & Aslan 2020).

2.4.8 *Pay allowance*

Pay allowance is one of the strategic factors that can be used to attract and retain female employees in the South African construction industry. Regular payment of allowances or increasing allowances can result in happy workers, increasing their propensity to stay (Kalema & Maina 2021).

In conclusion, retaining female employees in the South African construction industry is crucial for promoting diversity and equal opportunities. Employers should consider the above motivational factors to promote a sense of belonging, empowerment, recognition, and job security, which can lead to higher productivity and job satisfaction.

3 RESEARCH METHODOLOGY

A questionnaire survey study was conducted among female professionals of the Master Builder Association (MBA) of KwaZulu-Natal, Durban. The sample population consisted of 68 individuals from the MBA, Durban, and due to its small size, the entire sample was surveyed (Leedy & Omrod 2013). Therefore, the convenience sampling method was employed. The research instrument was administered electronically. A total of 50 responses were received, representing a 73.5% response rate. These responses were obtained from professionals with work experience ranging between 3 and 15 years, and the average age of respondents was 25 years. Respondents have worked for both private and public organizations. The least represented qualification among the respondents is a National Diploma in Building, accounting for 25% of the sample. The majority, 68%, hold a B. Tech qualification, while the remaining 7% hold other qualifications.

Simple statistical analyses were utilized in this study to extract meaningful insights from the collected survey data. The use of these statistical tools was carefully selected to address the research objectives and enhance the validity of the findings. To begin with, descriptive statistics such as mean scores and standard deviation were employed. These measures allowed for a concise summary of the survey responses, providing an overview of the central tendency and dispersion of the data. By calculating the mean scores, we were able to determine the average response for each item, enabling us to understand the overall perception of the participants regarding the constructs under investigation. The standard deviation, on the other hand, provided information about the variability or spread of responses, giving insights into the consensus or divergence among the participants and enables the ranking of the factors in each construct.

Moreover, the survey instruments underwent a rigorous validation process involving experts from both industry and academia. This validation aimed to assess the reliability and validity of the research instruments. The involvement of these experts, well-versed in the discipline, ensured that the instruments effectively measured the intended design and captured the constructs of interest accurately. Furthermore, the questionnaire items were thoughtfully crafted to provide a comprehensive description of each construct, enabling a thorough assessment of the respective underlying phenomena.

4 DISCUSSION

Based on Table 1, Discrimination has the highest mean score (MS = 4.38) among factors that contribute to female talent shortages in the KwaZulu Natal construction industry. Discrimination takes varied forms such as the belief that the construction industry involves physically demanding tasks that females may not be capable of performing. Discrimination is also based on the expectation that females may not be willing to work long hours or have family responsibilities that could limit their availability for work.

Male-dominated culture and environment (MS = 3.82) is the next factor that contributes to the shortage of female talents in the industry. This is due to the nature of construction activities which are considered difficult, hard, and masculine-oriented, leading to a perception that females may not be suited for such work. Harsh working conditions (MS = 3.04) are another factor that contributes to the shortage of female talents in the industry. The nature of construction activities being carried out in the open makes them prone to extreme temperatures, which may not be suitable for pregnant women.

Finally, the image of the industry (MS = 1.78) is considered a contributing factor to the shortage of female talent. The construction industry is perceived as high-paying but difficult, and the nature of the work may not appeal to females.

Overall, the table suggests that discriminatory attitudes and cultural perceptions towards females in the construction industry, as well as harsh working conditions, are significant factors contributing to the shortage of female talents in the KwaZulu-Natal construction industry. However, the analysis could be improved by providing specific examples of discriminatory attitudes towards females in the industry, the impact of harsh working conditions on pregnant women, and more information on how the industry's image may affect females' decisions to pursue careers in construction.

The data provided in Table 2 highlights some important factors that can enhance female equality in the construction industry of KwaZulu Natal province. Economic prosperity (MS = 4.44) is rightly identified as the most important factor. Equal pay for equal work is critical to

Table 1. Factors contributing to the shortage of female talents in the industry.

S/No	Factor	Mean	Average	SD	Ranking
1.1	Discrimination	4.38	3.2	0.167	1
1.2	Male-dominated culture attitude	3.82		0.088	2
1.3	Sexist attitude	3.40		0.028	3
1.4	Harsh working conditions	3.04		0.201	4
1.5	Encouragement from young	2.84		0.051	5
1.6	Career Knowledge	2.8		0.057	6
1.7	Unsociable working hours	2.54		0.093	7
1.8	Image of the industry	1.78		0.023	8

attracting and retaining female talent in the industry. Additionally, it allows women to fulfil their financial obligations, support their families, and invest in their personal and professional growth.

Growing competencies in women (MS = 4.26) is also an important factor that can enhance female equality in the construction industry. Providing training and development opportunities for women not only improves their skills and productivity but also boosts their confidence and self-esteem. This can have a positive impact on their career progression and overall job satisfaction. Attracting more females to the industry (MS = 4.18) is another critical factor to enhance female equality. Special incentives, such as scholarships for children's education, house loans, and flexible working hours, can encourage more women to consider careers in the construction industry. Such incentives can help bridge the gender gap and create a more diverse and inclusive workplace.

Table 2 identifies gender imbalances in housework (MS = 2.34) as the least important factor contributing to enhancing female equality in the industry. While it is true that women have responsibilities at home, and work-life balance is crucial, it is important to note that this factor should not be overlooked. Providing flexible working arrangements, such as flexible hours or work-from-home options, can help female workers manage their responsibilities at home while pursuing their careers in the industry. This can create a more supportive and inclusive work environment for women.

In conclusion, Table 2 ranks the factors that enhance females' equality in the construction industry of the KwaZulu Natal province of South Africa. Economic prosperity (MS = 4.44) is the most important factor, followed by growing competencies in women (MS = 4.26), and attracting more females to the industry (MS = 4.18). These top three factors reflect the need for fair compensation, skill development, and creating incentives for women to enter and stay in the industry. Health and happiness (MS = 4.08) and addressing gender equality (MS = 4.06) also rank highly, showing the importance of creating a safe and equitable workplace for women. Human rights (MS = 3.9) and diversity in leadership roles (MS = 3.28) are also identified as factors that enhance female equality in the construction industry, highlighting the need for a culture that values and respects all employees regardless of gender.

Table 2. Factors enhancing females' Equality in the construction industry of the KwaZulu Natal province of South Africa.

S/No	Factor	Mean	Average	SD	Ranking
2.1	Economic prosperity	4.44	3.7	0.048	1
2.2	Growing competencies in women	4.26		0.069	2
2.3	Attracting more females to the industry	4.18		0.204	3
2.4	Health & Happiness	4.08		0.099	4
2.5	Addressing gender equality	4.06		0.045	5
2.6	Human rights	3.9		0.023	6
2.7	Diversity in leadership roles	3.28		0.065	7
2.8	Gender Imbalances in Housework	2.34		0.198	8

Finally, gender imbalances in housework (MS = 2.34) are the least important factor, indicating that it has a negligible effect on female equality in the KwaZulu Natal construction industry.

Table 3 presents the motivational factors that are relative to retaining females in the construction industry of KwaZulu Natal province in South Africa. The most significant factor in retaining females in the industry is equitable rewards for relative input to the project (MS = 4.84). This means that when females are paid in proportion to the input they make to a project, it encourages them to stay and contribute their best. It is important to note that women are often paid less than their male counterparts in the industry, and ensuring equitable rewards helps to address this disparity.

The second most influential factor in retaining females in the construction industry is the exercise of power (MS = 4.64). When women are given decision-making authority without having to seek approval from colleagues, it indicates a sense of equality and enhances their retention in the industry. This also improves their performance as they are empowered to take ownership of their work. Recognition of contribution made (MS = 4.36) is the third most important factor in retaining females in the industry. This means that acknowledging and valuing the contributions of female workers in the construction industry is crucial in keeping them motivated and engaged. Job security (MS = 4.24) is the fourth most significant factor in retaining females in the construction industry. Women need to feel secure in their jobs to be able to focus on their work and contribute effectively. This can be achieved through the provision of clear job descriptions and contracts and a stable working environment.

A sense of belonging and identification with the project team (MS = 4.00) and achievement from meeting complex tasks (MS = 4.00) are ranked equally as the fifth most important factor in retaining females in the construction industry. These factors are linked to a positive workplace culture where women feel valued and appreciated, and they are given challenging tasks that help them to develop their skills. Opportunity to extend skills and experience (MS = 3.64) is the seventh most significant factor in retaining females in the industry. This means that providing training and development opportunities to female workers is important in keeping them engaged and motivated. Finally, pay allowance (MS = 3.62) is the least influential factor in retaining females in the construction industry. While pay is important, it is only one of many factors that contribute to motivation and retention, and it should not be the only focus of efforts to retain female workers.

Overall, this table highlights the key factors that are important in retaining female workers in the construction industry of KwaZulu Natal province in South Africa. By addressing these factors, organizations can improve their retention rates of female workers and create a more diverse and inclusive workplace.

Table 3. Motivational factors relative to retaining females in the industry.

S/No	Factor	Mean	Average	SD	Ranking
1	Equitable rewards to relative input to the project	4.84	4.23	0.086	1
2	Exercise of power	4.64		0.058	2
3	Recognition of contribution made	4.36		0.018	3
4	Job security	4.24		0.007	4
5	A sense of belonging and identification with the project team	4.00		0.033	5
6	Achievement from meeting complex tasks	4.00		0.033	5
7	Opportunity to extend skills and experience	3.64		0.084	7
8	Pay allowance	3.62		0.086	8

The conclusion regarding the causes of the shortage of female talent in the construction industry in the KwaZulu Natal province of South Africa aligns with the findings of a study conducted by Eige (2016). To maximize the contribution of women in the construction industry, it is beneficial to place them in areas where they can excel, leveraging their strengths in soft skills and other domains. Economic prosperity and the development of women's competencies are crucial factors for promoting gender equality in the construction industry. Economic prosperity ensures that women can earn a living, while training programs help them acquire skills that are advantageous in a male-dominated field. This viewpoint is supported by McKinsey & Company's study (2015). In terms of retaining women in the construction industry, equitable pay and rewards have the greatest impact. When women are not treated as inferior and are recognized as valued members of the system, they develop a sense of belonging. This finding is consistent with a study conducted by Nkomo *et al.* (2021) on the same topic.

5 CONCLUSIONS AND RECOMMENDATIONS

5.1 *Conclusions*

The findings of this study shed light on the factors contributing to the shortage of female talent in the construction industry of KwaZulu Natal province, South Africa. Discrimination, male-dominated culture, harsh working conditions, and the industry's image were identified as significant factors affecting the participation of women in the industry. The study also highlighted several factors that enhance female equality in the construction industry as well as motivational factors that can contribute to the retention of female workers in the industry.

5.1.1 *Implications of study to literature, policy development & industry*

The findings contribute to the existing literature on gender disparities in the Kwazulu-Natal Construction Industry, confirming discrimination, male-dominated culture, and harsh working conditions as barriers to women's participation. The study provides specific insights into the experiences of female professionals in the KwaZulu-Natal construction industry, adding to knowledge on gender equality and diversity in the workplace. These findings have important implications for policy development in the construction industry. Policies should focus on creating a supportive and inclusive work environment, ensuring equal pay and opportunities for women, promoting diversity in leadership roles, improving working conditions, and addressing the industry's image to attract and retain female talent. Implementing these policies will foster a more gender-inclusive industry. For the construction industry in KwaZulu-Natal, the study's findings offer practical implications. Understanding the factors contributing to the shortage of female talent allows industry stakeholders to take proactive measures. Creating awareness about gender biases, promoting inclusivity, providing equal opportunities for women in terms of compensation, career advancement, and decision-making, investing in women's competencies through training and development, and ensuring job security and work-life balance is essential. These strategies will harness the potential of female professionals and create a more diverse and prosperous workforce.

5.2 *Recommendations*

- It is recommended that equitable pay and rewards be given to female counterparts to retain them and make them feel a sense of belonging within the company and project team.
- There should be no discrimination against females in the appointment of top management positions, with decisions based solely on merit.
- Ensuring job security can help retain workers in an organization for longer periods. By providing job security, opportunities can be equalized for females in the industry, promoting their development on the job.

By implementing the recommendations outlined in this study, stakeholders can pave the way for a more diverse, prosperous, and gender-balanced construction industry in KwaZulu Natal.

REFERENCES

Abdullahi, U., & Adeniran, O. A. (2020). Gender Discrimination and Workplace Performance: Evidence from Nigerian Banks. *International Journal of Business and Management*, 15(11), 112–125.

Agumba, J. N., & Kihumba, E. N. (2021). An Investigation into the Challenges Facing Women in the South African Construction Industry. *Journal of Construction in Developing Countries*, 26(1), 15–32.

Chinomona, R., Musengi, R., & Muzenda, E. (2020). Factors Influencing the Participation of Women in Construction. *Sustainability*, 12(3), 1288. https://doi.org/10.3390/su12031288

Cilliers, E., & Pretorius, L. (2016). Women in Construction: South Africa's Experience. *South African Journal of Human Resource Management*, 14(1), 1–10.

Deeb, A., & Alam, M. (2021). The Impact of Work Environment on Construction Labour Productivity in South Africa. *Journal of Construction Business and Management*, 5(1), 1–13. https://doi.org/10.1080/24683524.2020.1855874

Eige, M. (2016). *The Economic Advantages of Gender Equality in the Workplace*. UN Women. https://www.unwomen.org/-/media/headquarters/attachments/sections/library/publications/2016/economic-advantages-gender-equality-workplace-en.pdf?la=en&vs=4315

Ergen, E. & Aslan, D. (2020). The Role of Job Satisfaction, Organizational Commitment, and Work-Life Balance in Women's Intention to Stay in the Construction Industry: *A Study in Turkey. Sustainability*, 12(21), 9021.

Govere, I. M., Odumosu, T. O., & Oyedele, L. O. (2020). Assessing the Effectiveness of Gender Mainstreaming Initiatives in the South African Construction Industry. *Journal of Construction in Developing Countries*, 25(1), 1–15.

Kalema, B. M. & Maina, S.W. (2021). Impact of Employee Motivation on Organizational Performance in the Construction Industry in South Africa. *Journal of Construction Business and Management*, 5(1), 1–12

Kekana, M. (2021). Exploring the Experiences of Women in the South African Construction Industry: A Qualitative Study. *Construction Economics and Building*, 21(3), 1–15.

Khumalo, B., Radingoana, T. N., & Choenyane, O. (2021). Gender Transformation in the South African Construction Industry: A Desktop Study of the State of the Industry. *African Journal of Science, Technology, Innovation and Development*, 13(3), 408–417.

Leedy, P. D. and Omrod, J. E. (2013) *Practical Research Planning and Design*. Tenth Edition, Pearson New International Edition.

Madonsela, N., & Matolweni, P. (2019). Women's Participation in the Construction Industry in South Africa: An Analysis of Issues and Challenges. *Journal of Gender, Information and Development in Africa*, 8(2), 57–76. https://doi.org/10.31920/2050-4292/2019/v8n2a4

Master Builder Association, *Master Builders KwaZulu-Natal*, Durban.

McKinsey & Company. (2015). *Women Matter Africa: A South African Perspective*. https://www.mckinsey.com/~/media/McKinsey/Featured%20Insights/Employ

Mkhize, S., & Ngwenya, S. (2021). Female Participation in the South African Construction Industry: Challenges and Opportunities. *South African Journal of Science*, 117(7/8), 1–6.

Mphahlele, T. N., Nkuna, K. K., & Mathebula, L. N. (2020). An Assessment of Women's Participation in the South African Construction Industry. *International Journal of Construction Management*, 20(3), 253–265.

Muthathi, F., Managa, T.G., & Nkosi, G. (2017). Factors Affecting Retention of Professional Women in the South African Construction Industry. *Journal of African Construction and Building Technology*, 4(1), 74–87.

Nkomo, S., Kekana, K., & Mphahlele, R. (2021). Motivational Factors for Retaining Female Employees in the South African Construction Industry. *Journal of Economics and Behavioral Studies*, 13(2), 126–137.

Sikakana, B. N., Mpekoa, T. A., & Maphalala, M. C. (2021). Enhancing the Participation of Women in Construction: Opportunities and Challenges in South Africa. *Journal of Construction in Developing Countries*, 26(1), 29–45.

Stats SA. (2018). *Quarterly Labour Force Survey*. Statistics South Africa. Retrieved from https://www.statssa.gov.za/publications/P0211/P02112ndQuarter2018.pdf

Tenurial arrangements to encourage growth and development in urban villages in Nigeria

A.N. Abdullahi
Kaduna Polytechnic, Kaduna, Nigeria

G.O. Udo
University of Uyo, Nigeria
University of Johannesburg, South Africa

F.P. Udoudoh & J. Udo
University of Uyo, Nigeria

C.S. Okoro
University of Johannesburg, South Africa

ABSTRACT: This study examines urban villages as transformative symbols in Nigerian cities, providing affordable housing to low-income rural migrants and serving as a vital labor force for manufacturing and low-end services. However, limited property rights hinder the socio-economic progress of urban village inhabitants, preventing effective utilization and transfer for economic purposes. The paper proposes solutions by advocating for liberalizing the Nigerian Land Use Act (LUA) to establish a continuum of land rights encompassing formal and informal arrangements. Amendments to the LUA, particularly addressing urban villages, should consider customary laws and informal markets, ensuring permanent statutory right of occupancy under Section 8. Suspending or removing Section 22 would facilitate access to mortgage loans. These measures strengthen tenurial arrangements, promoting equitable and efficient land resource utilization. By endorsing a broader spectrum of land rights, the LUA can contribute to the economic development of urban villages and enhance the socio-economic well-being of their residents.

Keywords: Affordable housing, low income, urban villages, development

1 INTRODUCTION

Rapid urbanization has led to the proliferation of urban villages in most cities of developing countries in Asia and Africa. It is characterized by mixed tenure and poses challenges to global development objectives, such as inclusive development, therefore necessitating a need for investigation. Through complex dynamics, urbanization tends to irreversibly change everything in its path including institutions, customs, and lifestyle (Berrisford & MacAuslan 2017). In line with this assertion, Hong (2015) postulates that urbanization leads to the expansion of urban built areas, without firstly providing for basic facilities (roads, electricity, pipe borne water) into formerly rural villages, hence, creating urban villages with their predominant customary/informal tenurial arrangements within most cities. Udoudoh & Ofem (2016) showed how these urban villages were rural settlements that were later taken over and integrated into urban areas because of urban expansion into the peripheral regions.

Zhao (2020) observed that a fallout of increasing urbanization is the exacerbation of the gap between urban and rural areas. The rapid urbanization coupled with the region's burgeoning population and upsurge of internally displaced persons (because of insecurity from the Boko Haram, armed banditry, and farmers/herders' clashes) are causing the increasing outflow of rural residents seeking employments and better livelihoods to nearby cities. These immigrants are offered limited alternatives and end up in the proliferating urban villages. Unfortunately, the rising inequalities and exclusion because of non-recognition of urban villages by the various State governors has exacerbated the living conditions in the settlements in Nigeria's North-West geopolitical zone.

Other countries like South Africa have devised strategies such as bottom-up "community based" approach; sustainable livelihoods programmes; participatory development strategy amongst others that are recording some success in the upgrading of settlements (Daniel et al. 2015). In Nigeria, the government has experimented with approaches which include benign neglect, repressive options, resettlement or relocation, slum upgrading and regularization programmes, which have not been sustainable (ibid.). This study explores the impact of alternative tenurial arrangements on the socio-economic development of urban villages.

The phenomenon of urban villages proliferating in our cities arises due to an accelerated influx of migrants from rural settlements into urban areas has been described as worrisome (Kamalipour 2017). These urban villages though not empowered with legal and institutional rights, meet the housing needs of urban refugees, asylum seekers, internally displaced persons, new rural-urban migrants as well as the cities indigent population excluded from the urban housing systems (Udoudoh & Ofem 2016; UN Habitat 2016). Despite this Vij et al. (2018) observe that there is prevailing legal and institutional orthodoxies that give privileges to cities and their needs over the surrounding urban villages. In the last decade the topic of urban villages has been described extensively from different perspectives by scholars. They are described as villages that create new urban and non-state spaces, producing a new urban mosaic (Hong 2015). Others have focused on property rights (Yani 2017), the housing market, construction work and sanitary conditions (Wang et al. 2009), and the challenges of managing physical infrastructure in urbanized villages in Nigeria. Most of these studies promote urban villages as a positive aspect of urbanization and as a solution rather than a problem even though urban villages share some important characteristics with informal settlements, including precarious tenurial security (Wang et al. 2009). There is a dearth of empirical data on how the predominant customary/informal tenures have impacted the socio-economic development of these urban spatial phenomenon particularly in North-West Nigeria. This paper seeks to discover the power behind that phenomenon.

Therefore, this research explored the characteristics of urban villages towards developing strategies to improve the socio-cultural and economic status of inhabitants through an ideology of inclusive development. The specific objectives were to examine the tenurial arrangements existing in urban villages of the study area and the impact of existing tenurial arrangements on the socio-economic development of urban villages of the study area. The remaining sections of the paper presents an overview of the concept of urban villages, the methods employed to conduct the study including the study area, and the results of the study. The discussion of the findings and conclusion follow.

2 CONCEPT OF URBAN VILLAGE AND THEIR TENURIAL ARRANGEMENTS

The concept of urban villages can be viewed from two perspectives: as a planned development, and a product of urbanization (Wu & Zhang 2022). In the first scenario, an urban village is a well planned development concept in which the population resides in the surrounding areas that fulfill and provide their basic needs and wants (Brindley 2003). The second scenario, which is the focus of this study, is a settlement inhabited by an urban society that lives and resides in a village abosorbed into a city due to the rapid urbanization process. Urban villages, also known as urbanized villages (Ikurekong 2007) or villages in the city (Cong 2017) are defined as rural settlements that have been taken over and integrated into urban areas as a result of urban expansion into the peripheral rural regions (Udoudoh & Ofem 2016). This implies that large cities fuelled by rapid urbanization continuously engulf more and more communities and villages on its path through a process termed "quiet encroachment"; thus, forming this type of spatiality. Zhang (2012) believes that urban villages are a unique form of slum developed from rural settlements, representing an existing conflict in the allocation of public resources to different social groups. Public resources such as infrastructures and social services are poorly provided, the society is binded by deep social networks (affinity, geography, clan, beliefs and customs), and the administrative status and land use remain unchanged as formal urban development simply leapfrogs the village settlement component (Hao et al. 2011). Therefore, urban villages are traditional settlements transformed by the rapid urbanization and industrialization of the urban center contiguous to it. These envelope and takes over the village, its surrounding farmlands for urban purposes and its administrative functions. These towns also have varying degrees of impact on the economic, social and cultural life, but the villages still maintain their largely rural traits. The tenurial arrangements that exist show that they are a transitional evolving neighborhood characterized by a mixture of rural and urban features.

The modes of allocation of title and rights to land in any given society is an important indicator of the type, character and organization of that society, since land rights can be held to reflect rights in other areas of public life. In Nigeria, the primary tenure is a statutory ownership in line with the provisions of the LUA which subsists for every part of the country. Also prevalent in the urban villages of the North-West region is the customary tenure, which predates the colonial era, but is gradually being eroded by the operations of the LUA and neo-liberal economic practices. Nevertheless, customary tenure constitutes over 80% of the entire landholding operating in most rural communities and determines ownership, possession, access, use and transfer (Onyebueke et al. 2020). It represents evolving uncodified laws, principles and norms that prevail despite the LUA. Under the customary regime, land is considered to be owned by a supreme being, held by a community, and administered for the benefit of the community by the village head, chief, or oba (head chief) with access being achieved through kinship/membership of a group/community (Maduekwe 2014).

In the North-West geopolitical zone of Nigeria, land use, transfer, development, and inheritance are traditionally managed according to communal needs rather than through payment. However, it is being replaced by a combination of reinterpreted and innovative customary practices with other informal and formal practices (neo-customary land tenure), which make it easier to acquire property or secure access to land, blend pre-colonial land management procedures and informal settlements in the North-West geopolitical zone (with their own actors and procedures). The contracts are usually not registered, and they are responsible for most of the multiple land sales disputes, the spatial development and environmental problems within the metropolis. To avoid these problems, various governments have embarked upon land registration programmes to ensure that formal rules and procedures are followed in the land market in line with the provisions of the LUA.

In urban villages within the cities of Kaduna, Kano, Katsina, Zamfara and Sokoto, a wide range of primarily de facto tenure categories, with varying degrees of legality and limitation on development, are observed. The impact of these tenurial arrangements on the overall development of urban village is spatial, social and economic. Onyebueke et al. (2020) examined clashes between customary tenure and statutory practices and how stakeholders are appropriating them to promote and resist displacement/eviction in a community in Enugu, Nigeria. The study discovered that peri-urbanisation is consequential to displacement risks confronting customary landholders and communities. Legal reforms that endorse a 'continuum of land rights', ranging from documented occupancies to others that are less formalized, were recommended.

Shi (2018) examined land ownership in Chinese cities and revealed that collective tenure enabled rural villages to create self-governance mechanisms that allowed transformation of individual and collective assets into vibrant, well-serviced, and mixed-use neighborhoods. It would be interesting to see how such property rights regimes can support equitable and holistic notions of urban resilience in other parts of the world. Further, Yani (2017) examined how landed property rights regime in China affects village-led land development behaviour and spatial outcomes in urban villages and revealed that various dynamic land development strategies are adopted by villagers, contributing to the coexistence of sub-optimal industrial and high-quality housing developments in urban villages. This is in response to the changing market environment and internal economic conditions in the dynamic urbanization process. Similarly, Nagpal (2018) examined the effect of property titles on housing investment decisions (improving quality and amount invested in improvements) of 1376 low-income urban households in three Indian resettlement localities, comparing officially resettled and unregistered households. The study discovered that more secure property titles are not associated with a higher probability of engaging in housing investment, with higher levels of housing investment, or indeed with measures of housing quality. The results suggest that urban property titling programs in developing countries can be more successful if they also expand access to complementary inputs, such as the formal financial sector.

Lasisi et al. (2017) studied the city expansion and agricultural land loss in Osun, Nigeria, to determine the rate, pattern, and effects of uncontrolled spatial expansion and found that a perceived risk of losing farmlands to encroachment within 20 years particularly given that formal registration of property titles was low. Protection of agricultural land rights, proper compensation for forced eviction, encroachment, and conversion of farmlands into housing development, organized relocation and provision of supporting infrastructure were recommended. Similar views were expressed in Valencia (2016) who assessed the different peri-urban settlements in Columbia including agriculture-based, informal, formalized, and state-subsidized housing. The study found that inhabitants of agriculture-based settlements are the most legally and physically marginalized as their key resources are increasingly degraded; informal settlers are often exposed to a variety of social and environmental stressors and poor access to basic services (after building new homes), and the rights of informal dwellers are increasingly recognized through formalization policies, improved access to basic social and physical services, and the introduction of subsidized housing to counteract informality and housing deficits. Concurring with these views, Udoudoh & Ofem (2016) advocated that stakeholders including the community should be involved in providing and managing physical infrastructure in urbanized villages.

Therefore, tenure in urban villages is a burning issue that has been extensively studied. However, its effect on the socio-economic development in Nigeria's North-West geopolitical zone has not been fully explored; a gap which this study fills.

3 METHODOLOGY

The study employed survey research design to collect quantitative data through household survey. However, a pragmatic stance guided by the literature and conceptual framework was taken using epistemological perspective of critical realism. This enabled descriptions, interpretations, and possible explanations to be made in the interest of recommending positive change strategies for the settlements. Therefore, based on the "what" and "how" questions, the tool included pre-coded response and open-ended questions to obtain participants' opinions and perceptions.

The population used for the study were the inhabitants of the urban villages in the North-West geopolitical zones in Nigeria. The region, dominated by the Hausa-Fulani ethnic stock, is in the northern part of Nigeria. It comprises seven States including Kano, Kaduna, Katsina, Sokoto, Zamfara, Jigawa and Kebbi, representing 25.75% of Nigeria's landmass with 216,065 km^2. The region is generally lowland of about 100–300m located in Sudan savannah and home to vast forest reserves and Fulani herders (International Crises Group 2020). It is drained by rivers Kaduna, Rima, Sokoto and Zamfara. Agriculture is the dominant occupation employing approximately two-thirds of the population in the area, with subsistence (millet, guinea-corn, rice, maize and yams) and cash crops (peanuts, cotton, tobacco and gum Arabic). Trading is the second most prominent occupation, with Kano serving as the southernmost point of the famous Trans-Saharan Trade Routes (TSTRs) and well connected with cities in North Africa and Southern Europe. There is low level of literacy attributed to apathy towards and inadequate investment in formal education and this has heightened poverty, disease, squalor, malnutrition, and poor shelter. As of 2019, all seven

States in the zone had poverty level above the national average of 40.1% led by Sokoto (87.7%), Jigawa (87%) and Zamfara (74%) (National Bureau of Statistics 2020). This has led to the zone lagging other geopolitical zones. The region is characterized by deep poverty, sluggish economic growth and limited access to basic services and infrastructures and a profusion of urban villages which serves as haven for many migrant populations.

The research adopted multi–stage or cluster sampling technique in obtaining information from the heads of households and purposive sampling method was used for village heads and key informants. The urban villages were first divided into clusters consisting of District Wards (Ungwa). The second stage division used streets (Layi). Data from 2006 housing census from five of the states including Kaduna (1,115,968), Kano (1,603,335), Katsina (1,066,316), Sokoto (688,648), and Zamfara (591,446) were used. Due to the large population, a sample size was determined using Yamane formular, $n = N/1 + N(e)2$, where n signifies the sample size, N = the population under study, e signifies the margin error/ level of significance. A $\pm$ 5% error margin at 95% confidence level for population above 100,000 was applied. A sample size of 398 rounded up to 400 each state (i.e., 2000) was estimated. This was further spread evenly across the four selected settlements in each state. Thus, a total of 100 questionnaires were administered in each selected urban village settlement. These selected settlements are at the periphery on the four geographical points (North, South, East and West) along the entry/exit point into the selected state capitals. Households were surveyed in selected streets and buildings using simple random sampling technique. Household heads and key informants were approached.

The responses from the household surveys resulted in both quantitative and qualitative data. Of the 2000 copies of questionnaire distributed across the 20 settlement areas covered in the study, 1796 were retrieved and adjudged good enough for analysis, representing about 89.8% of the total distributed. The quantitative data gathered through questionnaire was analysed using descriptive and inferential (factor analysis) statistics. Factor analysis was done to determine the nature of the construct influencing tenurial arrangements and to achieve the validity of data purposes (Latif et al. 2019). The rotated component matrix was the main output from the factor analysis. It showed the most important factors and estimates of the correlations between the variables (principal components). The cut-off value of 0.5 was used for the factor loadings. The open-ended reponses were analysed using content analysis and integrated in the discussion of findings. The factor analysis was also used as a face validity technique as independently loading variables were extracted through rotation.

4 RESULTS

4.1 *Descriptive analysis*

To determine the tenurial arrangements existing in urban villages, household heads were asked to indicate the channel utilized to access property, the type of title/document that they possessed which supports their ownership claim, and for those without title, the reasons for lack of titles. Analysis of data showed that, the most common form of property acquisition was inheritance, as 753 household heads (42%) indicated that this was the channel used to gain access to their property. This is followed by purchase (30 %), customary allocation (18%). Surprisingly, only 6% got access to property through statutory allocation.

The most common document supporting ownership possessed by household heads is agreement certified by community head, with a frequency of 970 respondents (54%). This was followed by the possession of certificate of occupancy (C of O) (19%) and sales agreement (14%). Those that did not possess any form of documentation to support their ownership claim were 5% of the household heads. This indicated that apart from the 335 respondents (19%) that possess C of O, the remaining 1461 respondents do not have formal documents to legalise their ownershop of land. The major reasons given were the prohibitive cost (43%), the lack of knowledge/awareness of the importance C of O (19%), lack of interest in acquiring C of O (15%), and laborious bureaucracy (14%).

The study further inquired into the property rights available to households, the strength of property rights and the property use types by household heads. The dominant property rights were the rights to inherit property from relatives (43%), occupy (36%), and dispose of property (3%). On the strength of their property rights, 79% of the respondents indicated that they had the right to be compensated in the event of compulsory acquisition, 10% had the right to exclusive occupation, while 1% had the right of the living/to alienate. Most inhabitants of urban villages solely use their property for residential purpose only (67%), for dual purpose of residential and commercial (22%), and residential cum service industry (5%).

Household heads were asked if they had at any time used their property as collateral for accessing credits. The data showed that 86% (1538 respondents) had never used their property to secure funding, while 14% affirmed to the use of their property to secure funding (of which 69% obtained the loans from friends and relatives, 14% from credit and thrift societies, and 9% from corporate bodies such as banks). Household heads prefer informal sources for reasons that include ease of access (43%), no interest rate (30%), little interest (13%) and convenient repayment plan (13%).

4.2 Factor analysis results

The use of factor analysis to measure the variables in tenurial arrangements resulted in seven underlying components for tenurial arrangements in urban villages of North-West Nigeria. Details of the factor loadings are shown in Table 1. The communality values were all above 0.6. However, some factors including customary allocation, purchase, no document and high cost of funds cross-loaded on two or more factors (above 0.45) and were removed. The emerged components were named as follows: 1 (difficulty in acquisition of formal rights), 2 (cost of documentation), 3 (local inheritance), 4 (awareness of security of rights), 5 (extra illegal occupation), 6 (customary rights) and 7 (formal acquisition).

Table 1. Factor loadings for tenurial arrangements variables.

	\multicolumn{7}{c}{Rotated Component Matrix}	Communalities
	1	2	3	4	5	6	7	
Trespass	.950							.952
Not aware of implications	.939							.932
Govt allocation	.919							.933
SROO	.906							.942
Statutory allocation	.767							.971
Laborious bureaucracy		.904						.917
CROO		.861						.921
Purchase		-.851						.921
Indigene			.954					.975
Settlers			-.939					.959
Inheritance			.626					.956
Not interested				.865				.842
Not sure of process				.754				.687
Gift				.694				.823
Grant					.857			.770
Squatter					.734			.787
Agreement cert. by comm. head						-.838		.986
Sales/deed of Assignment						.810		.914
Pledge/foreclosure							.836	.794
Tenancy agreement R of O							.630	.822
C of O							.506	.767

5 DISCUSSION OF FINDINGS

5.1 The tenurial arrangements existing in urban villages of the study area

The most common channel for property acquisition is inheritance, as 42% of household heads indicated that this was the channel through which they achieved access. This supports findings of Eboiyehi & Akinyemi (2016) that inheritance is a major vehicle for access for individuals. Results from factor analysis categorised inheritance in relation to settlers and indigenes in the third component. as shown in Component 3. This suggests that although security of property titles does not provide a higher probability of engaging in housing investment, as Nagpal (2018) argued, there is a possibility of inheritance among dwellers, whether they are natives or otherwise. Acquisition by purchase ranked (second) and loaded highly (0.85). Although the LUA does not allow sales of land without authorization, this practice still continues in urban villages. Acquisition of rights through communal allocation was low, probably as a result of dwindling communal reserves due to urbanization.

Memorandum of land transfer (agreement) certified by the community head was the predominant document possessed, which also loaded strongly on the sixth component (0.84), customary rights. This is because most of the respondents do not see the importance of legal documentation of title, as they explained in the open-ended responses. For most of the household heads, land transactions take place between the buyer and the seller in consultation with the local Village Head, who witnesses the transaction and issues a signed and stamped 'paper' that connotes approval of the sale, in return for a payment or token. This local approval is then taken to the next chiefdom level, the District Head, who also appends a seal of approval to the transaction and again takes a small commission. This document presumes secured ownership claim to land. Onyebueke & Ikejiofor (2017) and Chimhowu (2019) believe that urban communities are witnessing innovative neo-customary practices with assorted socio-economic and institutional transformation. The recognition of the customary land delivery system by parties involved in the transaction and the communities provides sufficient tenure security for the lands and buildings in these areas. These locally issued land exchange documents do not have the same powers as an official C of O issued by Government under the Land Tenure Laws and LUA, when presented to commercial banks and mortgage agencies as collaterals for a loan.

Furthermore, even with the touted advantages of C of O, as an instrument for security of tenure and its use as collateral for securing finance, it is observed that a high percentage of the inhabitants of urban villages do not possess this valuable document due to reasons of costs (as agreed by which 43%, and), Although only 14% agreed on bureaucracy being a reason for non-possession of a valuable dosucment, this variable loaded strongly on the second component (costs of documentation), with as high as 0.90, indicating that it is relevant. In addition, because of the high perception of de facto rights that the villagers have over the land, their current priority is not on possession of land titles but rather on communities' modernization in terms of infrastructure, social amenities and extension of utility services to their settlements.

The finding that the dominant property right available to household heads is the right to inherit property from relatives, could also be related to the third component (local inheritance). Unsurprisingly the right to alienate property is negligible as only 3% of household heads claim to have such right. This is because of the embargo on the disposal of customary land. Household heads are very confident of their rights as majority of the respondents claimed to possess the highest rights possible (the right to be compensated in the event of compulsory acquisition). The least right claimed was the statutory right to alienate, probably because of the laborious bureaucracy involved in getting the Governor's consent as stipulated in the LUA.

5.2 The impact of existing tenurial arrangements on the socio-economic development of urban villages of the study area

The analysis of data revealed that only 14% of household heads affirmed to the use of their property to secure funding. This is largely due to customary restriction on the sales and use of communal/family property as collateral for accessing credit facilities and limited access to formal (banking) credit source. Households that used their property to access credit stated that the most preferred credit source is that of friends and relatives. This supports assertions of Moahid & Maharjan (2020) on the relevance of informal credit sources such as these which are usually cost-free. Additionally, the predominant low income status of inhabitants of urban villages, coupled with irregular income, affects their ability to access loan from formal financial institutions. They then resort to family and friends; these sources may be unreliable as they also struggle or have limited financial capability to offer such help. About 14% of the respondents of the study reported that sourcing for credit from thrift societies is their preferable source of credit facilities. This is despite the observed disadvantage of accessing loan from informal sources. The findings also showed that most inhabitants of urban villages (67%) solely use their property for residential purpose. This is in line with findings of Hao et al. (2011). Further, some household heads reported using their property for dual purposes of residential and commercial, and some for residential cum service industry, and others residential cum manufacturing. These findings correspond with views shared by Hao et al. (2011) who opined that formal land developments leapfrogs in urban villages and dwellers tend to be binded in social networks and thus it is possible to operate or engage in multiple land uses to support socio-economic sustenance. The economically active inhabitants of urban villages operate under de facto tenure arrangement and although it provides high level security of tenure to meet their livelihood needs, it has also been discovered to be the antithesis for socio-economic developments of urban villages as it inhibits the use and transfer of properties for purposes of alienation and securing of consociate funds and thus contributes to their exclusion from the urban economy. To encourage urban villages to thrive, the government needs to amend the LUA to endorse a continuum of land rights that includes documented and verified de jure rights and de facto less formalized, individualized undocumented rights and provide basic infrastructure and services to support formal investments and informal economy.

6 CONCLUSION

The study aimed to establish tenurial arrangements and their impacts in urban villages in the North-West of Nigeria. The study found, through descriptive and inferential statistics, that tenurial arrangements are underlain by seven key factors including difficulty in acquisition of formal rights, cost of documentation, local inheritance, awareness of security of rights, extra illegal occupation, customary rights and formal acquisition.

Urban villages epitomize the changing societal and economic form of Nigerian cities and thus play a crucial role as they offer a large scale affordable accommodation to shelter very large numbers of low-income rural migrants in urban areas. These constitute the cheap labour force for both manufacturing and low end services. To encourage this segment of the society, the government needs to endorse a continuum of land rights that includes documented and verified de jure rights and de facto less formalized, individualized undocumented rights. To achieve this, the LUA could be amended to capture the prevailing realities surrounding customary laws and informal markets. This should include the amendment of Section 8 of the Act to make the grant of statutory right of occupancy permanent; eliminating Section 22 of the Act to remove the right of obtaining government approval for transaction, which will reduce delay and ease the obtaining of credit facilities using land as collateral; amendment of Section 28 to remove revocation on grounds of alienation by land owner without governor's approval; amendment of Section 29 to make room for compensation for loss of land, at market value not only for unexhausted improvements. These measures will strengthen tenurial arrangements for effective and equitable utilization of land and land resources for economic development.

REFERENCES

Berrisford, S. & MacAuslan, P. (2017) *Reforming Urban Laws in Africa: A Practical Guide.*
Brindley, T. (2003). *The Social Dimension of the Urban Village: A Comparison of Models for Sustainable Urban Development.* Urban Design International.

Chimhowu, A (2019) The New African Customary Land Tenure. Characteristic, Features and Policy Implications of a New Paradigm. *Land Use Policy.* 81: 897–903.

Cong, H. (2017). *An Assessment of Urban Village Redevelopment in China: A Case Study of Medium Sized City of Weihai.* (Doctoral thesis). Heriot-Watt University.

Daniel, M. M., Wapwera, S. D., Akande, E. M., Musa, C. C. & Aliyu, A. A. (2015). Slum Housing Conditions and Eradication Practices in Some Selected Nigerian Cities. *Journal of Sustainable Development,* 8(2), 230–242.

Hao, P., Sliuzas, R., Geertman, S. (2011). The Development and Redevelopment of Urban Villages in Shenzhen. *Habitat International,* 35(2): 214–224.

Hong, Q. (2015). *Stakeholders' Interaction in the Redevelopment of Urban Villages: A Case Study in Xiamen.* (Doctoral dissertation) University of Sheffield. http:etheses.whiterose.ac.uk/10762/

Ikurekong, E. A. (2007). An Assessment of Urbanized Villages for Housing Development Projects in Uyo, Akwa Ibom State. Being a Paper Delivered During the *Annual Seminar by Faculty of Environmental Studies,* Uniuyo. April, 18.

International Crises Group [ICG] (2020). *Violence in Nigeria's Northwest: Rolling Back the Mayhem. No. 288 African Report.* Abuja/ Brussels.

Kamalipour, H. (2017). *Urban Informalogy: The Morphologies and Incremental Transformation of Informal Settlements.* (Degree thesis). University of Melbourne.

Land Use Act 1978 CAP L5. *Laws of Federation of Nigeria 2004.*

Lasisi, M., Popoola A., Adediji A., Adedeji, O., and Babalola, K. (2017). City Expansion and Agricultural Land Loss within the Peri-Urban Area of Osun State, Nigeria. *Ghana Journal of Geography* Vol. 9(3), 2017 pages 132–163.

Latif, N. A. A., Abidin, I. M. Z., Azaman, N., Jamaludin, N. & Mokhtar, A. A. (2019). A Feature Extraction Technique Based on Factor Analysis for Pulsed Eddy Current Defects Categorization. *IOP Conf. Series: Materials Science and Engineering* 554 (2019) 012001.

Maduekwe, N. (2014). *The Land Tenure System Under the Customary Law.* Available SSRN 2813056. 1–20.

Moahid, M & Maharjan, K. L. (2020). Factors Affecting Farmers' Access to Formal and Informal Credit: Evidence from Rural Afghanistan. *Sustainability,* 12, 1268.

Nagpal, K. (2018) *Essays on Urbanisation and Governance in Developing Countries* (Doctoral Dissertation) Jesus College (University of Oxford).

National Bureau of Statistics (2020). *2019 Poverty and Inequality in Nigeria: Executive Summary.*

Onyebueke, V., Walker, J., Lipietz, B., Ujah, O. and Ibezim-Ohaeri, V., (2020) Urbanisation-Induced Displacement in Peri-urban Areas: Clashes Between Customary Tenure and Statutory Practices in Ugbo-Okonkwo Community in Enugu, Nigeria. *Elservier Land Use Policy* 99(2020) 104884.

Onyebueke, V.U., Ikejiofor, C (2017) Neo-customary Land Delivery Systems and the Rise of Community Mediated Settlements in Peri-urban Enugu, Nigeria. *International Development Plan Rreview.* 39(3)319–340

Shi, L. (2018), *The Promises and Perils of Collective Land Tenure in Promoting Urban Resilience: Learning from China's Urban Villages.* Habitat International. 77.

Udoudoh, F. P. & Ofem B. I. (2016). The Stakeholders and Challenges of Managing Physical Infrastructure in Urbanized Villages in Nigeria. *The International Journal of Social Sciences and Humanities Invention,* 3(11), 3015–3034.

United Nations Habitat (2016) *Habitat Country Programme Document Nigeria (2017–2021).* London: Earthscan.

Valencia, S., (2016) *Caught Between Spaces: Socio-economic Vulnerability in Formal and Informal Peri-urban Bogota and Soacha Colombia* (Doctoral Dissertation) Lund University – Sweden.

Vij, S., Narain, V., Karpouzoglou, T., Mishra, P., (2018) From the Core to the Pheriphery: Conflicts and Cooperation Over Land and Water in Peri-urban Guragaon, India. *Land Use Policy* 76 (382–390).

Wang Y.P., Wang Y. & Wu, J. (2009). Urbanisation and Informal Development in China: Urban Villages in Shenzhen. *International Journal of Urban and Regional Research,* 33(4), 957–973.

Wu, Y. & Zhang, Y. (2022). Formal and Informal Planning-Dominated Urban Village Development: A Comparative Study of Luojiazhuang and Yangjiapailou in Hangzhou, China. *Land, 11*(4), 546.

Yani, L. (2017) The Role of Urban Villages in the Urbanization Process in China: The Case of Shenzhen. *International Conference on Construction and Real Estate Management.* Edmonton – Canada.

Zhang, X. (2012). *The Making of Public Open Space Accessible to Underserved Populations in Urban Village.* Unpublished Master's Dissertation. Columbia University in the City of New York, United States.

Zhao, X. (2020). *Living Heritage and Place Revitalization: A Case Study of Place Identity of Lili, a Rural Historic Canal Town in China.* (Doctoral dissertation). University of Queensland. http://espace.library.uq.edu.au/view/uq:98

Environment, climate change and shock events impact and response and water resources

Strategies to respond to climate change effects in the Zimbabwean construction industry

M. James
Faculty of the Built Environment, National University of Science and Technology, Bulawayo, Zimbabwe

T. Moyo
Department of Quantity Surveying, Nelson Mandela University, Port Elizabeth, South Africa

ABSTRACT: The Zimbabwean construction industry is vulnerable to impacts of climate change, but lacks effective responses and policies to strengthen climate change resilience. The effects are characterized by destruction of infrastructure, construction workers' health and safety concerns, among others, therefore, probing the study. Qualitative methods were incorporated to gather insights from purposively selected seven key informants. Face-to-face interview data, analyzed through thematic analysis revealed common themes including the need to; integrate construction-related policies, improve enforcement of existing regulations, provide adequate funding for climate change requirements, facilitate climate change literacy for construction professionals, and promulgate building by-laws that support construction activities. Findings encompassed the need to deal with climate change throughout the construction process. However, consensus on these issues needs to be made apparent. The major limitation was data collection from only key informants; nevertheless, this augers well with exploratory studies. Future studies should further interrogate identified strategies to establish appropriate interventions.

Keywords: Climate Change, Policy, Construction, Sustainability, Developing countries

1 INTRODUCTION

Climate change has been universally recognized as a global problem (Rashid *et al.* 2014), and its burden is huge for poor and developing countries (Eshete *et al.* 2020). The effects have led to climate hazards which directly impact the environment (Zieba *et al.* 2020), project life cycle, operation of buildings and engineering structures, repair and reconstruction, destruction, utilization, and recycling (Kryvomaz 2021). In Zimbabwe, the recurrence of climate hazards and effects (World Bank 2021) supports scholarly scrutiny in terms of policy and also indicates the need to strengthen climate change resilient strategies. Thus, the effects of climate change in the construction industry have been amplified due to the lack of implementable strategies to respond to the challenge. Simpson *et al.* (2021) bemoan that hundreds of millions of people across Africa lack climate risk knowledge for adaptation to climate change. This is synonymous with the Zimbabwean construction industry despite the United Nations Environment Programme (2022) supporting the adaption in terms of policy implementation and encouraging all African countries to decarbonize the carbon stock by making the building codes mandatory during massive construction, investing in building capacity, resources, and supply chain to promote energy efficient designs and low carbon sustainable construction. Although previous studies in Zimbabwe have indicated the effects of Climate Change on Millennium Development Goals (MDGs) and Sustainable Development Goals (SDGs) (Brazier 2015; Brown *et al.* 2012; United Nations 2023), there is limited focus on the resilient strategies on the Zimbabwean construction sector. Therefore, there is a need to explore the strategies that can be employed to respond to climate change's effects on the construction industry in Zimbabwe. The study will explore climate change measures that can be integrated into construction-related policies. The remedies enhance the achievement of one of the prominent Sustainable Development Goals (SDG) of Climate Action – Take Urgent Action to combat climate change and its impacts (United Nations Development Programme 2020). In a study undertaken by Pravin *et al.* (2017), it was noted that climate change is on the leading edge of construction industry discussions because of the impacts it imposes on the construction project's life cycle and on the built environment. Therefore, the construction sector has to be more effective and tackle challenges. United Nations Environmental Programme (UNEP) (2022) also comments that the construction industry and the built environment have a great potential to vastly reduce greenhouse emissions since it is counted as one of the major sectors contributing to them. This calls for a quick response through research and development so that problems of the exact nature are not continuously experienced or adaptive strategies are employed to strengthen resilience in the construction industry regarding project management and the related professionals involved. For clients, project managers and contractors, this study will sensitize them to be more conscious of the effects of climate change so that they are well prepared during project execution.

The following sections review the contrasting perspectives and existing theories on climate change and construction policies. Strategies that have been employed internationally and locally are discussed. The identified gaps are also discussed. The methodology and description of the study area are presented. The empirical findings from the in-depth interviews are articulated. The paper ends with a discussion of findings within the context of literature and conclusions on the implication of findings on construction policy.

2 CLIMATE CHANGE

The concept of Climate change adaptation and mitigation has evolved over the years. Climate change adaptation has shifted from a single-dimension to an integrative approach that aligns with vulnerability and resilience concepts (Tu Dam 2021). Adaptation planning is similarly guided by hazard-based, vulnerability-based, and urban resilience frameworks (ibid). Al-so, Schweizer *et al.* (2013) validated a theoretical framework for place-based climate change engagement based on place attachment theory, place-based education, free-choice learning, and norm activation theory principles. This framework demonstrates the power of involving locals in action-based learning at physical, material locations that also serve as symbolic sites for inspiring political action and learning about the effects of climate change. However, there are conceptual and methodological challenges in defining an adaptation goal, as well as mixed evidence on what effective adaptation looks like and how to enable it. (Singh *et al.* 2022). Singh *et al.* (2021) demonstrated different normative views on adaptation outcomes due to different epistemological and disciplinary entry points and how they can lead to different interpretations of adaptation effectiveness. However, using this theoretical framework for the current study is relevant for three reasons. Firstly, the context of tracking and monitoring (monitoring and evaluation) adaptation is one of the critical areas to ensure effectiveness in strategies to resolve climate change impact on the construction sector, as Smith *et al.* (2019) recommended. Secondly, climate change adaptation and mitigation are mostly characterized by a dynamic environment, including clients who wish to complete the project at minimum cost (Highman & Thompson 2015), contractors who want to maximize profits, politicians who are involved in policy making (UNEP 2022), professionals who manage the construction process from initiation to close-out (Pravin *et al.* 2017). Thirdly, the current study enhances theoretical consideration of climate change with empirical evidence via the domain of the construction industry.

3 CLIMATE CHANGE AND RELATED CONSTRUCTION POLICIES

Climate change response in Zimbabwe can be traced back to 1992 when Zimbabwe ratified the United Nations Framework and consented to the Kyoto Protocol in 2009 and the Paris Agreement of 2015 – Convention on Climate Change (UNFCC) (GOZ 2016). It is against this background that the government of Zimbabwe developed relevant policies in which construction issues were incorporated within major regulations and policies governed by common law (DLA Piper 2023). The supreme law, the Constitution of Zimbabwe, gives everyone the right to an environment that is not harmful to their health or well-being and an environment protected for the benefit of the future generation (Akesson et al. 2016). Specifications for local authorities, there are stipulated building standards through model building by-laws for the design and carrying out of building works. In support, the construction sector has requirements for licenses and permits, which include approvals from the Factories and Works Act [14:08], Regional Town and Country Planning Act, and Model Building By-laws. Most of these approvals lack information specific to climate change-related strategies (DLA Piper 2023). In addition, construction Health and Safety rules are governed by a number of statutes that also recognizes environmental rights and human rights and ensure health and safety in the construction sector. These include the National Social Security Authority (NSSA) Act [17:04], the Environmental Management Agency (EMA) Act [20:27], Pneumoconiosis Act [15:08] (ibid). However, the provisions in these policies are just general and are not specific to health hazards encountered as a result of climate hazards. According to DLA Piper (2023), only two types of legislation (the Environmental Management (Effluents and Solid Waste Disposal) Regulations, 2007 and the Environmental Management Act [Chapter 20:27] and the Environmental Management (Environmental Impact Assessment & Ecosystems Protection) Regulations, SI 7 of 2007) exists which deals with environmental issues affecting building works and with promoting sustainable developments. The missing link is regulations or policies that speak directly to the construction sector.

According to Murombo et al. (2019), in light of the Katowice Partnership for E-Mobility, the transport sector is one of the critical categories of Zimbabwe's NDCs in which electrification of the modes of transport is advocated for. It is, however, noted that electrification of construction equipment has been omitted, which could be essential in the reduction of greenhouse gas emissions since Zimbabwe is a developing country. Massive developments in terms of construction are yet to take place. In addition, the Low Emission Development Strategy (LEDS) specific approaches are selected as mitigation measures, including feedstock improvement, conservation agriculture, reduction of deforestation, fruit tree planting, and commercial forestry (GOZ 2021). Yet, these did not relate directly to the construction sector.

However, Akesson et al. (2016) argue that Zimbabwean policy frameworks are quite robust. However, the challenge is implementing and enforcing existing legislation and policies derailed by a lack of good governance, transparency, accountability, and political will to enforce the laws, insufficient capacity, and coordination among law enforcement agencies. Forsyth (2019) attempts to assist developing countries with climate adaptation and mitigation strategies but asserts that development aid and financial assistance are easily siphoned off by corrupt institutions. This is also supported by Anderson (2021), who concluded that other climate change mitigative solutions are weakened by corruption as some issues that are already high on the anticorruption agenda are closely connected to climate change. However, in trying to deal with these barriers, monitoring, and evaluation mechanisms are recommended.

Nevertheless, in practice, ambitions regarding the quality of monitoring and evaluation processes and frameworks sometimes correlate with the resources allocated to the activities (Christiansen et al. 2016). Climate finance, financial resilience, and financial inclusion are equally substantial for adaptation. There is a need to prioritize and link these to climate change adaptation and mitigation to allow societies and economies to adapt to the adverse effects and reduce the impacts of climate change (Hussain et al. 2021; UNEP 2023). Also, Usman et al. (2022) hinted at the need for government and policy-makers to stimulate income levels to achieve comprehensive and effective environmental policies in Africa. Further, in research by Highman & Thomson (2015), it was observed that construction professionals needed to display the required level of climate literacy. In support, Walshe et al. (2017) indicated that even though climate change is a human-induced problem and a threat to society, priority is given to other factors, like poverty, livelihoods, and food security, over climate change. However, According to Biesbroek (2021), policy integration and climate change adaptation are still in their infancy. In South Africa, it is also argued challenges to the implementation of climate change mitigation policies involve issues of the current status of emissions data and reporting regulations and inadequate credible data for effective policy formulation (Trollip & Boulle 2017).

Scholars have suggested different views in responding to climate change's effects on the construction sector. Among these include, Marichova (2020), who analyzed the specifics of the problem of sustainability in the construction market and the role and influence of the government in the development of sustainable construction. Similarly, Yeukai & Gumbu (2021) suggest that public policy requires a combination of professional and political approaches. Usman et al. (2022) also hinted at the need for government and policy-makers to stimulate income levels as a prerequisite for achieving sound and effective environmental policies in Africa. In support, Murombo et al. (2019) highlight the need for a climate change Act in Zimbabwe in-order to manage climate change risks and set a clear policy pathway that supports the Paris Agreement to reduce greenhouse emissions.

On the other hand, Venugopal et al. (2022) promote in-depth assessments because they can develop scientifically sound preventative and protective labour policies to avert adverse occupational health and productivity consequences for workers globally. Furthermore, Marjanac & Patton (2018) recommend the detection of future extreme weather events for valuable information about the future risks of such events to emergency managers, regional planners, and policy-makers at all levels of government, and this is likely to have implications for the planning and management of building codes, land use, water, health and food management, insurance, and transportation networks. However, these studies needed a comprehensive approach as they show fragmentation in approaches.

4 METHODOLOGY

The study was qualitative, and data was collected using in-depth face-to-face interviews with purposively selected key informants, and 30 minutes length interviews were conducted in their respective offices with the support of recording devices upon request. The aim was to get in-depth knowledge, perceptions, ideas, and advice on the measures that can be integrated into policies and regulations in response to climate change effects in the Zimbabwean construction industry. Respondents included relevant government department heads from government ministries, departments, and agencies (MDAs) who, in most cases, represent contractors, construction professionals, climate change academics, and sustainability professionals in Zimbabwe. More so, the targeted respondents are highly involved in issues to do with policy-making, and most of these have headquarters in Harare, Zimbabwe. Expert sampling was used to obtain expert knowledge (Frost 2023). This allowed the researcher the ability to clarify issues as they arise through probing. The semi-structured interview guide consisted of 2 parts. The first part consisted of Interviewee's profile, designation, highest academic qualification, and experience, and another part dealt with questions on climate change measures that can be integrated into construction-related policies. To ensure validity, the researcher combined negative case analysis and member checking (Birt et al. 2016). An audit trail was also kept by monitoring and recording all research-related activities and data

(Carcary 2020). Reliability was ensured through carefully and honestly carrying out the research (Cypress 2022). Thematic analysis was used to provide brief descriptions and interpretations in terms of themes and patterns from a data set. It was also used to further interpret essential aspects of the research topic and hence, describe the research data in an organized and rich form (Majumdar 2022). The researcher used six steps, according to Scharp & Sanders (2019), in which one first understands the data in general, categorizes and gives subcategories to the data, creates themes in the data, evaluates the articles that have been made, labeled existing pieces, and identified them as a whole. The study is limited to MDAs, and stakeholders like the Construction Industry Federation of Zimbabwe (CIFOZ), Architects Institute of Zimbabwe (AIZ), Zimbabwe Association of Consulting Engineers (ZACE), and Small to Medium Enterprise (SMEs) are significant.

5 FINDINGS AND DISCUSSIONS

The findings were presented and discussed in this section

5.1 *Profile of interviewees*

Table 1. Respondents' profile.

Respondent	Designation	Organization	Experience	Highest Qualification
R1	Director of Urban Planning and Development-	Ministry of National Housing and Social Amenities	15 years	Master of Science Degree
R2	Director – EMA	Environmental Management Agency	10 years	Master of Science Degree
R3	University Lecturer Consultant for the Government of Zimbabwe on Climate change	University of Zimbabwe	17 years	Master of Science Degree
R4	Director Public Works	Ministry of Local Government and Public Works	16 years	Master of Science Degree
R5	Deputy Director Climate Change Mitigation	Ministry of Environment, Climate, Tourism, and Hospitality Industry	15 + years	Master of Science Degree + Doctor of Philosophy Degree candidate
R6	Director	National Social Security Authority	18 years	Master of Science Degree
R7	Director Estates Development and Maintenance	Ministry of National Housing and Social Amenities	15 years	Master of Science Degree in Urban Design

* Source: Survey findings (2023).

5.2 *Strategies to resolve climate change effects – major themes generated*

Themes were generated from the qualitative data analysis. These included the need to coordinate institutions and integrate sector policies coupled with monitoring and evaluation mechanisms, climate change financing and political will, enhanced awareness and climate change literacy, stricter or stringent clauses, review of building codes and health and safety regulations, and timeliness in reviews. The themes are discussed hereafter.

5.2.1 Coordination of institutions and integration of sector policies coupled with monitoring and evaluation mechanisms

Having policies is one thing, and integration of these is another thing. It was noted that the success of climate-related policies is dependent on how they are integrated with sectoral policies, policies from other levels of government, civil society, institutions, and within themselves (integration of mitigation–adaptation policies), as supported by (De Olivera 2009; Murombo et al. 2019). According to Biesbroek (2021), policy integration and climate change adaptation are still in their infancy, as supported by respondents 1, 2, 4, 5, and 6. For instance R1 and R2 remarked as follows;

R1 *"The policy of using solar geysers for water heating is not resident in housing but is implemented by Ministry of Energy, so by-laws cannot be enforced where there is no common ground with the electricity department."*

R2 *"The construction industry needs to overlay their proposed developmental projects with other sector policies and plans; despite EMA having a wetlands map, construction is still taking place on wetlands which increases vulnerability to the construction sector."*

It was similarly observed that respondents agreed that when it comes to implementation, several barriers emerge as respondent 1, from the ministry, argued that some green building codes are managed by another ministry which makes it difficult to implement. This is supported by Macheka (2021), when the author reported that existing environmental regulations are fragmented and difficult to enforce. Integration also requires a certain degree of literacy, as indicated by Respondent 2:

R2 *"There is a need to understand what mainstreaming and what integration means".*

This supports the need for improving awareness concerning climate change so that integration is successful. Ferreira (2022) emphasizes the importance of involving key stakeholders in climate change issues. Furthermore, other respondents argued that awareness needs to begin from a tender age and progress to tertiary education, which is supported by Kuthe et al. (2020). Respondent 1's comments exemplify this:

R1 *"We need to introduce climate change education from the grassroots by including it in the Primary or ECD level curriculum; why should we teach vanasekuru (older people) climate change?*

This is consistent with UN reports that concluded that young people can be taught about the effects of global warming and how to adapt to it in the classroom (UN 2023). Furthermore, UNDP report expresses that coordinated policy-making would help governments navigate these tradeoffs transparently and equitably and ensure that climate change solution leaves no one behind (UNDP 2020). This, however, touches on the need for inclusivity in policy-making as strongly suggested by respondents that climate literacy with regards to construction should also extend to rural or urban construction professionals as per the comment:

R2 *"literacy needs not to be confined to these big construction guys but should extend to the rural area builder who builds those rural schools and year in and year out we are losing a lot of infrastructure, so if we don't make sure they are not sensitized on these types of materials which are climate change resilient we will face the same problem."*

R6 *"literacy needs to extend to the rural folk who is a builder and must be sensitized on proper health and safety procedures during construction, especially in the face of climate change. I don't know if the employers are ignoring occupational health and safety or they are ignorant."*

In addition, inclusivity can go along with enhancing awareness campaigns for all construction industry players. This notion also supports the writings of Highman & Thompson (2015), wherein construction professionals faced client resistance regarding materials used during construction. The importance of material selection is mostly affected by clients' preferences, who commonly aim to reduce the cost of construction or are unwilling to compromise on quality, budget, scope, or schedule at the expense of climate change mitigation solutions. This calls for continuous awareness and the need for continuous research and development for all. As Revel et al. (2009) recommended, inclusivity is also key so that everyone is on the same page and speaks the same language regarding climate change and the construction industry.

5.2.2 Climate change financing and political will

It is apparent that policy integration, literacy, and awareness are inadequate to improve resilience to climate change hazards in the construction sector. As supported by Yeukai & Gumbu (2021), public policy requires a combination of professional and political approaches, and this calls for the need to determine ways in which policies can improve the resilience of this sector to climate change hazards. Some responses indicated the need for adequate funding for climate change requirements, as alluded to by respondents 3, 4,5, 6, and 7. According to Hussain *et al.* (2021), this supports the literature, which recommends that financial inclusion, financial resilience, and climate change resilience are linked.

R3 *"Those who are involved in the monitoring of climate policy regulations need support and resources to do their work. If there are no resources, then monitors cannot reach point B, and as a result, they will end up doing desk research or mining the computer for information".*

R4 *"The policies are there but are not backed with enough resources for them to be implementable; the inspector needs resources to carry out his / her duties; otherwise, he will end up approved by the office. Otherwise he cannot board a combi to go for an inspection."*

R6 *"We are required to monitor construction sites in terms of occupational health and safety, but mmm, the resources to go around all sites limit us."*

Usman *et al.* (2022) support this by recommending the need for government and policymakers to stimulate income levels as a prerequisite for achieving sound and effective environmental policies in Africa. The respondents further commented:

R5 *"At a policy level, there is a need for budgeting for climate change".*

"Ministry of Finance needs to uphold their stance of making it a requirement that Ministries/Departments/Agencies (MDAs) demonstrate mainstreaming of climate change when applying for yearly budget allocation."

Based on these comments, it is clear that it is critical to strengthen political will in policymaking, as supported by Gumbie *et al.* (2023) so that climate change adaptation and mitigation are prioritized. Creating policies necessitates both political will and financial support. However, R3 contends that policy-makers should be able to visualize climate change issues through environmental lenses rather than political lenses, ideologies, and worldviews. As a result, as previously stated by respondents, they require training, awareness, and education to investigate climate change issues. It is therefore noted that if the construction sector is not explicitly regarded as a key sector, ensuring its implementation might be difficult. This is also evidenced by the absence of the construction sector as a key priority area in the Climate Change Policy (GOZ 2015). Therefore, prioritization is also regarded as a key area in ensuring policy implementation.

5.2.3 Stricter or stringent clauses

The respondents kept on buttressing the importance of putting a stricter penalty for offenders but also mentioned the importance of a political will to back this. More so, all respondents concurred on the need for improvement in the enforcement of existing regulations combined with political will, as exemplified by respondents' comments:

R3 *"Policies cannot improve resilience when they are not implemented, not binding, and have no arresting powers. Policies should not be for window dressing purposes but should be functional, integrated, and inclusive".*

R2 *"Law enforcement needs political will and support; otherwise, right now, we are big vicious barking toothless bulldogs because we have no arresting powers."*

R4 *"There is a need for stronger punitive measures for the would-be offenders."*

These responses showed the need to enhance our policies to be implementable, taking it from the EMA Act. However, literature has concluded that the polluter pays principle needs to be revised as it focuses on curing damages rather than on the process that halts the environmental problems (Ingwani 2010). In addition, most construction companies calculate the cost of paying to EMA vis a vis the profit, and they rather pay for pollution or degradation. It is against this background that the respondents feel that there is a need for a more painful penalty other than putting a price tag on issues to do with the environment (ibid).

However, R3 had a different perspective. The respondent advocated that there is a need to be more stringent on policy enforcement so that everyone knows and understands climate change issues in fear of being arrested based on the notion that 'ignorance of the law is no bliss'.

5.2.4 *Review of building codes, occupational health and safety regulations, and timeliness*

All respondents acknowledged the need for the promulgation of building by-laws that support construction activities related to conserving the planet, as exemplified by 4 respondents (R1, R2, R3, R6, R7) who categorically stated that:

R1 "There is a need to revisit by-laws to speak to climate change."

R2 "We need to review the building codes so that they also integrate climate codes."

R3 "We need green building codes to be added to the building by-laws and create a matrix that can be ticked during inspection."

R6 "The existing health and safety regulations must be reviewed in line with climate change issues."

Kryvomaz & Savchenko (2021) also indicate the importance of investing in a green building approach as an adaptation measure against climate change. This supports the move by UNEP (2022) of the need to support adaptation in terms of policy implementation and encourage all African countries to decarbonize the carbon stock by making building codes mandatory during massive constructions. In addition, R 5 bemoaned that the reviews take very long such that, at times, the amendments are overtaken by events and hence calls for the need for set timelines for the completion of policy reviews, even formulation.

5.3 *Conclusion*

The effects of climate change in the construction industry have been amplified due to the lack of implementable strategies to respond to the challenge. Therefore, this study sought to interrogate strategies that can be implemented to resolve the challenge. All the key informants confirmed that political will and support are critical. The findings also indicate the need for policy to ensure construction players be climate change literate through awareness and education. In order to enable players in the built environment to exercise authority in terms of monitoring and implementation and activate legal mandate to arrest offenders, enhanced enforcement through the introduction of a climate change Act is also recommended. Furthermore, it is recommended to introduce industry-specific policies so that policies work in harmony. Policy-makers should now regard the construction sector as an independent sector and not fall under other key sectors. Integrated implementation and integrated monitoring and evaluation of these existing policies are encouraged. Research and development in the built environment are highly recommended to inform policy-makers to prioritize the construction industry and make well-informed decisions when coming up with the aforementioned proposed policies. Stakeholders need to provide adequate funding for climate change initiatives to enhance awareness. Building codes need to include how to deal with climate change hazards during each stage of the construction project process. Future studies should consider views from other construction players in terms of CIFOZ, AIZ, ZACE, artisans, small to medium enterprises, and clients. The qualitative nature of the study is a limitation, as insights from other stakeholders are pertinent.

REFERENCES

Akesson, U., Wingqvist, G. Ö., Ek, G., & César, E. 2016. Environmental and Climate Change Policy Brief Zimbabwe. Gothenburg. *Sida's Helpdesk for Environment and Climate Change.*

Anderson, J. 2021. https://blogs.worldbank.org/climatechange/tackle-climate-change-take-corruption. [Accessed 21 February 2023]

Biesbroek, R. 2021. Policy Integration and Climate Change Adaptation. *Current Opinion in Environmental*

Birt, L., Scott, S., Cavers, D., Campbell C. & Walter F. 2016. Member Checking: A Tool to Enhance Trustworthiness or Merely a Nod to Validation? *Qual Health Res.* 2016 Nov;26(13):1802–1811. doi:10.1177/1049732316654870. Epub 2016 Jul 10. PMID: 27340178.

Brazier, A. 2015. *Climate Change in Zimbabwe Facts for Planners and Decision Makers.* Publishers Konrad-Adenauer-Stiftung 26 Sandringham Drive, Alexandra Park, Harare

Brown, D., Chanakira, R., Chatiza, K., Dhliwayo, M., Dodman, D., Masiiwa, M., Muchadenyika, D., Prisca Mugabe, P. & Zvigadza, S. 2012. Climate Change Impacts, Vulnerability, and Adaptation in Zimbabwe. *IIED Climate Change Working Paper No. 3, Climate Change Working Paper Series.*

Carcary, M. 2020. The Research Audit Trail: Methodological Guidance for Application in Practice. *The Electronic Journal of Business Research Methods, 18.*

Christiansen, L; Schaer, C. Naswa, P.L. 2016. *Monitoring & Evaluation for Climate Change Adaptation. A Summary of Key Challenges and Emerging Practice. Understanding, Discussing, and Exemplifying the Key Challenges of M&E for Adaptation UNEP DTU Partnership Working Papers series;* Climate Resilient Development Programme.

Cypress, B. S. 2017. Rigour or Reliability and Validity in Qualitative Research: Perspectives, Strategies, Reconceptualization, and Recommendations. *Dimens Crit Care Nurse.* 2017 Jul/Aug;36(4):253–263. doi: 10.1097/DCC.0000000000000253. PMID: 28570380.

De Oliveira, J. A. P. 2009. The Implementation of Climate Change Related Policies at the Subnational Level: An Analysis of Three Countries, *Habitat International*, 33(3): 253–259, ISSN 0197-3975, https://doi.org/10.1016/j.habitatint.2008.10.006. [Accessed 19 January 2023]

DLA Piper. 2023. *Real World Law Construction.* https://www.dlapiperrealworld.com/law/index.html?t=construction&s=legal-framework&c=ZW. [Accessed 24 May 2023]

Eshete, Z.S., Mulatu, D.W. and Gatiso, T.G. 2020. "CO_2 Emissions, Agricultural Productivity and Welfare in Ethiopia", *International Journal of Climate Change Strategies and Management*, 12(5): 687–704.

Ferreira, V., Barreira, A., Loures, L., Antunes, D., & Panagopoulos, T. (2020). Stakeholders' Engagement on Nature-Based Solutions: A Systematic Literature Review. *Sustainability* 12(2): 640. https://doi.org/10.3390/su12020640 [Accessed 10 March 2023]

Forsyth, T., McDermott, C. L., & Dhakal, R. 2022. What is Equitable About Equitable Resilience? Dynamic Risks and Subjectivities in Nepal. *World Development* 159, 106020.

Frost, J. 2023. *Purposive Sampling: Definition & Examples*: https://statisticsbyjim.com/basics/purposive-sampling/ [Accessed 09 March 2023]

Government of Zimbabwe. 2016. *Zimbabwe National Climate Policy Report.* [Accessed 20 March 203]

Government of Zimbabwe. 2021. *Zimbabwe Long Term Low Greenhouse Emission.* https://unfccc.int/sites/default/files/resource/Zimbabwe%20LEDS_08Nov2022.pdf [Accessed 20 March 203]

Highman, A. P., & Thomson, C. 2015. An Evaluation of Construction Professionals' Sustainability Literacy in North West England. In *ARCOM Conference* (pp. 417–426). Association of Researchers in Construction Management.

Hussain, B., Islam, M., Ahmed, K., Islam, M. & Atiqul Haq, S. M. 2021. *Financial Inclusion, Financial Resilience, and Climate Change Resilience.*

ICLG.com. 2022. *Construction & Engineering Laws and Regulations Report 2022–2023 Zimbabwe (iclg.com)* https://iclg.com/practice-areas/construction-and-engineering-law-laws-and-regulations/zimbabwe. [Accessed 23 May 2023]

Ingwani, E., Gondo, T., & Gumbo, T. 2010. The Polluter Pay Principle and the Damage Done: Controversies for Sustainable Development. *Economia. Seria Management.* 13. 53–60.

Kryvomaz T. I., & Savchenko, A. M. 2021. Reducing the Impact of the Construction Industry on Climate Change by Introducing the Principles of Green Building. *Environmental Safety and Environmental Management* 37(1): 55–68. https://doi.org/10.32347/2411-4049.2021.1.55-68. [Accessed 23 March 2023]

Kuthe, A., Körfgen, A., Stötter, J., & Keller, L. 2020. Strengthening Their Climate Change Literacy: A Case Study Addressing the Weaknesses in Young People's Climate Change Awareness. *Applied Environmental Education & Communication*, 19(4): 375–388, DOI: 10.1080/1533015X.2019.1597661

Macheka, M. T., Hardman, M. (ed) 2021. Environmental Management and Practises in Zimbabwe's Chivi District: A Political Ecology Analysis. *Cogent Social Sciences*, 7:1, DOI: 10.1080/23311886.2021.2000569

Majumdar, A. 2022. *Thematic Analysis in Qualitative Research.* 10.4018/978-1-6684-3881-7.ch031.

Marichova, A. 2020. Role of the Government for Development of Sustainable Construction. *Ovidius University Annals of Constanta – Series Civil Engineering.* 22. 53–62. 10.2478/ouacsce-2020-0006.

Marjanac, S., & Patton, L. 2018. Extreme Weather Event Attribution Science and Climate Change Litigation: An Essential Step in the Causal Chain? *Journal of Energy & Natural Resources Law*, 36(3): 265–298.

Murombo, T., Dhliwayo M., Dhlakama T. 2019. *Climate Change Law in Zimbabwe: Concept and Insights. The Konrad Adenauer Foundation.* 26 Sandringham Drive, Alexandra Park.

Pravin, S. N. K., Murali, K., & Shanmugapriyan, R. 2017. Review on Climate Change and Its Effects on Construction Industry. *International Research Journal of Engineering and Technology IRJET*, 4(11).

Rashid, M., Begum, R., Mokhtar, M. and Pereira, J. 2014. Physical Development Framework for Climate Change Adaptation in Malaysian Construction Industry – Current Scenario and Way to Improve. *Current World Environment*, 9(3): 552–560.

Scharp, K. M., Sanders, M.L. 2019. What is a Theme? Teaching Thematic Analysis in Qualitative Communication Research Methods. *Communication Teacher*, 33 (2):117–121

Schweizer, S., Davis, S. &Thompson, J.L. 2013. Changing the Conversation About Climate Change: A Theoretical Framework for Place-Based Climate Change Engagement *Environmental Communication*, 7(1): 4262. http://dx.doi.org/10.1080/17524032.2012.753634, [Accessed 17 March 2023]

Singh, C., Iyer, S., Mark G. N., Few, R., Kuchimanchi, B., Segnon, A. C. & Morchain, D. 2022. Interrogating 'Effectiveness' in Climate Change Adaptation: 11 Guiding Principles for Adaptation Research and Practice, *Climate and Development*, 14(7): 650–664, DOI: 10.1080/17565529.2021.1964937 [Accessed 27 March 2023]

Singh, C., Iyer, S., New, MG., Few, R., Kuchimanchi., B., Segnon, A.C. & Morchain, D. 2021. Interrogating 'Effectiveness' in Climate Change Adaptation: 11 Guiding Principles for Adaptation Research and Practice. *Climate and Development*. https://hdl.handle.net/10568/114779 [Accessed 27 March 2023]

Smith, B., Rai, N., D'Errico, S., Argon, I. & Brooks, N. 2019. *Monitoring and Evaluation of Adaptation – an Introduction Deutsche Gesellschaft für Internationale Zusammenarbeit (GIZ) GmbH Sustainability* 52:75–81, ISSN 1877–3435, https://doi.org/10.1016/j.cosust.2021.07.003.(https://www.sciencedirect.com/science/article/pii/S1877343521000890) [Accessed 20 February 203]

Trollip, H., & Boulle, M. 2017. *Challenges Associated with Implementing Climate Change Mitigation Policy in South Africa.*

Tu Dam Ngoc Le 2023. Theoretical Frameworks in Climate Change Adaptation Planning: A Comparative Study in Coastal Cities of Developing Countries1, *Journal of Environmental Planning and Management*, 66 (2): 424–444, DOI: 10.1080/09640568.2021.1990028 [Accessed 27 March 2023]

UNDP Zimbabwe. 2020. *Zimbabwe Progress Review Report of Sustainable Development Goals (SDGs), December.* Progress report.

United Nations Environment Programme. 2022. *2022 Global Status Report for Buildings and Construction: Towards a Zero-Emission, Efficient and Resilient Buildings, and Construction Sector.* https://wedocs.unep.org/20.500.11822/41133 [Accessed: 21 November 2022]

United Nations Zimbabwe 2023, https://zimbabwe.un.org/en/sdgs/13

United Nations. 2023. https://www.un.org/en/climatechange/climate-solutions/education-key-addressing-climate-change. [Accessed 27 March 2023]

Usman, M., Balsalobre-Lorente, D., Jahangir, A., & Ahmad, P. 2022. Pollution Concern During Globalization Mode in Financially Resource-rich Countries: Do Financial Development, Natural Resources, and Renewable Energy Consumption Matter? *Renewable Energy*, 183: 90–102.

World Bank, 2022. https://www.worldbank.org/en/news/feature/2022/12/16/climate-action-in-2022-a-look-back-at-the-world-bank-s-climate-work-for-the-past-year#:~:text=2022%20saw%20the%20first%20of,while%20achieving%20critical%20development%20goals. Accessed 20 march 2023]

Yeukai, L. & Gumbu, Y. 2021. The Efficacy of Regional Organizations in Maintaining Peace and Security in the Region. The Case of SADC. *European Academic Research*. IX. 1388–1402.

Zieba, M., Durst S. & Hinteregger. 2020. *The Impact of Knowledge Risk Management on Sustainability.* Emerald Publishing Limited.

Critical factors influencing environmental management best practices in the developing countries real estate industry: A systematic review

H. Adjarko, I.C. Anugwo & A.O. Aiyetan
Durban University of Technology (DUT), South Africa

ABSTRACT: This study examines the influential factors in environmental management practices within the Ghanaian real estate industry, particularly in developing countries. The integration of effective environmental management practices in the industry is vital for achieving sustainable and competitive advantages while minimizing negative impacts on the natural environment. However, environmental management goals have received limited attention in Ghana's real estate sector. Through a systematic review of 96 papers using Google Scholar and ScienceDirect, 20 relevant papers were critically reviewed. A total of 65 success factors were identified and classified into ten themes: institutional/organizational, stakeholder influence, professional relations, mimicry from other companies, construction team, government pressure, social considerations, economic considerations, market forces, and social considerations. This comprehensive review fills a gap in the literature, offering insights for developing interventions that promote environmental sustainability and reshape national and organizational culture within the real estate industry.

Keywords: Best practices, critical factors, environmental management, real estate, sustainable construction

1 INTRODUCTION

Ghana has been dubbed the "best place for doing business" following the discovery of oil in 2007 to boost the economic (World Bank Group 2019). As such, Ghana's real estate sector has gained considerable momentum over the past few years due to urbanization and the influx of foreign investors. For example, in 2019, the real estate sector contributed 8.8 billion to Ghana's Gross Domestic Product, and by the end of the first quarter of 2019, real estate growth stood at 9.1 per cent (Ghana Investment Promotion Centre 2020). Ghana's real estate sector is one of the fastest growing of the 21 sub-sectors of the economy and employs almost 3% of the labour workforce. Real estate refers to tangible and investment asset that can be seen and touched (Abdulai & Awuah 2022; Abdulai & Hammond 2010). It can be described under four (4) broad categories: Residential Real Estate: that is undeveloped land, houses, multi-family house, apartments, and Cooperative housing; Commercial Real Estate: office buildings, warehouses and retail store buildings; Industrial Real Estate: factories, mines and farms; Agriculture Real Estate: tea garden, rubber & cotton palm oil plantation, Timberland and farmland.

Ghana Investment Promotion Centre (2020), reports a deficit of about 2 million residential/housing units as of 2019 in Ghana. The affordable housing schemes, such as the Sanglemi Housing Project (1,500 units), 10,000 affordable housing units by Solin, a private Hungarian company, and 100,000 affordable housing by the United Nations Office for Project Services (UNOPS) units etc are efforts to reduce the deficit. (Darko *et al.* 2018; Ghana Investment Promotion Centre 2020). Governments are urged to achieve and lead the sustainable development goals by making cities and human settlements inclusive, safe, resilient and sustainable. The real estate sector must promote inclusive and sustainable industrialization and foster innovations that promote environmental protection (Adinyira et al. 2018). What are the factors influencing environmental management best practices in the real estate sector in Ghana? While studies show that the property sector can play such a major role in the reduction of energy use and carbon emissions, it is worthwhile to note that the real estate sector has been found to have little impact on social and governance dimensions of Environmental, Social and Corporate Governance (ESG) policies (Kok et al. 2010).

Similarly, according to Sindhi & Kumar (2012) there is a growing awareness among organizations of ways to conserve and utilize natural resources to gain a competitive advantage. In this regard, a growing body of environmental management literature suggests that firms can gain sustainable competitive advantages by reducing the adverse impacts of their operations on the natural environment (Clarkson et al. 2011). Enforcement by governmental agencies has been found to be relatively loose, however, organizations are gradually moving toward proactive environmental approaches. The issue of environmental management is largely defined and serviced by non-construction related parties and few construction professionals are contributing to or influencing the direction of environmental research (Pasquire 1999). This research seeks to fill that gap. Darko et al. (2020) note that the adoption of Environmental management in the real estate sector in the country has been slow. According to Darko et al. (2017), there are currently no government policies and regulations that mandate environmental management adoption in the real estate industry in Ghana although the government is desirous of promoting sustainable buildings (Abdulai & Awuah 2022).

Some Ghanaian construction firms have voluntarily adopted sustainability-rating systems like: Leadership in Energy and Environmental Design (LEED), Excellence in Design for Greater Efficiencies (EDGE) and South Africa's (SA's) Green Star certification (Green Star SA). Sustainable buildings in Ghana are measured and certified by these foreign rating systems. DiMaggio & Powell, (1983) cited in Darnall & Pavlichev (2004), in explaining Institutional theory stated that an organizational actions are shaped by coercive, mimetic and normative pressures. Coercive pressures come from Institutions and their stakeholders. Normative pressures come from professional relationships such as industry associations. Mimicry pressure comes from copying or modeling other organizations. The inability of a firm to promote eco-innovation negatively affect its productivity and business performance because this places the firm at a disadvantaged position with respect to competitors (Lööf & Heshmati 2002 cited in De la Torre Ruiz et al. (2012)). The paper aimed to identify factors influencing environmental management best practices in the real estate industry in research publications guided by Hoffman's expansion of the Institutional theory. 20 available scientific literature dealing with theoretical aspects related to the topic was analyzed thoroughly. The paper is structured into five sections: the introduction; a systematic analysis regarding environmental management theories and critical factors influencing environmental management best practices in the real estate sector; the materials and methods; the results and discussion and future research directions; and the conclusions.

2 MATERIALS AND METHODS

Figure 1 shows the research process followed. The research was conducted through a systematic review of literature with keywords used, date and database used presented in Table 1.

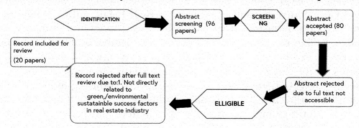

Figure 1. Showing flowchart of how papers were selected for the systematic review.

Table 1. Database for research.

Database	Search date	Keywords	Limit
1 Google scholar and science direct	From 12th Dec. 2022 to 31st December 2022	environmental management; sustainable construction; success factors, real estate, sustainability indicators; and housing development	All publications from 2014 to 2022, English publications

Searching two widely accepted databases ensured that relevant publications in Science Direct and Google Scholar and outside this range of year was omitted (see Table 1). Two databases are enough to offer adequate insight into the subject, complemented by the citation approach and with the help of Keenious research tool. In the citation approach, the references of materials (e.g., books and journal articles) were searched to find related materials. This resulted in identifying additional related papers dating before 2014 and were used depending on their relevance. Author such as (Umeokafor & Okoro 2020) have adopted this approach to complete the systematic literature search. The Keenious research explorer tool was downloaded and installed as an add-in to Microsoft word and utilized in identifying further related papers by scanning through the manuscript.

3 RESULTS AND DISCUSSION

3.1 Summary of critical success factors Identified from the systematic literature review

Figure 2 shows the Conceptual framework or categorizing literature review findings on identified success factors to enhance environmental best practices in the real estate industry. In all ten (10) broad categorizations were made based on the similarities in the factors identified and the institutional and eco-innovations theories.

Figure 2. Conceptual framework for categorizing literature review findings on identified success factors to best practices to environmental management in the real estate industry.

3.2 Success factors in the real estate industry

According to the reviewed literature, it is imperative to clearly identify, adopt and apply environmental management best practices within the real estate and built environment. Thus, effective application of environmental management best practices in Ghana real estate industry would justify positive and productive benefits to the built environment professionals and clients. The thorough literature review identified environmental management best practices and success factors which are based on their similarities were grouped under ten (10) main themes discussed below:

1. The Institution/Organization: it was found that organizations must take the initiative to access product innovation available, process innovation available and making use of organizational

environmental innovations (Murat & Mohammed 2020). According to Darko *et al.* (2018), thirteen product innovation factors include: application of energy-efficient systems (lighting, windows, HVAC systems, appliances (e.g., energy-efficient refrigerators), solar technology, rooftop wind turbines, natural lighting, solar water heating technology, solar shading devices, ground source heat pump technology);Control systems technology (HVAC, Security, Audio visual, Occupancy/motion sensors); and water efficiency technologies (quality of potable water, alternative water resources, and water conservation techniques) ((Darko *et al.* 2018; Owusu & Asumadu-Sarkodie 2016; Tupenaite & Geipele 2017). Others are installation of water-efficient appliances and fixtures (e.g., low-flow toilets), rainwater harvesting technology, grey water reclaiming and reuse technology (Darko *et al.* (2018). Innovation in operational and design process considerations include: embracing innovation in design, environmental friendly design, quality of facilities, architectural heritage considerations, architectural functionality, flexibility and adaptability (Tupenaite & Geipele 2017). Organizations' environmental innovations include access to general & environmental resources, access to external support, prior management system experience, basic environmental management capabilities, prior accounting & innovation expertise, EMS development, environmental investments & innovation etc. (Tupenaite & Geipele 2017).

2. Environmental Concerns: According to Abu *et al.* (2010) some environmental factors include: achieving energy efficiency, water efficiency/conservation, reduction of greenhouse gas emissions, waste management/recycling, material efficiency, pollution prevention– noise, water, air, optimization & conservation of land, protecting and enhancing biodiversity, reduction of car dependency (Dobias & Macek 2014). The gap identified was that only environmental related factors were considered by the researchers. Other environment related factors is found in Table 2 (Tupenaite & Geipele 2017). Atmosphere considerations include energy and lighting efficiency of housing, renewable energy use, and reduced greenhouse gas emission (CO_2) (Tupenaite & Geipele 2017). Other factors identified are: reduced use of natural resources (water, gas and electricity); reduced waste and increased recycling; enhanced building occupant health, comfort, and safety; production of renewable resources; collection of water for potable and non-potable uses; recycling and treatment of sewage and waste water; and improved indoor environmental quality (Darko *et al.* 2018; Tupenaite & Geipele 2017; Warren-Myers 2012).

3. Construction Team: The systematic review revealed construction team as the most critical theme affecting environmental management best practices. Murat & Mohammed (2020) identified seven critical factors for sustainable construction project management namely: project-related, work-related, client-related, project management, design-team-related, contractor-related, and project-manager-related factors. Written environmental policies, External and internal environmental accounting and audits, benchmarking environmental performance, public environmental reporting, and environmental performance goals were identified by Darnall & Pavlichev (2004). Other factors include effectiveness of leadership, supervision and decision making of the project manager, effectiveness of subcontractors, consultancy services, schedules, design, coordination and communication among project participants, level of compliance with government laws, regulations, quality, health and safety requirements etc. (Pasquire 1999; Sang & Yao 2019; Shen *et al.* 2018).

4. Economic Considerations: Achieving best environmental practices in the real estate sector comes with a cost. Accuracy of cost estimation, effective cost control and marketing, regular reporting, savings, success of sales/renting of real estate, financial resources, accurate financing plans, credibility, accuracy of risk assessment, proper analysis of financial and operational risks, and correctness of feasibility study are grouped under this category (Roshani *et al.* 2018). Value adding factors include level of project employee competency and effectiveness in the headquarters and on site, level of contribution to business value, extent the project adds value to the company through awards or recognitions, level of innovativeness of the project, new ideas, methods, and technology employed in the project (Roshani *et al.* 2018).

5. Professional Relations: Network relations of different professionals with different experiences in the implementation of environmental management best practices was found to be a vital factor (Roshani et al. 2018). Access to External Support through associations was found to be a major factor. In a bid to protect the sanctity of the profession, most professional organizations provide resources to members to maintain standards and high-quality products. Access to General & Environmental Resources are usually made available to members of professional bodies (Darko et al. 2018).
6. Stakeholders Influence: The external stakeholders whose influence affect decisions in the real estate industry include Government authorities, Markets (shareholders, investors, banks, buyers, suppliers, household consumers), Society (environmental groups, neighborhood groups, trade associations, unions) (Satankar & Jain 2015). Level of satisfaction of corporate, management, non-management employees, as internal factors affect best practices in the real estate industry. Satisfaction of clients were identified as an external factor. This satisfaction during or after the project are influenced by functionality of the space design, ergonomic solutions, attractiveness of project, location (preferred district by the customers), environmental friendliness, low environmental effect; sustainable designs (Satankar & Jain 2015).
7. Mimicry from Other Companies: The manufacturing industry has well-regulated systems of implementing and monitoring their environmental performance. Sanctions are easy to implement in the manufacturing industry. Lessons could be drawn from these industries and implemented in the real estate industry (Zhang et al. 2015).
8. Government Pressure: The existence of environmental regulations in the industry is a catalyst to developing and implementing eco-innovations that have better performance, higher-quality, safer, lower product costs (material substitution or less packaging)and in avoiding penalties (De la Torre Ruiz et al. 2012). Compliance can lead to products with higher resale or scrap value (ease in recycling or disassembly) and lower disposal cost. High environmental regulatory costs is an incentive for organizations to improve efficiency (De la Torre Ruiz et al. 2012). Effective regulations lead to monitoring and maintenance, material savings (substitution, reuse or recycling of production inputs), better utilization of by-products, lower energy consumption, reduced material storage and handling costs, conversion of waste into valuable forms, reduced waste disposal costs, safer workplace conditions. Darnall & Pavlichev (2004) identified: Traditional Regulation, Incentive-based (Such as voluntary environmental programs, Information-based environmental requirements, Pollution trading, Emissions taxes), Policies Technical Support, Financial Support, and Voluntary Environmental Programs.
9. Market Forces: According to Shen et al. (2018), market demand and technological advancement are fundamental drivers for sustainable estates. Future policies focusing on increasing market demand and market-based incentives for sustainable products may go a long way to promote environmental performance of the estate industry. Due to perceived higher initial costs than conventional projects, promotion of sustainable buildings in the market still faces challenges. According to Shen et al. (2018) high upfront costs may reduce attraction to green estates, thus affecting market demand for sustainable buildings. Project stakeholders are urged to improve their competencies continuously in the area of affordable and sustainable products to the market (Shen et al. 2018).
10. Social Considerations: Social issues are issues that affect people both in and around the estate during and after construction. Shen et al. (2018) argues that social influence or norms can significantly affect attitude towards energy use, suggesting that normative-based intervention can be an effective way to educate the public to change their attitude toward sustainable estates. Promotion of health and safety among builders, occupants and the community, professional training and public education are important measures to raise industry and public awareness about the long-term economic, environmental, and social value of the adopting sustainable practices (Shen et al. 2018). Ghana's population is up above 32 million as at 2023 with over 50% of the population under the age of 25 years (Ali et al. 2021; World Bank Group 2023). Government could exploit current dynamics of green technologies and empower the youth with the necessary skill set for an all-inclusive green economy (Ali et al. 2021).

Table 2. Summary of Critical Success factors to enhance environmental best practices in the real estate industry.

No	Categorization	Frequency	Rank	Success Factors	References
1	Institution/ Organization (IO)	19	1st	External pollution, Innovation and design process considerations, Energy Efficiency, Control systems, use of structure, the strategic orientation of the firm, Competencies Corporate headquarters Communicate ambition early, Management employees, non-management employees Top Management Support	(Abdulai & Awuah (2022), (Abu Hassan Abu et al. 2010), (Adinyira et al. 2018), (Ali et al. 2021), (Darko et al. 2018), (Karakhan, (2016), (Magliocco 2019), (Maqsood et al. 2022), (Murat & Mohammed 2020), (Owusu & Asumadu-Sarkodie, (2016), (Pasquire 1999), (Roshani et al. 2018), (Sang & Yao 2019), (Satankar & Jain 2015), (Shen et al. 2018), (Tupenaite & Geipele 2017), (Volker 2011), (Zhang et al. 2015), (Zia 2020)
2	Environmental Concerns (EC)	18	2nd	Land use Considerations, sustainable site, Site limitation and location, Atmosphere/ climate considerations, Water efficiency Considerations, Materials and waste Management, Indoor environmental quality, /Daylight/ ventilation Conservation considerations	(Abdulai & Awuah (2022), (Abu Hassan Abu et al. 2010), (Adinyira et al. 2018), (Ali et al. 2021), (Darko et al. 2018), (Karakhan, (2016), (Magliocco 2019), (Maqsood et al. 2022), (Murat and Mohammed 2020), (Owusu & Asumadu-Sarkodie, (2016), (Pasquire 1999), (Roshani et al. 2018), (Sang & Yao 2019), (Satankar & Jain 2015), (Tupenaite & Geipele 2017), (Volker 2011), (Zhang et al. 2015), (Zia 2020)
3	Construction Team (CT)	14	3rd	Construction methods and materials, Services/ Quality, Creativity & enthusiasm, Early involvement, end user Integrated design & calculation, Authority of Maintain ambition level the Project Good management & leadership Manager/Leader Successful tender High ambition level Good collaboration Realistic Cost and Time Estimates, Competent Project Team Project mission /common goal, Monitor performance and feedback, Project Control, Problem Solving Abilities, Technical assessment, experience and executive records of project manager project strategic planning, time cost and quality management, health and environmental safety, user affordability, Design consideration, cost of unit's technology	(Magliocco 2019), (Maqsood et al. 2022), (Murat and Mohammed 2020), (Owusu & Asumadu-Sarkodie, (2016), (Pasquire 1999), (Roshani et al. 2018), (Sang and Yao 2019), (Satankar & Jain 2015), (Shen et al. 2017), (Shen et al. 2018), (Tupenaite & Geipele 2017), (Volker 2011)
4	Economic Factors (EF)	15	4th	Financial, Value adding, Transportation	(Magliocco 2019), (Murat & Mohammed 2020), (Owusu & Asumadu-Sarkodie, (2016), (Pasquire 1999), (Roshani et al. 2018), (Sang & Yao 2019), (Satankar & Jain 2015), (Shen et al. 2017), (Shen et al. 2018), (Tupenaite & Geipele 2017), (Volker 2011)

(continued)

Table 2. Continued

No	Categorization	Frequency	Rank	Success Factors	References
5	Professional Relations (PR)	13	5th	Network relations, Access to External Support, Access to General & Environmental Resources	(Pasquire 1999) (Magliocco 2019), (Maqsood et al. 2022), (Murat and Mohammed 2020), (Owusu & Asumadu-Sarkodie, (2016), (Pasquire 1999), (Roshani et al. 2018), (Sang & Yao 2019), (Satankar & Jain 2015), (Shen et al. 2017), (Shen et al. 2018), (Tupenaite & Geipele 2017), (Volker 2011)
6	Stakeholders Influence (SI)	11	6th	Household consumers, Commercial buyers, Suppliers of goods and services Shareholders and investment funds Banks and other lenders Labor unions, Industry or trade associations Environmental groups or organizations Neighborhood/community groups & organizations Client Involvement	(Volker 2011) (Roshani et al. 2018), (Satankar & Jain 2015), (Magliocco 2019), (Maqsood et al. 2022), (Murat and Mohammed 2020), (Owusu & Asumadu-Sarkodie, (2016), (Sang & Yao 2019), (Shen et al. 2017), (Shen et al. 2018), (Tupenaite & Geipele 2017)
7	Mimicry From Other Companies (MC)	11	6th	Lessons from Implementation, monitoring and sanctions implemented in the manufacturing industry	Zhang et al. 2015) (Zia 2020), (Volker 2011) (Roshani et al. 2018), (Satankar & Jain 2015), (Magliocco 2019), (Maqsood et al. 2022), (Murat & Mohammed 2020), (Owusu & Asumadu-Sarkodie, (2016), (Sang & Yao 2019), (Shen et al. 2017)
8	Government Pressure (Gp)	4	8th	Public authorities (government, state, municipal) Environmental regulation	(Abu Hassan Abu et al. 2010), (Ali et al. 2021) (Abdulai & Awuah (2022), (Abu Hassan Abu et al. 2010)
9	Market Forces (MF)	3	9th	Green procurement, energy performance contracting	(Shen et al. 2018), (Zhang et al. 2015) (Zia 2020)
10	Social Considerations (SO)	2	10th	Health & Comfort, Local and international support for green economy, High awareness and understanding of environmental protection	(Ali et al. 2021), (Maqsood et al. 2022)

Table 2 below shows the summary of success factors that were identified in the systematic literature review as discussed above. Authors have been identified and the categorization of the factors have been represented in the table for easy understanding. It also shows the prioritization of critical success factors (CSFs) in accordance with the rate of frequency appeared in the literature reviewed. In all the ten (10) broad categorizations, environmental concerns were ranked first as the most talked about success factor for environmental management in the real estate industry.

3.3 Future research

Many of the critical success factors have been limited to the construction industry. It is recommended that future research is conducted empirically using both qualitative and quantitative research methods to ascertain the critical success factors for the real estate industry for developing countries.

4 CONCLUSION AND RECOMMENDATIONS

Many of the critical success factors identified by researchers have focused on the construction industry. However, this paper has extended the factors to include those related to the real estate sector. This research is highly relevant for future research works in the real estate industry, due to the low level of environmental management related research works in the real estate sector as compared to the construction sector. This would help improve understanding of how to incorporate environmental management into the real estate industry. The literature review helped to identify critical success factors to environmental management best practices in the real estate industry. The ten (10) broad themes identified include: Institutional/ Organizational, Stakeholders Influence, Professional Relations, Mimicry, Construction Team, Government Pressure, Social Considerations, Economic Considerations, and Market forces (Pasquire 1999). The two types of industries, construction, and real estate, though similar differ in ways, such as the scope of technical definition, and the type of stakeholders. The latter is highly relevant for future research, due to the low level of environmental management related works in the real estate sector. Furthermore, an in-depth qualitative analysis of the articles is recommended. These would help improve the incorporation of environmental management into the real estate industry. The study concludes that more research is needed to narrow the issues more clearly, provide qualitative and quantitative data and develop a model for incorporating of environmental management into the real estate industry in Ghana.

REFERENCES

Abdulai, R. T., & Awuah, K. G. B. (2022). *Real Estate (RE) and Sustainable Development Goals (SDGs) in Ghana* [Https://perlego.mention-me.com/m/ol/xo4dy-harold-adjarko].

Abdulai, R. T., & Hammond, F. N. (2010). Landed Property Market Information Management and Access to Finance. *Property Management, 28*((4)), 228–244.

Abu Hassan Abu Bakar, A. A. R., Shardy Abdullah, Aidah Awang, Vasanthi Perumal. (2010). Critical Success Factors for Sustainable Housing: A Framework from the Project Management View. *Asian Journal of Management Research*.

Adinyira, E.; Kwofie, T. E.; Quarcoo, F. (2018) Stakeholder Requirements for Building Energy Efficiency in Mass Housing Delivery: The House of Quality Approach Environment, *Development and Sustainability* Vol. 20 Issue 3, pp. 1115–1131.

Adinyira, E., &, E. B., & Kwofie, T. E. (2012). Determining Critical Project Success Criteria for Public Housing Building Projects (PHBPS) in Ghana. *Engineering Management Research, Vol. 1* (No. 2). https://doi.org/DOI: 10.5539/emr.v1n2p122

Ali, Ernest Baba; Valery Pavlovich Anufriev; Bismark Amfo, (2021), *Green Economy Implementation in Ghana as a Road Map for a Sustainable Development Drive: A Review*

Clarkson, P. M., Li, Y., R.D., G., & Vasvari, F. P. (2011). Does it Really Pay to be Green? Determinants and Consequences of Proactive Environmental Strategies. *Journal of Account. Public Policy, Vol. 30*(No. 2), pp. 122–144.

Darko, A., Chan, A. P. C., & Owusu., E. K. (2018). What are the Green Technologies for Sustainable Housing Development? An Empirical Study in Ghana *Business Strategy and Development*.

Darnall, N., & Pavlichev, A. (2004). Environmental Policy Tools & Firm-level Management practices IN the United States. *SSRN Electronic Journal*. https://doi.org/DOI: 10.2139/ssrn.1030609

De la Torre Ruiz, J. M., Ferrón Vilchez, V., Aguilera Caracuel, J., & Martín Rojas, R. (2012). Advanced Environmental Management and Innovation: A Theoretical Framework. In C. Kuei & C. Madu (Eds.), *Handbook of Sustainability Management* (pp. 421–440). World Scientific.

Din, A., Hoesli, M., & Bender, A. (2001). *Environmental Variables and Real Estate Prices*. [Record #30 is using a reference type undefined in this output style.]

Ghana Investment Promotion Centre, G. I. P. (2020). <Property-Development-Sector-Profile.pdf>.

Gunduz, Murat; Almuajebh, Mohammed (2020), Critical Success Factors for Sustainable Construction Project Management Sustainability Vol. 12 Issue 5, p. 1990.

Karakhan, Ali (2016), A LEED-Certified Projects: Green or Sustainable? *Journal of Management in Engineering*.Vol. 32 Issue 5.

Kok, N., Eichholtz, P., Bauer, R., & Peneda, P. (2010). Environmental Performance, A Global Perspective on Commercial Real Estate The *European Centre for Corporate Engagement Maastricht University School of Business and Economics Netherlands*.

Kusku, F. (2007). From Necessity to Responsibility: Evidence for Corporate Environmentalcitizenship Activities from a Developing Country Perspective. *Corporate Social Responsibility and Environmental Management, Vol. 14*(No. 2), pp. 74–87.

Magliocco, Adriano, (2018) *Vertical Greening Systems: Social and Aesthetic Aspects, Nature Based Strategies for Urban an Building Sustainability*, Butterworth-Heinemann, Pages 263–271, ISBN 9780128121504,

Maqsood, S., Yan, Z., Xintong, L., Shuai, H., Ihsan, J., & Khurram, S. (2022). Critical Success Factors for Adopting Green Supply Chain Management and Clean Innovation Technology in the Small and Medium-sized Enterprises: A Structural Equation Modeling Approach. *Frontiers in Psychology*, 13:1008982. https://doi.org/doi:10.3389/fpsyg.2022.1008982

Murat, Gunduz and Mohammed, Almuajebh (2020) Critical Success Factors for Sustainable Construction Project Management, *Sustainability*, 12, doi:10.3390/su12051990 www.mdpi.com/journal/sustainability

Owusu, P. A., & Asumadu-Sarkodie, S. (2016). A Review of Renewable Energy Sources, Sustainability Issues and Climate Change Mitigation. *Cogent Engineering, Vol. 3*(No. 1), pp. 1–14. https://doi.org/doi:10.1080/23311916.2016.1167990.

Pasquire, C. (1999). "The Implications of Environmental Issues on UK Construction Management" *Engineering, Construction and Architectural Management, Vol. 6*(Iss 3), pp. 276–286.

Roshani, Alireza; Gerami, Mohsen; Rezaeifar, Omid (2018) New Rethinking on Managers' Competency Criteria and Success Factors in Airport Construction Projects *Civil Engineering Journal*

Sang, Peidong; Yao, Haona (2019), Exploring Critical Success Factors for Green Housing Projects: An Empirical Survey of Urban Areas in China *Advances in Civil Engineering* Vol., pp. 1–13, 2019.

Satankar, P. P., & Jain, A. (2015). Study of Success Factors for Real Estate Construction Projects. *International Research Journal of Engineering and Technology, 2*(4), 804–808.

Shen, Wenxin; Tang, Wenzhe; Siripanan, Atthaset; Lei, Zhen; Duffield, Colin; Hui, Felix (2018), Understanding the Green Technical Capabilities and Barriers to Green Buildings in Developing Countries: A Case Study of Thailand, *Sustainability*, Vol. 10 Issue 10, p. 3585.

Shen, Wenxin; Tang, Wenzhe; Siripanan, Atthaset; Lei, Zhen; Duffield, Colin F.; Wilson, David; Hui, Felix Kin Peng; Wei, Yongping (2017) Critical Success Factors in Thailand's Green Building Industry, *Journal of Asian Architecture and Building Engineering*, Vol. 16 Issue 2, pp. 317–324.

Sindhi, S., & Kumar, N. (2012). "Corporate Environmental Responsibility – Transitional and Evolving". *Management of Environmental Quality: An International Journal, Vol. 23*(Iss: 6), pp. 640–657.

Tupenaite, L. I., Lill; &, & Geipele, I. N., Jurga. (2017). Ranking of Sustainability Indicators for Assessment of the New Housing Development Projects: Case of the Baltic States. *Resourses* www.mdpi.com/journal/resources. https://doi.org/10.3390/resources6040055

Umeokafor, N., & Okoro, C. (2020). Barriers to Social Support in the Mental Health and Wellbeing of Construction Workers in Emerging and Developing Economies: A Systematic Review. *Proceedings of the Joint CIB W099 & TG59 International Web-Conference 2020: Good Health, Wellbeing & Decent Work*.

Volker, L. (2011). *Success And Fail Factors In Sustainable Real Estate Renovation Projects Management and Innovation for a Sustainable Built Environment*.

Warren-Myers, G. (2012). Sustainable Management of Real Estate: Is It Really Sustainability? *JOSRE Vol. 4* (No. 1).

World Bank Group, (2019), *Doing Business Survey 2019: Training for Reform – Economy Profile Ghana*.

Zhang, Xiaoling; Wu, Zezhou; Feng, Yong; Xu, Pengpeng (2015), "Turning Green Into Gold": A Framework for Energy Performance Contracting (EPC) in China's Real Estate Industry *Journal of Cleaner Production* Vol. 109, pp. 166–173.

Zia, M. N. (2020). A Review Paper on Identification of Critical Success Factors (CSFs) for Successful Project Management of Construction Projects *Journal of Management Practices, Humanities and Social Sciences, Vol 4* (Issue 2), pp. 29–36. https://doi.org/https://doi.org/10.33152/jmphss-4.2.2

Author index

Abdullahi, A.N. 268
Adeniran, A.A. 12
Adewunmi, Y. 21
Aiyetan, A.O. 250, 259, 289
Allen, C.J. 225
Anugwo, I.C. 289
Adjarko, H. 289

Beene, M. 203
Bhila, N. 54
Bremer, T. 233

du Toit, B. 125
David, A.B. 250, 259
Dumba, S. 29

Gumbo, T. 29, 67, 74, 119, 242

James, M. 279
Jayeola, F.C. 191

Kabundu, E. 12
Kajimo-Shakantu, K. 125, 134, 152, 161, 182, 203, 213
Kalaoane, R.C. 119
Kangwa, K. 90

Kanyembo, A. 213
Kibangou, A.Y. 119
Kumalo, L. 37
Kuotcha, W. 143

Lungu, A. 203

Mahachi, J. 37, 54
Makashini, L. 21
Mateza, W. 173
Mathane, T.P. 67, 74
Mbanga, S.L. 12
Mndzebele, M.G. 242
Mntu, A. 125, 134
Mogorosi, T. 134
Monoametsi, S.C. 233
Morejele-zwane, L. 29
Moyo, T. 82, 109, 173, 279
Mukumba, C.P. 152
Muleya, F. 182, 203, 213
Mulokwe, R. 46
Munshifwa, E.K. 21
Musakwa, W. 119
Musonda, I. 82, 109, 119
Muzioreva, H. 82, 109
Mwakatobe, B.F. 143

Mwanaumo, E. 99
Mwanaumo, E.M. 191
Mwiya, B. 90, 99

Nengovhela, H. 182
Ngema, N.N. 12
Ngoma, I. 143
Nkambule, N.P. 3
Ntakana, K. 3

Ogungbile, A.J. 191
Oke, A.E. 191
Okoro, C.S. 268
Onososen, A.O. 82, 109

Risimati, B. 46
Le Roux, L. 161

Tembo, C.K. 203, 213
Tsegay, F.G. 99

Udo, G.O. 268
Udo, J. 268
Udoudoh, F.P. 268

Xulaba, S. 225

Zulu, S. 143